# *Extraction of Metals from Soils and Waters*

# MODERN INORGANIC CHEMISTRY

Series Editor: John P. Fackler, Jr., *Texas A&M University*

Recent volumes in the series:

CARBON-FUNCTIONAL ORGANOSILICON COMPOUNDS
Edited by Václav Chvalovský and Jon M. Bellama

COMPUTATIONAL METHODS FOR THE DETERMINATION OF FORMATION CONSTANTS
Edited by David J. Leggett

COOPERATIVE PHENOMENA IN JAHN–TELLER CRYSTALS
Michael D. Kaplan and Benjamin G. Vekhter

EXTRACTION OF METALS FROM SOILS AND WATERS
D. Max Roundhill

GAS PHASE INORGANIC CHEMISTRY
Edited by David H. Russell

HOMOGENEOUS CATALYSIS WITH METAL PHOSPHINE COMPLEXES
Edited by Louis H. Pignolet

INORGANOMETALLIC CHEMISTRY
Edited by Thomas P. Fehlner

THE JAHN–TELLER EFFECT AND VIBRONIC INTERACTIONS IN MODERN CHEMISTRY
I. B. Bersuker

METAL COMPLEXES IN AQUEOUS SOLUTIONS
Arthur E. Martell and Robert D. Hancock

METAL DIHYDROGEN AND $\sigma$-BOND COMPLEXES
Gregory J. Kubas

MÖSSBAUER SPECTROSCOPY APPLIED TO INORGANIC CHEMISTRY
*Volumes 1 and 2* • Edited by Gary J. Long
*Volume 3* • Edited by Gary J. Long and Fernande Grandjean

MÖSSBAUER SPECTROSCOPY APPLIED TO MAGNETISM AND MATERIALS SCIENCE
*Volumes 1 and 2* • Edited by Gary J. Long and Fernande Grandjean

OPTOELECTRONIC PROPERTIES OF INORGANIC COMPOUNDS
Edited by D. Max Roundhill and John P. Fackler, Jr.

ORGANOMETALLIC CHEMISTRY OF THE TRANSITION ELEMENTS
Florian P. Pruchnik
Translated from Polish by Stan A. Duraj

PHOTOCHEMISTRY AND PHOTOPHYSICS OF METAL COMPLEXES
D. Max Roundhill

# *Extraction of Metals from Soils and Waters*

**D. Max Roundhill**
*Texas Tech University*
*Lubbock, Texas*

**Kluwer Academic/Plenum Publishers**
**New York, Boston, Dordrecht, London, Moscow**

Library of Congress Cataloging-in-Publication Data

Extraction of metals from soils and waters/edited by D. Max Roundhill.
p. cm. — (Modern inorganic chemistry)
Includes bibliographical references and index.
ISBN 0-306-46722-4
1. Metals—Environmental aspects. 2. Extraction (Chemistry) 3. Soil remediation. 4. Water—Purification. I. Roundhill, D. M. II. Series.

TD879.M47 .E94 2001
628.5′5—dc21

2001053916

ISBN 0-306-46722-4


233 Spring Street, New York, New York 10013

http://www.wkap.nl/

10 9 8 7 6 5 4 3 2 1

A C.I.P. record for this book is available from the Library of Congress



Printed in the United States of America

For Stephanie, Brian, Ian, and David

# PREFACE

This book is intended as a reference source to guide chemists into the subject of metal extraction. The field of metal coordination chemistry has reached a level of maturity where it can move into the applications phase. This book fits that profile. Much of the focus is on the use of complexants for the extraction of metals from either waters or soils.

The field of metal extraction is far too broad to be covered comprehensively in this volume. Since this series focuses on Inorganic Chemistry, the main theme of this book falls within this direction. The emphasis therefore is on the particular properties of the metals and complexants that result in them being chosen for individual applications. The metals upon which this book focuses are those for which environmental remediation is important. This usually means metals that are toxic because of either their chemical properties or because they are present in the environment as long-lived radioactive isotopes.

The book is divided into chapters that cover both techniques and individual metals. The techniques include soil washing and electrokinetic extraction, the metals include the alkali metals with a particular emphasis on cesium, heavy metals such as mercury, lead, cadmium, copper, silver and gold, and lanthanides and actinides. In a later chapter the newly emerging fields of bioremediation and phytoremediation are covered. These fields have now reached the level of maturity where they are being applied in the field, and where studies are being carried out to determine the binding sites and chemical processes that occur during their operation. In the final chapter the development and use of selective sensors for metal ions is covered. As legislation continues to reduce the allowable amounts of metals that can be released into the environment, monitoring systems will need to be put in place to ensure that facilities conform to these requirements.

I wish to acknowledge my research students and postdoctoral associates who have worked with me in the field of metal extraction, with particular appreciation to Alex Yordanov, Nick Wolf, Fred Koch, Mathew Falana, and Jinyu Shen. I thank the U.S. Department of Energy, through the Pacific Northwest National Laboratory, and the U.S. Army

Research Office, for supporting my own research work in this area. I thank Dr. J. L. Gardea-Torresdey for assistance with the material for Chapter 12. I thank Nicole Johns-Street for assistance in putting together the final version. I also thank my wife, Stephanie, for seeing me through the writing of this book, my third one in this series.

# Contents


1. **Introduction**

1. Occurrence and Remediation Strategies ................................ 1
   1.1. Heavy Metals ................................ 2
   1.2. Precious Metals ................................ 4
   1.3. Strategies for Obtaining Metal Ion Recognition ................................ 5

2. **Phase Transfer Extraction and Adsorption Methods**

1. Introduction ................................ 9
2. Liquid-Liquid Systems ................................ 11
   2.1. Equilibria ................................ 11
   2.2. Surfactants ................................ 12
   2.3. Ion Hydration ................................ 12
   2.4. Aqueous Biphasic Systems ................................ 15
   2.5. Liquid Membranes ................................ 16
   2.6. Picrate Method of Analysis ................................ 19
3. Adsorption Methods ................................ 20
   3.1 Ion Exchange Methods ................................ 20
   3.2 Adsorbent and Membrane Methods ................................ 23
   3.3 Polymer Filtration Methods ................................ 24
   3.4 Uncharged Solid Phase Adsorbent Materials ................................ 25
      3.4.1 Metals ................................ 25
      3.4.2 Carbon and Silicon ................................ 26
      3.4.3 Metal Oxides ................................ 27
   3.5 Silicates and Zeolites ................................ 28
   3.6 Polymers and Biopolymers ................................ 29

3. **Soil Washing Methods and *In Situ* Stabilization Methods**

1. Introduction ................................ 35

2. Soil Washing and Flooding ........ 36
  2.1 Soil Washing ........ 37
  2.2 *In Situ* Stabilization ........ 39
3. Analytical Techniques ........ 41
4. Extraction from Soils ........ 43
5. Complexant Design for Soil Extraction ........ 44

**4. Electrokinetic Extraction of Metals from Soils**

1. Introduction ........ 47
2. Electrokinetic Extraction ........ 47
  2.1. Electrokinetic Transport Model ........ 50
  2.2. Experimental Method ........ 52
  2.3. Electrokinetic Extraction of Copper, Lead, Cadmium, and Mercury ........ 54
  2.4. Electrokinetic Extraction of Uranium and the Other Radionuclides ........ 57
  2.5. Electrokinetic Extraction of Chromium ........ 58

**5. Selective Extraction with Chelates, Macrocycles, and Calixarenes**

1. Introduction ........ 63
2. Multidentate Ligands ........ 65
  2.1. Dithio- and Diselenocarbamates ........ 66
  2.2. Aminopolycarboxylic Acids, Amides, and Aminimides ........ 68
  2.3. Oximes and Hydroxyaromatic Derivatives ........ 72
  2.4. Others ........ 75
  2.5. Extractants for Supercritical Carbon Dioxide ........ 80
3. Macrocyclic Ligands ........ 81
  3.1. Oxygen Macrocycles ........ 82
  3.2. Azamacrocycles and Thiamacrocycles ........ 84
4. Calixarenes ........ 95
5. Limitations of Chelate, Macrocycle, and Calixarene Extractants ........ 100

**6. Phase Transfer Extraction of Mercury**

1. Background ........ 105
  1.1. Toxicity and Occurrence ........ 106

2. Complexants ........ 107
2.1. Oxygen Donors ........ 108
2.2. Nitrogen Donors ........ 109
2.3. Sulfur Donors ........ 109
3. Chelates and Multidentates ........ 109
3.1. Sulfur and Nitrogen Donor Chelates ........ 110
3.1.1. Dithiocarbamates ........ 111
3.1.2. Thiocarbazone ........ 112
3.1.3. Sulfur Containing Metal Complexes ........ 112
4. Extraction and Detection Techniques ........ 113
5. Column Materials for Mercury Separation ........ 114
5.1. Organic Adsorbents ........ 115
5.2. Inorganic Adsorbents ........ 115
6. Macrocycles and Open Chain Analogs ........ 116
6.1. Open Chain Analogs ........ 116
6.2. N-Donor Macrocycles ........ 117
6.3. S-Donor Macrocycles ........ 117
6.4. O-Donor Macrocycles ........ 118
6.5. N, O-Donor Macrocycles ........ 119
6.6. Chromogenic and Fluorogenic Macrocycles ........ 121
7. Porphyrins ........ 121
8. Calixarenes ........ 122
8.1. P and S-Donor Calixarenes ........ 122
8.2. N-Donor Calixarenes ........ 124
9. Elemental Mercury ........ 125

**7. Phase Transfer Extraction of Lead and Cadmium**

1. Lead ........ 129
1.1. Toxicity and Occurrence ........ 129
1.2. Complexants ........ 130
1.2.1. Halides and Acids ........ 130
1.2.2. Poly(ethylene glycol) ........ 132
1.3. Chelates and Macrocycles ........ 133
1.3.1. Chelating Agents for *In Vivo* Extraction Therapy ........ 133
1.3.2. Oxygen and Nitrogen Donor Extractants ........ 134
1.3.3. Sulfur Donor Dithiocarbamate Extractants ........ 135

1.3.4. Analytical Methods for Lead ................ 136
1.3.5. Other Sulfur Donor Extractants ............ 137
1.4. Macrocycles ........................................ 139
1.4.1. Oxygen Donors .................................... 140
1.4.2. Nitrogen Donors .................................. 141
1.4.3. Sulfur Donors ...................................... 143
1.5. Calixarenes ......................................... 145
1.6. Liquid Carbon Dioxide as the Non-Aqueous Phase ................................................... 146
1.7. Biological ........................................... 148
1.8. Ingestion from Paper ........................... 148
1.9. Analytical Methods for Lead ................ 148
1.9.1. Biological Samples .............................. 149
1.9.2. Sediment Samples ............................... 149
1.9.3. Microextraction .................................... 150
2. Cadmium ................................................ 151
2.1. Toxicity and Occurrence ...................... 151
2.2. Chelates and Macrocycles ................... 152
2.2.1. Chelating Agents for Extraction Therapy ................................................ 152
2.2.2. Nitrogen and Oxygen Donor Extractants ............................................ 152
2.3. Sulfur Donors ...................................... 154
2.3.1. Dithiocarbamates ................................. 155
2.3.2. Other Sulfur Donors ............................ 156
2.4. Cyclodextrins and Carbohydrates ....... 157
2.5. Cadmium Incorporation into the Food Chain ...... 158

**8. Phase Transfer Extraction of Copper, Silver and Gold**
1. Copper. ................................................... 164
1.1. Monodentate Ligands .......................... 164
1.1.1. Thiocyanate ......................................... 164
1.1.2. Cyanide ............................................... 164
1.1.3. Mineral Acids ...................................... 165
1.1.4. Amines ................................................ 165
1.1.5. Micelles and Vesicles .......................... 165
1.2. Chelate Ligands ................................... 166
1.2.1. Oxygen Donors .................................... 166
1.2.2. Oximes ................................................ 168
1.2.3. Aliphatic and Aromatic Nitrogen Chelates ............................................... 172

1.2.4. Other N,O Chelates ............ 173
1.2.5. Sulfur-Donor Chelates ............ 174
1.3. Macrocycles and Calixarenes ............ 175
1.3.1. Azamacrocycles ............ 175
1.3.2. Thiamacrocycles ............ 176
1.4. Calixarenes ............ 177
1.5. Porphyrins ............ 177
1.6. Carbon Dioxide ............ 177
1.7. Others ............ 178
2. Silver and Gold ............ 179
2.1. Monodentates ............ 179
2.1.1. Cyanide ............ 179
2.1.2. Amines ............ 180
2.1.3 Oxygen Donors ............ 180
2.1.4. Particles ............ 180
2.1.5. Detergents and Salts ............ 181
2.2. Chelates ............ 181
2.2.1. N, O-Chelates ............ 181
2.2.2. S-Chelates ............ 183
2.3. Macrocycles ............ 184
2.3.1. Azamacrocycles and Thiamacrocycles ............ 184
2.3.2. Thiolariat Ethers ............ 187
2.3.3. Hydrocarbons ............ 187
2.4. Calixarenes ............ 187

**9. Extraction of Actinides and Lanthanides**

1. Actinides ............ 193
1.1. Introduction ............ 193
1.2. Monodentates ............ 193
1.2.1. Carboxylic Acids ............ 194
1.2.2. Phosphates and Phosphine Oxides ............ 194
1.2.3. Amines and Ketones ............ 196
1.3. Multidentates ............ 197
1.3.1. N, O Chelates ............ 197
1.3.2. Malonamides ............ 198
1.3.3. Chemical Detection and Analysis ............ 202
1.3.4. Stereognostic Coordination ............ 203
1.3.5. Carbamoyl Phosphonates ............ 204
1.3.6. Crown Ethers and Cavitands ............ 204

1.3.7. Calixarenes ... 205
2. Lanthanides ... 206
2.1. Monodentates ... 207
2.2.1. Carboxylates ... 207
2.2.2. Phosphates ... 209
2.2.3. Ketones and Phosphine Oxides ... 210
2.2.4. Crown Ethers ... 212
2.2.5. Nitrogen Donors ... 214
2.2.6. EDTA and Analogues ... 217
2.2.7. Aromatic Chelates ... 219
2.2.8. Phosphoramides ... 222
2.2.9. Miscellaneous Methods ... 225

**10. Extraction of Anions and Oxyanions**

1. Introduction ... 231
2. Toxicological Effects ... 233
3. Analytical Techniques ... 234
4. Extraction Methods ... 234
5. Complexant Design ... 237
5.1. Acyclic Ligating Sites ... 237
5.2. Amines ... 238
5.2.1 Ferrocene Derivatives ... 239
5.3. Calixarenes ... 240
5.3.1. Calixarenes Amines and Amides ... 240
5.4. Poly(ethylene glycol) ... 243
6. Extraction from Soils ... 243

**11. Extraction of Alkali and Alkaline Earth Metals**

1. Introduction ... 247
2. Crown Ethers ... 248
2.1. Crown and *Bis*-Crowns ... 248
2.2. Capped Cleft Molecules ... 253
2.3. Biphasic Media ... 254
2.4. Transport Phenomena ... 255
2.5. Carboxylate Crowns ... 261
2.6. Phosphonate Crowns ... 261
2.7. Lariat Crowns ... 263
2.8. Amino Crowns ... 264
2.9. Aza Crowns and Photocrowns ... 266

3. Specialty Crowns ........ 270
4. Cavitands and Cryptands ........ 273
5. Other Oxygen Donors ........ 276
6. Metal Oxides ........ 279
7. Physical Methods ........ 279
8. Cesium ........ 281

**12. Phytoremediation and Bioremediation of Soils and Waters**

1. Introduction ........ 289
2. Phytoremediation ........ 290
   2.1. Chromium (VI) ........ 292
   2.2. Cadmium, Lead, Copper, and Zinc ........ 293
   2.3 Mercury ........ 296
   2.4. Radionuclides ........ 296
3. Bioremediation ........ 297
   3.1. Mercury ........ 298
   3.2. Cadmium, Copper, and Lead ........ 299
   3.3. Actinides ........ 300
   3.4. Chromium ........ 301

**13. Optical and Redox Sensors with Metal Ions**

1. Introduction ........ 307
2. Optical Reporter Molecules ........ 308
3. Redox Reporter Molecules ........ 310
4. Chemiluminescence ........ 310
5. Chelate Complexants ........ 311
   5.1. Nitrogen and Oxygen Donors ........ 311
   5.2. Conformational and Steric Effects ........ 316
   5.3. Sulfur Donors ........ 317
   5.4. Ruthenium Bipyridyl Complexes as Sensors ........ 319
   5.5. Calixarene Complexants ........ 320
   5.6. Optical Reporters on Calixarenes ........ 320
   5.7. Redox Reporters on Calixarenes ........ 326
   5.8. Lanthanide Reporters on Calixarenes ........ 326
6. Macrocyclic Complexants ........ 327
   6.1. N, O, and S-Donors ........ 328
   6.2. Azamacrocycles ........ 332
   6.3. Cryptands ........ 334
   6.4. Porphyrins and Pyrroles ........ 336

7. Crown Ethers and Cryptands for Alkali and Alkaline Earth Chemosensors ... 338
7.1. Naphthalene and Anthracene Reporters ... 339
7.2. Iron Sulfur Cluster Reporters ... 344
7.3. Structural Features ... 344
8. Sensors for Biological Applications ... 345
8.1. Sodium ... 346
8.2. Potassium ... 347
8.3. Calcium ... 347
8.4. Zinc ... 349
8.5. Lanthanides ... 351
8.6. Signal Transmission ... 352
9. Solid State Sensors ... 353
10. Sensors for Anions ... 355
10.1. Acyclic Ligating Sites ... 355
10.2. Calixarenes ... 357
11. Molecular Computation ... 359

**Index** ... 365

# 1

# INTRODUCTION

## 1. OCCURRENCE AND REMEDIATION STRATEGIES

Heavy metals have found industrial, agricultural and military uses for several centuries of time. As a result they are now widely dispersed in a range of different forms, and there are environmental problems resulting from their mining, extraction, and purification. In many cases these metals are present as mixtures rather than as a metal residue in pure form. As a result, if these metals are to be recovered in pure form, it will be necessary for selective extraction processes to be developed. Because many heavy metals are toxic, they cannot be left unrecovered. However, since these metals have significant commercial value, it is realistic that economically justifiable methods can be developed for their separation and recovery from waste residues.

For actinides and radioactive elements the situation is somewhat different. In this case these materials are stored in a limited number of sites. Nevertheless, these materials are often present as mixtures in solutions that have high salt concentrations, and may have high acidities. These materials also have a high priority for recovery or disposal because they are often the result of weapons production, and their storage facilities are unsuitable for long term use. Furthermore the international and political aspects of this problem require that urgent consideration be given now to finding a solution to it.

Heavy metals can be present in both soils and waters. In soils they are usually tightly bound because the cationic metal center strongly associates with the anionic structures of soils. In waters, heavy metals can be present under conditions of high acidity or basicity, or in the presence of high concentrations of other salts. Special conditions also prevail when the metal is radioactive because the extractant used may be decomposed by the nuclear decay of the element.

In choosing a remediation approach, several considerations must be applied. One of these is that the chosen method does not leave toxic residues that must themselves be subsequently removed. This can be avoided by eliminating the use of toxic compounds, or by ensuring that they undergo complete biodegradation to non-toxic compounds. Another method of achieving this goal with soils is to use electrokinetic or phytoremediation methods where no new chemical species are added during the removal process.

This book focuses on compounds and methods that are used for the extraction and subsequent recovery from the extractant solution or adsorbent. The two groups of elements that are included are those that we define as heavy metals or oxyions. These elements are chosen because they are frequently targeted for extraction or adsorption because they are toxic, costly to replace, or are long-lived radioisotopes. The coordination chemistry of these species has already been well studied, and because there is a need to develop economically viable methods of extracting, separating, and recovering them from waste sites, there is considerable interest in discovering new and better methods. This book outlines possible strategies that can be used to achieve these goals. The book also covers the available techniques and methods that are presently available for their selective extraction and recovery.

## 1.1. Heavy Metals

In order for a metal to be given high priority for remediation it needs to meet several criteria. One is that the metal should be present in a site in sufficient quantity that it poses a serious health threat. A second criterion that is often taken into consideration is that the metal has sufficient commercial value that its subsequent resale in pure form can recover the costs of the extraction process. When both criteria are met there is a strong impetus to develop extractants that are selective for the individual metals. Radioactive elements are an exception, however, since they often result from weapons production and the responsibility for their removal is a governmental one.

A broadly defined group of elements classified as heavy metals contain a number of members that are toxic. This toxicity may be by virtue of their interaction with enzymes, their tendency to bind strongly with sulfhydryl (-SH) groups on proteins, or other *in vivo* effects.[1] Several of these elements are either transition metal or post-transition metals (cadmium, mercury, lead, copper, zinc, chromium), while others are main group elements (selenium, arsenic, thallium). Inorganic forms of most heavy metals form strong bonds with proteins and other biological tissue, thus increasing bioaccumulation, and their inhibiting excretion. Tissues often exhibit a significant selectivity when binding these metals. Lead, for example, tends to accumulate in bone tissue, while cadmium and mercury predominantly accumulate in the kidneys. The donor groups in amino acid that are most available for *in vivo* binding to metal ions are amino and carboxyl groups, but binding to thiol (sulfhydryl) or thiolate groups is especially strong for many heavy metals. This feature is particularly significant because -SH groups are a common component of the active sites of many important enzymes, including those involved in energy output and oxygen transport. The combination of a widespread possibility for human exposure, coupled with an extremely high toxicity, leads to certain heavy metals being a particular concern with regard to their toxic effects. Among these metals are cadmium, mercury, lead, copper, and chromium in its hexavalent state. Extraction and recovery of these metals from soils and waters is therefore an important process.

A result of decades of industrial production that has involved their use, heavy metal contamination of soils and aquifers is becoming increasingly common at many hazardous waste sites. Furthermore, solid wastes generated through the mining and processing of mineral ores also contain residual metals along with other substances.[2] Improperly disposed mineral waste piles can be subject to severe wind and water erosion, resulting in transport of the contaminated material inherent in the wastes. Moreover, leakage generated from waste piles can pollute nearby groundwater systems and surface streams, thereby rendering valuable water supply sources unsuitable for use. Other activities contributing to heavy metal contamination of soil and aquifers include vehicle emission, smelting, metal plating and finishing, battery production recycling, agricultural and industrial chemical application, and incineration processes. Once released into the soil matrix, these positively charged metals are then strongly retained by the anionic zeolites present. As a result, the adverse impact of these metals both on environmental quality and human health often persists for substantial periods of time. As a result, studies focused on metal ion removal from both aqueous solutions and soils, either for pollution control or for raw material separation and recovery, have become of increasing importance in recent years. Although this is particularly true for metals that have both a high toxicity and a substantial commercial value if they can be recovered in pure form, there are locations where water is of sufficiently high value that metal removal is important if it results in water of sufficient purity that it can be used for agricultural use or human consumption.

## 1.2. Precious Metals

Another area where a strong demand for new methods of metal extraction and separation exists in the commercial sector is in the separation and recovery of precious metals. Precious metals such as copper, silver, gold, palladium and platinum posses similar chemical properties and are of commercial importance in the catalyst, jewelry and electronics industries. Many traditional processes for the extraction and purification of precious metals from their ores are pyrometallurgical.[3] These methods are most suited to the treatment of high-grade ores and concentrates. With the depletion of many high-grade ores, methods for effecting ore concentration have become increasingly important, and there is increased demand to process low-grade materials. There have accordingly been developments in the treatment of ores by both hydrometallurgical and solvent extraction processes. These processes employ aqueous or organic media at essentially ambient temperatures, so as a consequence they are cost effective and relatively non-polluting when compared with pyrometallurgical extraction methods.

In the analytical field, the determination of trace amounts of heavy and precious metals in various materials frequently requires their preliminary concentration and separation from large amounts of other metals. An emerging area therefore for the application of highly selective metal ion extracting agents is in the preconcentration of specific metals. This goal may place fewer demands on a complexant than does the recovery of a single metal because it may be acceptable to obtain a product that contains a mixture of selected metals. The challenge therefore is to find an extractant that is selective for precious metals. Since these metals generally prefer soft donor atom ligands there is a possibility of using such compound for their preconcentration from a solution containing a broader range of metals.

## 1.3 Strategies for Obtaining Metal Ion Recognition

The most commonly used treatment methods for heavy metal-containing waste include precipitation, solvent extraction, activated carbon adsorption, treatment with ion exchange resins, and bioremediation. Other methods that are less widely used include reverse osmosis, electrolysis, cementation, irradiation, zeolite adsorption, evaporation, membrane processes and ion flotation.[4]

Solvent extraction occurs when a metal ion associates with an organic complexant to form a coordination compound that is transferred from the aqueous to the organic phase in a two phase system. While much laboratory work has been carried out with chloroform as the organic phase, low-volatility aliphatic hydrocarbons are preferable for industrial applications because of their lower toxicity. Carbon dioxide is also becoming frequently used for the solid-liquid extraction of heavy metals from contaminated soils and waters. This fluid under supercritical conditions has the potential to become technologically important once its properties are better understood.

Two types of liquid-liquid extractions that are frequently encountered are those:

(1) where both the complexant and metal ion are soluble in the aqueous phase, and the extracted metal complex is soluble in the organic phase.

(2) where the complexant is insoluble in the aqueous phase, then the complexation with the metal ion occurs at the interphase surface, and the metal complex is transferred into the organic phase.

The complexant can be either negatively charged or neutral. The negatively charged complexants can be either organic or simple inorganic ions. The neutral complexants are usually organic compounds, and for both negatively charged and neutral monodentate complexants coordination to the metal ion is *via* a single site. Two groups of organic compounds with multiple binding sites to metal ions that are used for metal extractions are chelate and macrocyclic ligands.[5,6] These compounds are preferable because of their highly favored complexation due to the chelate and macrocyclic effects. The term "chelate effect" refers to the enhanced entropic stability of a complex formed between a metal ion and a ligand that contains multidentate binding sites as compared to an analogous ligand that contains only a single binding site. The term "macrocyclic effect" refers to the greater thermodynamic stability of a complex that has a cyclic polydentate ligand as compared to the analogous complex having an acyclic chelate ligand. The most efficient use of extracting agents requires that the following conditions are met:

(1) good ion uptake by the complexant that allows for fast binding with the metal ion.
(2) high stability against hydrolysis.
(3) selective ion complexation of heavy metals along with no affinity for alkali or alkaline earth ions whose concentrations are usually high in natural waters and soils.
(4) sufficiently high binding strength for the metal ions to be extracted.
(5) reversible complexation allowing for total recovery of the metal without significant ligand or resin destruction.[7]
(6) preference of the formed metal complex for the organic phase rather than the aqueous phase.

From studies of the interaction of transition metals and other heavy metals with chelate and macrocyclic ligands three major strategies emerge for achieving effective complexation:[8]

(1) use of ring size variation to maximize the thermodynamic stability of the complex by matching the radius of the metal ion to the bite angle of the chelate ligand and the hole size of the macrocyclic ligand.
(2) use of donor set variation to alter the affinities of the ligand systems towards particular metal ions.
(3) use of substituent variation to take advantage of the effect of appended substituents to the donor atoms of the parent chelate or macrocycle on metal ion discrimination.

Arland has proposed a classification according to which metal ions can be subdivided into three groups.[9] These are :

(1) elements that form the most stable complexes with the first element of their group in the periodic system, i.e., N, O, F.

(2) elements that form the most stable complexes with atoms of the second or the lower elements of their group (P, S, Cl).
(3) elements that display an intermediate behavior between these two options.

It is also possible that different valence states of a particular metal may behave in a different manner, because the "hardness" or "softness" of a metal ion is a reflection of its polarizability, and hence the formal charge on the metal center.

## REFERENCES

1. S. Manahan, *Toxicological Chemistry*, Lewis Publishers, Inc. 1992.
2. T. E. Clevenger, *Water, Air, and Soil Pollution*, **1990**, *50*, 241.
3. C. F. Bell, *Principles and Applications of Metal Chelation*, Clarendon Press, Oxford, 1977.
4. R. F. Hammen, D.C. Pang, L. S. Van Der Sluys, C. G. Cook, E. Loftsgarrden, EPD Congress 1993, *The Minerals, Metals & Materials Soc.*, J.P. Hager, ed., p. 3.
5. L. F. Lindoy, *Incl. Phenom. Mol. Recogn Plenum*, **1990**, 171.
6. J. J. R. Frausto da Silva, *J. Chem. Educ.*, **1983**, *60*, 390.
7. A. Deratani, B. Sebille, *Anal. Chem.*, **1981**, *53*, 1742.
8. L. F. Lindoy, *The Chemistry of Macrocyclic Ligand Complexes*, Cambridge University Press" 1989.
9. S. Arland, J. Chatt, N. Davies, *Quart. Rev.*, **1958**, *12*, 265.

# 2

# PHASE TRANSFER EXTRACTION AND ADSORPTION METHODS

## 1. INTRODUCTION

This chapter covers the principles involved in effecting the separation or preconcentration of metal ions from aqueous solution. For additional reading on the subject the reader should consult previous books and reviews that have been written on the subject.[1-9] Although there has been a long term interest in metal separation in the mining industry, the recent greater emphasis on environmental concerns has led to a much broader interest in developing new selective methods for lowering the concentrations of metals that are released, and recovering those that have been discharged. In addition, the change in the political climate has resulted in the need for fewer nuclear weapons, thereby placing a greater emphasis on the removal of nuclear wastes that have been generated over the past fifty years of weapons production.

The development of a useful system for the selective extraction of metal ions depends on a number of factors. The first of these is the identification of a complexant that is specific for the metal ion that is targeted. Subsequently it is then necessary to choose a method of using this compound in a manner that results in the effective extraction of the targeted metal ion. In making this choice it is necessary to develop a process that allows for the metal ion to be selectively recovered while producing a minimal volume of secondary waste. Two other preferred criteria are that the extractant has a high capacity for the metal ion, and that it can be recovered for further use. Several techniques are available for using the extractant for selective metal extraction. One of these is liquid-liquid extraction. This technique usually involves an immiscible mixture of water and an organic solvent where the metal ion is dissolved in the aqueous layer and the extractant in the organic layer. Another technique involves the use of liquid membranes. This procedure involves an arrangement involving a liquid membrane separating the source and receiving phases. A requirement of this procedure is that the liquid membrane is immiscible with both of these phases. In addition to liquid-liquid extraction, liquid-solid techniques can also be used. These liquid-solid techniques involve simple adsorption of the free or complexed metal ion onto a solid support. This support can act either as an adsorbent for an uncharged metal complex, or as an ion exchange resin for a cationic or anionic metal ion or complex. Another method for using a liquid-solid interface for the selective adsorption of a metal complex is to employ a porous membrane. For the cases of both a solid phase adsorbent and a porous membrane it is possible to incorporate complexants that are selective for the particular metal ions that are targeted. The advantages and disadvantages of these different techniques and methods will be discussed in this chapter.

A major research and development effort has been made in the area of liquid-liquid extraction for the removal of radioactive metals from nuclear waste materials.[10-12] Many of the considerations involved in choosing a system for this application is also applicable to other metals, but there is the additional factor that for nuclear waste treatment the system must be stable to nuclear irradiation. The most widely used liquid-liquid extraction process for the removal of both uranium and plutonium from the fission products is the PUREX process, which uses tributyl phosphate as the organic phase. In this process the actinide metals are extracted as their tetravalent cations from 2-3M aqueous nitric acid solution. The plutonium is subsequently removed by selectively reducing it to the trivalent state, which then transfers to the aqueous phase. A consequence of this process is that the residues from its application contain high concentrations of nitrate ion. Also, because the tanks used for its storage have metal liners, it has been necessary to neutralize the nitric acid with base before it can be stored. Any subsequent removal of material from these storage tanks must therefore involve the use of chemical processes that can tolerate high concentrations of nitrate salts. This can cause a significant problem because nitrate can act as an oxygen-donor ligand to actinide ions. In order to overcome this it is necessary therefore for an extractant to be used that is either present in very high concentration or is highly selective for the targeted actinide ions. Frequently the actinide complex that is extracted from such solutions has both nitrate and complexant bound within the coordination sphere, with the nitrate being either monodentate or bidentate.

# 2. LIQUID-LIQUID SYSTEMS

## 2.1. Equilibria

For liquid-liquid systems the most important consideration is the concentration of the metal ion in the two phases. Generally, one of the phases is an aqueous one, and the other is an organic solvent that is immiscible with water. The distribution ratio is given terms of metal concentration M in equation 1. Results obtained from extraction studies can be reported as distribution ratios or as

$$D = [M]_{org} / [M]_{aq} \quad (1)$$

percentages. For simplicity, the volume used for the phase ratio of the two liquids is usually set at 1:1. The metal species in the two layers depends on the type of complexant, the acidity of the aqueous layer, and the charge on the metal ion or complexant. What one is targeting for an efficient extractant is a situation where the hydrated metal ion in aqueous solution after its complexation becomes preferentially solvated by the organic layer, thereby becoming distributed into that phase. The complexation can occur at the phase boundary, or it can occur in the aqueous phase if the complexant has sufficient aqueous solubility that it is present in that phase. One group of complexants that satisfy these criteria are protonated chelates ($H_2L$) that have sufficient lipophilicity that they have some solubility in both the aqueous and organic phases. For a divalent ion $M^{2+}$, the system can be described by the equilibria shown in equations 2-4. The distribution ratio of such a system will therefore be given by equation 5, assuming only organic phase complexation. Aggregation can also occur leading to the formation of other charge-neutral complexes that can extract into the organic phase.

$$H_2L_{org} = H_2L_{aq} \quad (2)$$

$$H_2L_{aq} = H^+_{aq} + HL^-_{aq} \quad (3)$$

$$M^{2+}_{aq} + H_2L_{org} = ML_{org} + 2H^{+}_{aq} \quad (4)$$

$$D = [\ ML_{org}\ ] / [\ M^{2+}_{aq}] \quad (5)$$

Extraction of uncharged hydrophilic metal complexes can be achieved by use of an organic reagent (S) that solvates the complex, and converts it into a lipophilic aggregate that migrates into the organic phase. For a counterion $X^-$, the equilibria for a divalent metal ion are given by the equilibria shown in equations 6 and 7. If other complexes between $M^{2+}$, $X^-$ and S are formed, then these must also be included in the expression for the distribution equilibrium.

$$M^{2+}_{aq} + 2X^{-}_{aq} + S_n = [MX_2S_n\ ]_{org} \quad (6)$$

$$D = [MX_2S_n\ ]_{org} / M^{2+}_{aq} \quad (7)$$

## 2.2. Surfactants

Surfactants can also be used as extractants for transferring metal ions from an aqueous to an organic phase. These compounds are highly aggregated in the organic phase to form reverse micelles. The metal ions then become encapsulated in the hydrophilic inner region of the micelle. Since the aggregation of surfactants into micelles is a dynamic process, the cavity size can change to accommodate metal ions of different sizes. As a result, however, selectivity can be difficult to achieve. Since the outer part of the reverse micelle has bound protons or cations, ionization of these cations after metal ion complexation into the inner part of the micelle can lead to charge neutralization, and subsequent phase transfer of the metal ion micellar host into the organic phase.

## 2.3. Ion Hydration

Since the extraction of a metal cation from an aqueous phase to an organic phase is expected to result in a disruption of the hydration spheres about the aqueous phase metal ion, ion hydration is an important aspect of extraction from an aqueous into an organic phase. The three-dimensional hydrogen structure of water results in dissolved ions ordering the water structure over a large distance. As a consequence, ions have distinct first and second hydration spheres, and frequently ion-induced ordering of the water structure at even greater distances into the water lattice. Because of such effects, the presence of either hydrogen or hydroxide ions must be taken into consideration. Water-miscible liquids such as alcohols frequently reduce the hydration energies of the dissolved ions because of their involvement with the water hydrogen bonding. Cations having a high charge density induce order in the solution, while ones of low charge density have the opposite effect. For anions, the flouride ion induces order in aqueous solution. The halides that are lower in the periodic table, along with pseudohalide ions, have the opposite effect. Ions such as nitrate, phosphate and sulfate show an intermediate effect. The perchlorate ion reduces order in solution, thereby resulting in a net favorable entropy effect and an increased extraction coefficient. The hydration energies of a group of selected cations and anions are given in Tables 1, 2 and 3.

**Table 1. Enthalpies and Free Energies of Hydration of Selected Cations**

| Cation | $\Delta H_{hyd}$ (Kcal mol$^{-1}$) | $\Delta G_{hyd}$ (Kcal mol$^{-1}$) |
|---|---|---|
| $H^+$ | -267.9 | -259.2 |
| $Li^+$ | -130.3 | -120.8 |
| $Na^+$ | -104.1 | -97.0 |
| $K^+$ | -84.0 | -79.3 |
| $Cs^+$ | -70.1 | -66.5 |
| $Cu^{2+}$ | -516.4 | -494.9 |
| $Fe^{2+}$ | -473.4 | -450.5 |
| $Fe^{3+}$ | -1038.0 | -1002.0 |
| $Cr^{3+}$ | -1044.0 | -1006.0 |

| $La^{3+}$ | -806.3 | -777.5 |
|---|---|---|

From K. B. Harvey, G. B. Porter, *Introduction to Physical Inorganic Chemistry*, Addison-Wesley, Reading, MA, 1963, p. 326.

**Table 2. Enthalpies and Free Energies of Hydration of Selected Anions**

| Anion | $\Delta H_{hyd}$ (Kcal mol$^{-1}$) | $\Delta G_{hyd}$ (Kcal mol$^{-1}$) |
|---|---|---|
| $F^-$ | -113.2 | -103.1 |
| $Cl^-$ | -80.9 | -74.8 |
| $Br^-$ | -72.7 | -67.9 |
| $I^-$ | -62.4 | -59 |

From K. B. Harvey, G. B. Porter, *Introduction to Physical Inorganic Chemistry*, Addison-Wesley, Reading, MA, 1963, p. 326.

**Table 3. Free Energies (Kcal mol$^{-1}$) of Hydration of Selected Ions**

| Ion | $\Delta G_{hyd}$ | Ion | $\Delta G_{hyd}$ |
|---|---|---|---|
| $Ag^+$ | -102.8 | $OH^-$ | -102.8 |
| $NH_4^+$ | -68.1 | $CH_3CO_2^-$ | -87.2 |
| $Tl^+$ | -71.7 | $NO_3^-$ | -71.7 |
| $Pd^{2+}$ | -456.5 | $CN^-$ | -70.5 |
| $Cd^{2+}$ | -419.5 | $NO_2^-$ | -78.9 |
| $Hg^{2+}$ | -420.6 | $C1O_3^-$ | -66.9 |
| $Pb^{2+}$ | -340.6 | $H_2PO_4^-$ | -111.1 |
| $Ga^{3+}$ | 1079.1 | $SCN^-$ | -66.9 |

| $Au^{3+}$ | -1056.4 | $BF_4^-$ | -45.4 |
|---|---|---|---|
| $Eu^{3+}$ | -803.1 | $C1O_4^-$ | -102.8 |
| $Pu^{3+}$ | -773.2 | $ReO_4^-$ | -78.9 |
| $U^{3+}$ | -766.0 | $CO_3^{2-}$ | -314.3 |
| $Pu^{4+}$ | -1462.7 | $SO_3^{2-}$ | -309.5 |
| $U^{4+}$ | 1567.9 | $SO_4^{2-}$ | -258.1 |
| | | $CrO_4^{2-}$ | -227.1 |
| | | $PO_4^{3-}$ | 660.9 |

From Y. Marcus, *JCS, Farad. Trans.* **1991**, *87*, 2995.

Partial substitution of non-aqueous solvents for water lead to a decrease in the hydration of the cation, which subsequently reduces the energy required for its desolvation. As a consequence, in a liquid-liquid separation procedure it is advantageous to have an organic phase that has some solubility in water. This effect can be transferred to ion exchange separations where a solution of 20% ethanol in water is used rather than pure water itself.[13]

## 2.4. Aqueous Biphasic Systems

Aqueous biphasic systems have also been used for the extraction and separation of metal ions. These systems are formed when certain water-soluble polymers are combined either together, or with inorganic salts, over particular concentration ranges. A widely used system comprises of water and polyethylene glycol. In some cases the metal ion can be extracted into the polyethylene glycol-rich phase in the absence of a complexant, although in other cases a water-soluble organic complexant needs to be added. Advantages of an aqueous biphasic polyethylene glycol system are that they are inexpensive, non-flammable, and essentially non-toxic. With the aqueous polyethylene glycol system two immiscible aqueous phases can be obtained by the addition of salts to the solution. A wide range of salts have been used to obtain the phase-separated system. The more negative the free energy of hydration of the ions in the salt, the greater is its effect in causing this phase separation. The effect is additive for the cation and anion in the salt. A disadvantage of an aqueous biphasic polyethylene glycol based system is the difficulty of recovery of the organic-rich phase. In systems that use only an aqueous and an organic phase, the metal ions that are extracted into the organic phase can be readily recovered by extraction of the metal ion back into an aqueous phase. For aqueous polyethylene glycol this is more difficult to achieve because the phase separation step has been achieved by the addition of salts, and these salts cannot be readily removed and the polyethylene glycol subsequently recycled.

## 2.5. Liquid Membranes

Liquid membranes have also been used for the separation of metal ions. Liquids that are immiscible with both the source phase and the receiving phase are potentially useful as liquid membranes. High transport rates are obtained with such systems because diffusion is rapid between liquid phases. By adding complexants, high selectivity for the ions $M_A^+$ over $M_B^+$ can be achieved (Figure 1). Furthermore, since the liquid membrane

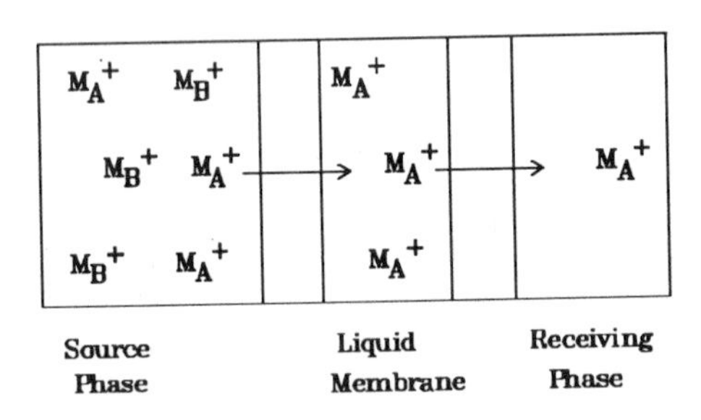


**Figure 1. Schematic Diagram for a Liquid Membrane**

occupies a small volume in the system, only small amounts of the complexant are required to achieve selective transport. These systems can be considered to be analogs of the biological membranes that provide a barrier between an intracellular and an extracellular aqueous environment.

Different configurations can be employed to achieve a liquid membrane system. Bulk liquid membranes comprise of an aqueous source phase and receiving phase separated by a water-immiscible organic liquid in a U-tube. An emulsion liquid membrane comprises of a dispersion of water-containing oil drops stirred in an aqueous source phase. Separation is rapid because of the high surface area of the organic layer. The source phase is the bulk liquid, and the receiving phase is the water trapped within the oil drop. A supported liquid membrane comprises an organic carrier phase immobilized in a porous polymer support. The support can be in the form of a flat sheet or a hollow fiber, with the latter being preferable because of its high surface area.

The majority of these membrane systems involve stirring or continuous flow of the source and receiving solutions, and in some cases of the liquid membrane. This action minimizes the time for diffusion of dissolved species through the membrane. In most cases diffusion, rather than complexation or decomplexation of the metal ion in the membrane phase, is the rate limiting step in the transport process. The Stokes-Einstein relation (equation 8) defines the diffusion coefficient, D as:

$$D = kT/6\pi\eta r \qquad (8)$$

where k is the mass transfer coefficient, T is the temperature, $\eta$ the solvent viscosity, and r is the molecular radius. For ions, other factors such as ionic strength and ionic charge must also be accounted for.

With respect to the solvent properties of the liquid membrane, the dielectric constant is an important factor since in some systems an increasing dielectric constant results in decreased transport. Other factors that have been considered are solvent donor number, viscosity, and dipole moment. Anion effects must also not be neglected when treating the simple transfer of metal ions between an aqueous and an organic phase.

A common type of emulsion liquid membrane has a water-oil-water interface system (Figure 2). These membranes are usually prepared by first

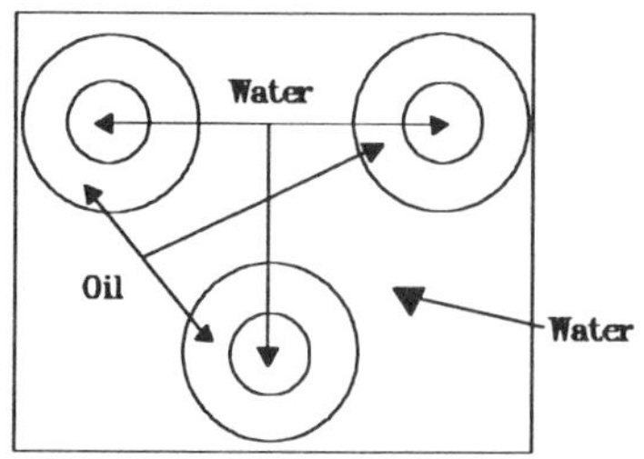


**Figure 2. Water-Oil-Emulsion Liquid Membrane**

forming an emulsion between two immiscible phases, and then dispersing the emulsion in a third phase by agitation for extraction purposes. The membrane phase is the oil phase at the boundaries. For metal salts, the metal ion migrates through the membrane layer into the aqueous phase within the emulsion drop, with protons migrating out of the drop to maintain charge balance (Figure 3). The anion remains in the bulk aqueous phase, and the

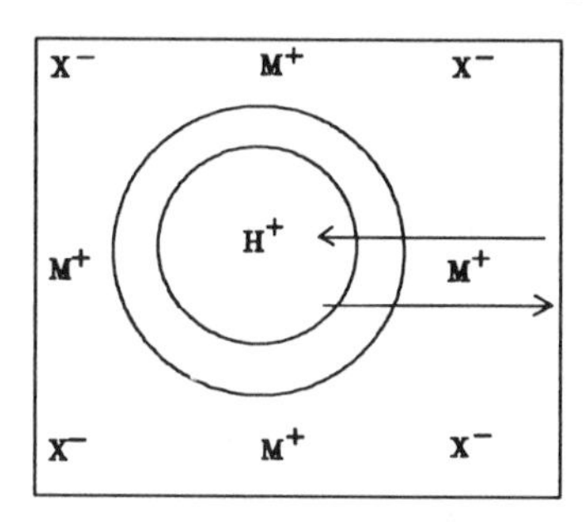


**Figure 3. Migration through a Membrane**

carrier is localized within the membrane (oil) layer. This system has been modeled to include diffusion of the carrier and its metal complex inside the emulsion, along with the reversible reactions at the external and internal interfaces. Mass transfer resistance has also been included in the treatment.[14,15] Alternatively the system can be modeled from the complexation-decomplexation rate constants and the equilibrium constants for the extraction and stripping across the membrane.[16] One of the problems associated with emulsion liquid membranes is maintaining the consistency of the emulsion. Swelling can occur when the feed stream enters the internal phase either by osmosis or the breakage and subsequent repair of the membrane wall. Swelling results in a lowering of the carrier reagent concentration, and a subsequent lowering in its stripping efficiency. Microemulsions have been considered as alternatives. For regular emulsions the drop size diameter is commonly 1μm, whereas for microemulsions the size is usually in the 0.005 to 0.1μm range. As a consequence, the internal phase in the microemulsion has a greater area per unit volume, and consequently a higher extraction rate into the receiving phase. The higher thermodynamic stability of microemulsions can be expected to lead to less breakage of the drops. Generally, microemulsions show extraction rates that are some ten-fold higher than regular emulsions. However, swelling remains a problem, and product recovery is more difficult with microemulsions.

Supported liquid membranes involving hollow fiber contactors are used, and they have certain advantages in metal ion separations. These membranes are obtained by impregnating the pore of a porous polymer with a carrier dissolved in a solvent. The impregnated membrane is then used as the interface between the source and receiving phases which are on each side of the membrane. In the extraction of metal ions, complexation occurs on the source side of the membrane, followed by decomplexation on the receiving side. The process may either involve co-transport of the metal cation and a counter-anion, or it may involve counter-transport where the metal cation that is removed from the source phase is replaced by a proton. In supported liquid membranes the polymeric support needs to have high porosity, small pore size, and good hydrophobicity. Among the materials that have been used are polypropylene, teflon, cellulose acetate, and silicones. Carrier loss can be a problem with these membranes, especially if the transport rate is high enough to carry it into the receiving phase. When choosing a polymeric support it is useful to have one that has the correct pore size to hold the carrier in place by capillary forces. It is also advantageous to use a polymeric support material that has good wetting properties with the liquid membrane phase. A contact angle close to zero is preferable. The operation of these supported liquid membranes can be modeled by transport equations involving diffusion and permeability.

## 2.6. Picrate Method of Analysis

Two recent papers have commented on the picrate method for analyzing for metal concentrations in liquid-liquid extractions. The method involves adding picrates, usually those of the alkali metals, in order that the metal ion is extracted into the organic layer as its picrate salt. One of these articles cautions against the use of free picric acid, especially with extractants such as amines, that can themselves be protonated and extracted into an organic solvent as the alkylammonium picrate salt.[17] A second paper investigates the association of picrate anions in solution. Molecular dynamics on the picrate-picrate anion pair shows that its behavior is strongly solvent dependent. In water, the intimately stacked pair is stable, while in acetonitrile it dissociates. Picrate anion is also a versatile chelating agent to metal ions, and there is evidence that more than one complexed form may be present in apolar solvents.[18] Since the change in the electronic spectrum of the picrate anion when a metal cation is extracted by a complexant is used to characterize the formation of a complex,[19] the stacking of picrate anions may contribute to the spectral shifts, and result in incorrect assays because the metal ion concentration in the solution is inferred from the intensity of the picrate absorption band.

Another factor that must be recognized when using picrate ion to estimate metal concentrations is the high affinity of picrate anions for the water/chloroform interface.[20] This behavior is similar to that observed with anionic surfactants. As a result, the picrate salt of the metal may be extracted better than the salt of another anion such as the chloride or nitrate. Since the latter anions are more likely to be the ones that are extracted from environmental sites, the use of the picrate analysis method my lead to a false estimate of the efficiency of a complexant as an extractant into the chosen solvent system.

## 3. ADSORPTION METHODS

Metal ions can be extracted and separated by adsorption from the liquid phase onto a solid support. These support materials can be ion exchange materials where attachment occurs by cation-anion attraction. Charge density is an important factor because the ions are subsequently separated by chromatography. Alternatively these support materials can be uncharged solids which extract metal ions by physical adsorption. Adsorbents do not have specific functional groups that bind the metal ions but instead involve surface-surface interactions such as van der Waals forces for ion removal from solution. The key parameters involved for adsorbents are pore size, pore volume, surface area, and particle size. Other approaches that have been used are polymer filtration and adsorption onto biological materials.

### 3.1. Ion Exchange Methods

Ion exchange resins are polymeric beads with particle sizes ranging from 0.3 to 1.2 mm that reversibly exchange ions from solutions. These resins typically have one of four functionalities: strong acid cations (SAC), weak acid cations (WAC), strong base anions (SBA), or weak base anions (WBA). The SAC exchange resins act as a sulfonic acid or its salts. They function well over a wide pH range, but typically need 200-300%

(stoichiometrically) of regenerant for effective regeneration. The WAC exchange resins act as a carboxylic acid or its salt. They have significantly higher capacity and are more easily regenerated (100% stoichiometry) than the SAC resins. The WAC resins, however, are only useful for solution that have a pH that is greater than 6. The WBA exchange resins act as tertiary amines with similar operating characteristics to WAC resins (higher capacity and better regeneration stoichiometry than SBA resins, but narrower pH ranges for operation). The SBA resins act as quaternary amines and have similar operating characteristics to SAC resins. The two main configurations of ion exchange resins are gellular and macroreticular. Gellular resins make use of micropores formed within the gel structure, and they are useful for ions of small size. Macroreticular resins are micropheres of resin assembled so as to produce mesopores and macropores between the resin spheres, and they work well with ions of large size.

Chelating resins are a type of macroreticular resin that is commonly used. New ligands continue to be immobilized onto polymer beads for specific applications, with post-functionalization being an effective technique for introducing ion-selective ligands. Nitrogen, oxygen and phosphorous functionalities continue to be studied for the preparations of new resins, and bifunctional resins are a useful way to combine ion selectivity with rapid complexation rates.[21]

Ion exchange materials are available for both cation and anion exchange, and both types are used for metal extraction and separation. Cation exchangers are useful for simple hydrated metal ions, and anion exchangers can be employed for negatively charged oxyanions. Conventional ion exchangers use sulfonated resins for cations, and alkylammonium substituted resins for anions. The principal application for cation exchange resins is in the softening of household and industrial water supplies by the removal of calcium ion. Another application is in the extraction and separation of lanthanide and actinide ions. The removal and separation of actinide ions presents a challenge to ion exchange methods since the species may be present in highly acidic or basic conditions, because there are often high concentrations of other salts present in the solution, and because the solutions contain highly radioactive elements. The development of new ion exchange resins for the separation of actinides and products of nuclear fission continues to be a research area of high significance.

Another important class of ion exchange materials are the Diphonix resins.[22] These materials have geminally substituted diphosphoric acid groups chemically bound to a polymer support. A widely used type contains these diphosphoric acid groups bonded to a sulfonated styrene-divinylbenzene matrix. Modifications include the addition of anion exchange groups such as *N*-methylpyridines. Another modification uses phenolic groups that after deprotonation can also act as cation exchange sites, but over a different pH range than the diphosphonate functionalities. Such bifunctional Diphonix resins are also frequently chosen because they couple an access mechanism that allows all ions to rapidly enter the resin structure with a recognition mechanism where the second ion type selectively complexes with the metal ion that is targeted for removal from the solution. The sulfonic acid groups provide the access function, and the diphosphoric acid groups the recognition function. An advantage of these Diphonix resins is that they can rapidly bind a broad range of metal ions from highly acidic source phase solutions. A further advantage of these resins is that they strongly bind iron(III), even from highly acidic solutions. The ability to bind iron(III) is important because it is a common component in nuclear waste materials. Other polymer supports have been used in the development of ion exchange resins. Among these are silica, cellulose, and a condensation polymer of the Bakelite type formed from phenol and formaldehyde. To be useful these materials need to be stable over a wide pH range, hydrolytically and thermally stable, available in different pore sizes, and have structures that can be readily functionalized for the incorporation of either cationic or anionic groups. Bifunctional resins involve intraligand cooperations. The access ligands allows ions to enter the matrix very rapidly where only certain ions are complexed. The second ligand leads to the high selectivity. Thus bifunctional resins combine metal ion selectivity and rapid complexation kinetics.[23] These resins can overcome the generally low metal ion selectivity that is observed for conventional ion exchange resins.

Inorganic materials have also been used as ion exchangers. The ones that are of particular interest are those having tunnel structures where the cavity size matches that of the metal ion that needs to be encapsulated. A requisite for these compounds is that the lattice is not disrupted by the presence of both acids or bases, which therefore eliminates the use of zeolites or phosphates for these ion exchange applications. One group of compounds with tunnel structures that have been used as cation exchangers are the silicates and germanates of titanium and zirconium, and the principles that have been developed to understand the function of these materials can be applied to other ion-exchangers of similar type. The silicates are exemplified by the compound $Na_2Ti_2O_3(SiO_4).2H_2O$, a material that is highly selective for cesium(I) and strontium(II) in the presence of large amounts of sodium(I). The affinity of the exchanger for alkali metal cations depends on the number and strength of the bonds formed by the framework oxygens and water molecules in the tunnels in the structure. Unlike the organic sulfonate ion exchangers, this material is quite rigid and symmetrical; therefore selectivity results from the metal ions having the correct size and coordination properties to match the hole sizes of the cavities. In some cases multiple sites occur within these ion exchangers, and selectivity for more than one cation may be observed.[24]

Natural zeolites, such as clinoptilolites, SIR-600, and manganese greensand, can be used for the selective removal of metal cations. Clinoptilolite is very selective for cesium(I), and manganese greensand can be used as a redox medium for the removal of iron and manganese. As with all zeolites, dissolution in aqueous solution occurs at very high or low pH.[25]

## 3.2. Adsorbent and Membrane Methods

Physical adsorption techniques for the removal of metal from aqueous solution can involve the use of either packed columns or membranes. Advantages of packed columns are that the loaded bed(s) can be easily regenerated, and the adsorbent is not susceptible to fouling or attack by the feed components. Technologies such as the adsorbent wheel can be used in conjunction with solid phase materials to make the system efficient as a metal concentrator. Advantages of membrane processes are that the membrane can be made resistant to fouling by the system components, and that they can be made energy efficient.

In general an adsorption process is a cyclic process in which a fluid is fractionated by selectively adsorbing one or more compounds from the fluid onto the adsorbent and then desorbing the adsorbed compounds, thereby restoring the adsorbent to its initial state. Several methods are available for carrying out the regeneration step. One method involves temperature-swing adsorption (TSA) whereby reduced bonding tenacity between the adsorbent and adsorbate occurs upon raising the temperature, resulting in desorption of the adsorbed metal ion or complex. Desorption can also be induced by displacement resulting from the addition of a second reagent. Under some conditions the adsorption is very strong, and restoration of the adsorbent and liberation of the metal may require removal and either chemical or thermal treatment of the adsorbent. Such a situation should be avoided. Selectivity of the adsorbent can be achieved by making use of various forces such as

Van der Waals interactions, hydrogen bonding, electrostatic forces, covalent bonding, and $\pi$-bonding. Selectivity can also be enhanced by the use of size separation methods with materials such as molecular sieves or by making use of differences in intra-particle diffusion rates. A challenge in the use of such systems is heat transfer, both in the removal of the heat of adsorption from the adsorbent, and in the supply of the heat for desorption of the adsorbent.

Among the operational advantages of membranes are that they are versatile, easy to operate, modular, operable over wide temperature and pressure ranges, chemically stable and compatible, and low cost. Functionalized microfiltration membranes can be used to remove heavy metals from water, examples being the supported liquid membranes.

## 3.3. Polymer Filtration Methods

Polymer Filtration™ is a process for the removal of metals from solution that involves polymeric binding followed by ultrafiltration. The process uses water-soluble metal-binding polymers with different types of complexants that are chosen to achieve the high selectivity and binding constants for the metals of interest. For solutions containing metal ions in concentrations that are $\leq$ 1,000 ppm, concentration factors ranging from 20 to several thousand have been achieved.[26] Polymer Filtration combines two operations into one hybrid process. In the first step, pH adjustments are made and water-soluble polymers are added to the solution containing the metal ions. The polymer concentration is typically 0.01 to 2% weight by volume. During the mixing step the metal ions bind to the polymer. The solution is then filtered through an ultrafiltration membrane. The metal-impregnated polymer is retained while the water and the free metal ions pass through the membrane. The polymer can be regenerated by lowering the pH of the retained material and recovering the metal ions by flushing with water. There are several advantages of the Polymer Filtration™ process. One advantage is that the metal ion binding occurs in a homogeneous solution, and high capacities and selectivities are attainable. Another is that no mechanical stability requirements for the polymers are required, and no organic solvents are used. Other advantages are that no fine particulates or colloids are formed, ambient temperatures and only slightly elevated pressures are used, and the treated water meets the legal limits for metals in the discharged water. The process can be used for the removal of actinides, precious metals, cations, anions, and metals from photofinishing and electroplating operations.

## 3.4. Uncharged Solid Phase Adsorbent Materials

Many adsorption materials have been considered for a wide range of applications. Adsorbents do not have specific functional groups that bind the metal ions but instead involve surface-surface interactions such as van der Waals forces for ion removal from solution. The key parameters involved for adsorbents are pore size, pore volume, surface area, and particle size. Among the adsorbents that have been used to remove metal ions from soils and waters are metal surfaces, metal oxides and non-metallic adsorbents such as carbon and silicon. The adsorption can be of the physical type, or direct chemical bonds may be formed between the adsorbent and adsorbate.

### 3.4.1. Metals

Although the use of metal surfaces for the removal of metals from the environment is not a common procedure, in certain cases it is a method that can be used to advantage. A case in point is the removal of elemental mercury. Although there are many methods available for the removal of mercury salts from water, there are very few options for removing elemental mercury itself. This is a major limitation since it is becoming increasingly apparent that airborne mercury is resulting in this metal becoming widely dispersed, even into areas where it is not being locally released. Among the metals that have been used to adsorb mercury are tungsten,[27] nickel,[28] gold,[29] and silver.[30] The processes that are involved in the adsorption of mercury involve amalgamation, and in some cases the formation of mercury-mercury bonds between the mercury atoms on the surface. Because of its relatively low boiling point, elemental lead can also be removed by vaporization and collect ion on a metal surface. Among the metals that have been used as adsorbents for lead are copper,[31-33] platinum,[34,35] gold,[36] and tungsten.[37,38] Dense island formation and surface alloying is observed. When a copper surface is used as adsorbent a structure having the composition $Pb_3Cu_4$ is observed. When a platinum surface is used, structures having both Pt-Pb and Pb-Pb bonds are observed. A problem with the use of metal surfaces as adsorbents, however, is that it is difficult to subsequently remove the metal. This difficulty is a result of the adsorbed metal being present as a bimetallic phase compounds usually have a high stability, and are therefore resistant to being separated back to the individual metals.

### 3.4.2. Carbon and Silicon

Carbon, an element that is available in several allotropic forms, has been used as an adsorbent for metals. An early review compares the use of activated carbon with other methods that have been used for the removal of mercury and lead.[39]

At ambient temperature mercury can be adsorbed onto carbon particles, and then desorbed at elevated temperature, thus showing that the mercury is either chemisorbed onto the particles or is contained in the form of mercury compounds.[40] Subsequent studies on activated carbon, and carbon impregnated with sulfur, confirm this chemisorption process. When these absorbents are used at higher temperature the mercury is preferentially adsorbed at the sulfur sites.[41-45] Copper(II), nickel(II) and lead(II) have been adsorbed from solution onto cloth materials containing activated carbon. The adsorption is believed to involve both adsorption onto the carbon particles and replacement of the protons on carboxylic acid groups on the cloth by metal ions.

Carbon materials have also been used for the electrochemical removal of heavy metals ions from an aqueous environment. Both electroplating and electrosorption have been carried out by using a packed bed of the carbon material in a flow-through electrochemical cell. The packed bed of conducting carbon is either in the form of carbon fibers or carbon foams. The large surface area minimizes the length of electrode material through which the solution must pass, with this decrease distance leading to reduced resistance to solution flow. Removal efficiencies above 90% are observed for cadmium, lead copper and nickel, as the levels are reduced from -1.0 V to -1.4 V, and times of up tp 72 hours were used. More negative potentials were not used in order to minimize the electrolysis of water.[46]

Silicon surfaces have also been used for the adsorption of mercury and lead. Mercury is strongly chemisorbed onto a silicon surface, likely into subsurface sites. Even after flashing the silicon crystal to 1250°C, some mercury atoms still remain on the silicon surface.[47] Surface studies on the deposition of lead on silicon verify strong interactions leading to phase transitions.[48] These lead atoms are highly mobile on the silicon surface, even at ambient temperature.[49]

### 3.4.3. Metal Oxides

A wide range of metal oxides and oxyanions have been used as adsorbents for metal ions. In some cases these compounds have been used as ion exchange resins, and in other cases as simple adsorbents. A review has been published of the kinetics and mechanisms of metal adsorption at the mineral-water interface which are critical in determining the mobility, speciation, and bioavailability of metal ions in aqueous and terrestrial environments. This review covers the non-equilibrium aspects of the metal adsorption at these interfaces, with an emphasis on molecular approaches and slow adsorption mechanisms.[50] Among the oxides and hydroxides that are widely used as adsorbents for metal ions are those of iron, silicon, aluminum and titanium. Hydrous ferric oxide adsorbs copper(II) and lead(II) from aqueous solution. The adsorption is greater from solutions at higher pH. The high capacity of the ferric oxide for these metals is suggestive of intercalation into the open permeable structure.[51] This postulate is supported when the reversibility of the process is considered. Although lowering the pH leads to desorption of the adsorbed metals, a fraction of the adsorbate is not easily desorbed, and this fraction increases when the metal ion is adsorbed from higher pH solutions that have longer contact times with the adsorbent.[52] Goethite, an iron oxide mineral comprised of $\alpha$-FeOOH, adsorbs divalent metal ions such as lead(II). The adsorption of lead(II) on this mineral is pH dependent, and maximizes at pH 7.5. X-ray photoelectron spectroscopy suggests the presence of $FeOPb^{+}$ complexes in the structure, and the presence of polynuclear complexes at high lead(II) concentrations.[53] Iron(III) hydroxides have also been used in conjunction with other metal hydroxides as adsorbents for metal ions. Examples are the use of a mixture of iron(II) hydroxide and chromium(III) hydroxide for the removal of lead(II),[54] and a mixture of iron(III) hydroxide and aluminum(III) hydroxide for heavy metal ions. This latter system has even increased adsorption capacity if a stoichiometric equivalent quantity of EDTA is added. At concentrations above stoichiometric, however, the adsorption capacity is decreased.[55] Silica in the form of quartz can be used as an adsorbent for metal ions. The adsorption of lead(II) follows a variable surface charge-variable surface potential model which considers the non-Nernstian dependence of surface potential on pH by accounting for the effect of absorbed metal cations on the Gouy potential at the plane of adsorption.[56] Titanium oxides have been used as adsorbents, especially for uranium(VI). This affinity is due to the formation of strong TiO-$UO_2$ bonds.[57] A homogeneously coprecipitated gel of silicon dioxide and titanium dioxide has been used for the extraction of uranium from seawater. More than 80% of the uranium can be recovered from the adsorbent by elution with ammonium carbonate.[58] Aluminum(III) oxide can also be used for the adsorption and desorption of uranium(VI). A kinetic study attributes the process to the interaction of $(UO_2)_3(OH)^{5+}$ with surface hydroxyl groups.[59] A later study of this system reveals the presence of a uranium(VI) complex that forms with both a single surface hydroxyl group and with two surface hydroxyls in the form of a bridged or bidentate complex. The adsorption of uranium(VI) is independent of the electrolyte concentration.[60]

Mercury vapor is adsorbed on silver- and copper-doped manganese (IV) dioxide, but not on aluminum(III) oxide. From the observed color changes it is proposed that oxidation of the mercury(0) occurs.[61]

## 3.5. Silicates and Zeolites

Silicates have also been used as extractants and adsorbents for metal ions, as well as acting as ion exchange materials. Glass,[62,63] and functionalized silicates are effective extractants for the removal of metal ions from solution. Chemically modified silicates are also effective adsorbents for metals. One approach is to introduce thiol or amine functionalities onto silicas or clays. The thiol derivatives are excellent adsorbents for mercury(II) and lead(II), and the amine derivatives are strong adsorbents for copper(II), nickel(II) and chromium(III). Zinc(II) is adsorbed by each of these derivatized materials.[64,65] Heavy metal ions can also be extracted using a silica-polyamine composite. This material is especially useful for copper(II), cobalt(II), and nickel(II), and it maintains its structural integrity after 3,000 cycles when used with copper(II) solutions at ambient temperature.[66] Zeolites can also be used for the extraction of metal ions from solution. Natural zeolites, such as clinoptilolites, SIR-600, and manganese greensand, can be used for the selective removal of metal cations. Clinoptilolites is very selective for cesium(I), and manganese greensand can be used as a redox medium for the removal of iron and manganese. As with all zeolites, dissolution in aqueous solution occurs at very high or low pH.[67] Among others that have been used are soils[68,69] and clay materials such as kaolinite,[70,71] montmorillonite,[70] clinoptilite,[72] biotite,[73] and hematite.[74,75] These materials have been used for the adsorption of uranium(VI), mercury(II), lead(II), and elemental mercury.

The effect of the siderophore desferrioxamine B on the adsorption of heavy metals on clays has been studied. Since siderophores are known to chelate divalent heavy metal ions, they have the potential to affect their adsorption and mobility in soils. In the absence of heavy metals, this particular siderophore is adsorbed onto montmorillonite much stronger than it is onto kaolinite. In montmorillonite suspensions, heavy metal adsorption is enhanced by the presence of this siderophore due to electrostatic interactions. In kaolinite, the addition of the siderophore diminishes the adsorption of heavy metals. The observed adsorption sequence is copper(II) > zinc(II) > cadmium(II), which correlates with their stability constants with the siderophore.[76]

A range of other miscellaneous materials have been used as extractants and adsorbents for metal ions. Among them are porous red pozzolan,[77] polyethyleneimine-modified wool fibers,[78] mercaptoethylated synthetic fibers,[79] cellulose,[80] fly ash,[81,82] aminoalkyl functionalized celluloses,[83,84] magnetic polyvinylbutyral-based microbeads containing Cibacron Blue,[85] starch,[86] and steamed hoof powder.[87]

## 3.6. Polymers and Biopolymers

Various polymeric adsorbents have been used for the extraction of metal ions from solution. The most effective ones are macroporous resins that are sufficiently hydrophilic that their surfaces are wetted by water. Among these are polymers having amidoxime,[88-90] acrylamide,[91] hydroxyapatite,[92] and diamine[93] functionalities appended. Among the metals that have been extracted with these materials are uranium(VI), copper(II), and lead(II).

Chitosan, a natural polysaccharide, has been used as a platform onto which functional groups can be appended to give metal ion selective extractants. Both chitosan itself and its functionalized form have been used for metal extraction. Chitosan has been used as an adsorbent for mercury(II), where intraparticle diffusion is the rate-limiting step. The data correlate well with the Langmuir isotherm equation, although at high solute concentrations a multilayer type of adsorption occurs.[94] Of the derivatized chitosans, functionalization with *N*-(2-pyridylmethyl) groups leads to materials that show selective adsorption of palladium(II), platinum(IV), and mercury(II),[95] or with polyamine groups for obtaining the selectivity order mercury(II) > uranium(VI) > cadmium(II) > zinc(II) > copper(II) > nickel(II),[96] and with dibenzo-crown functionalities for achieving high selectivities for lead(II) and copper(II).[97] Similarly, the preparation of a water insoluble sulfhydryl derivatized chitin can be used as an extractant for lead(II), copper(II), and cadmium(II). The performance of this material is much improved over the non-functionalized chitin.[98] Methylthio functionalities can also be appended to lignin to produce a material that has a lower aqueous solubility than lignin itself, and is a good extractant for lead(II), mercury(II), cadmium(II), chromium(III), and iron(III).[99]

# REFERENCES


1. *Principles and Practices of Solvent Extractions*, J. Rydberg, C. Musikas, G. Choppin, eds., Marcel Dekker, New York, N.Y., 1992.
2. *Solvent Extraction and Ion Exchange in the Nuclear Fuel Cycle*, D. H. Logsdail, A. L. Mills, eds., Ellis Horwood, Chichester, UK, 1985.
3. *Chemical Pretreatment of Nuclear Waste for Disposal*, W. W. Schulz, E. P. Horwitz, eds., Plenum, New York, N.Y, 1987.
4. *Handbook of Separation Process Technology*, R.W. Rousseau, ed., Wiley, New York, N.Y., 1987.
5. R. D. Noble, S. A. Stern, *Membrane Separation Technology: Principles and Applications*, Elsevier, New York, N. Y. , 1995.
6. *Metal Ion Separation and Preconcentration*, A. H. Bond, M. L. Dietz, R .D. Rogers, eds., ACS Symposium Ser., No. 716, American Chemical Society, Washington, D.C., 1999.
7. *Chemical Separations with Liquid Membranes*, R. A. Bartsch, J. D. Way, eds., ACS Symposium Ser., No. 642, American Chemical Society, Washington, D.C., 1996.
8. T. Sekine, Y. Hasegawa, *Solvent Extraction Chemistry; Fundamentals and Applications*, Marcel Dekker, New York, N.Y., 1977.
9. *Calixarenes for Separations*, G .J. Lumetta, R .D. Rogers, A. S. Gopalan, eds., ACS Symposium Ser., No. 757, American Chemical Society, Washington, D.C., 2000.
10. *New Separation Chemistry Techniques for Radioactive Waste and other Specific Applications*, L. Cecille, M. Casarci, L. Pietrelli, ed., Elsener, New York, N.Y., 1991.
11. *Chemical Pretreatment of Nuclear Waste for Disposal*, W. W. Schultz, E. P. Horwitz, eds., Plenum, New York, N.Y., 1994.
12. *Separation Techniques in Nuclear Waste Mangement*, T. E. Carleson, N. A.

Chipman, C. M. Wai, eds., CRC Press, New York, N.Y., 1995.
13. K. Street, Jr., G. T. Seaborg, *J. Am. Chem. Soc.*, **1950**, *72*, 2790.
14. M. Teramoto, T. Sakai, K. Yanagawa, M. Ohsuga, Y. Miyake, *Sep. Sci. Technol.*, **1983**, *18*, 735.
15. T. Kataoka, T. Nishiki, S. Kimura, Y. Tomioka, *J. Membr. Sci.*, **1989**, *46*, 67.
16. D. Lorbach, R. J. Marr, *Chem. Eng. Process*, **1987**, *21*, 83.
17. H. F. Koch, D. M. Roundhill, *Sep. Sci. Technol.*, **2000**, *35*, 779.
18. R. Abidi, J. M. Harrowfield, B. W. Skelton, J. Vicens, A. H. White, *J. Ind. Phenom. Mol. Recognit. Chem.*, **1997**, *27*, 291.
19. D. J. Cram, *Science*, **1998**, *240*, 760.
20. M. Lauterbach, E. Engler, N. Muzet, L. Troxler, G. Wipff, *J. Phys. Chem.*, **1998**, *102*, 245.
21. E. G. Isacoff, *EPA Metals Adsorption Workshop*, Cincinnati, OH, May 1998, pp. 10-11.
22. E. P. Horwitz, R. Chiarizia, S. D. Alexandratos, M. Gula, in *Metal Ion Separation and Preconcentration*, A. H. Bond, M. L. Dietz, R. D. Rogers, eds.,ACS Symposium Ser., No 716, American Chemical Society, Washington, D. C., 1999, Ch.14.
23. S. D. Alexandratos, *EPA Metals Adsorption Workshop*, Cincinnati, OH, May 1998, pp. 12-13.
24. A. Clearfield, D. M. Poojary, E. A. Behrens, R. A. Cohill, A. I. Bortun, L. N. Bortun in *Metal Ion Separation and Preconcentration*, A. H. Bond, M. L. Dietz, R. D. Rogers, eds.,ACS Symposium Ser., No 716, American Chemical Society, Washington, D. C., 1999, Ch. 11.
25. P. Meyers, *EPA Metals Adsorption Workshop*, Cincinnati, OH, May 1998, pp.29-32.
26. B. F. Smith, *EPA Metals Adsorption Workshop*, Cincinnati, OH, May 5-6, 1998, pp. 33-34
27. Y. B. Zhao, R. Gomer, *Surf. Sci.*, **1992**, *271*, 85.
28. N. K. Singh, R. G. Jones, *Surf. Sci.*, **1990**, *232*, 243.
29. M. A. Butler, A. J. Ricco, R. J. Baughman, *J. Appl. Phys.*, **1990**, *67*, 4320.
30. T. T. Mercer, *Anal. Chem.*, **1979**, *51*, 1026
31. F. Hofmann, V. Svenson, P. J. Toennies, *Surf. Sci.*, **1997**, *371*, 169.
32. S. Goapper, L. Barbier, B. Salanon, *Surf. Sci.* **1996**, *364*, 99.
33. S. Robert, S. Gauthier, F. Bocquet, S. Rousset, J. L. Dovault, J. Klein, *Surf. Sci.*, **1996**, *350*, 136.
34. J. McBreen, M. Sansone, *J. Electroanal. Chem.*, **1994**, *373*, 227.
35. M. Mazina-Ngokoudi, C. Argile, *Surf. Sci.*, **1992**, *262*, 307.
36. M. G. Barthes, G. E. Rhead, *Surf. Sci.*, **1979**, *85*, L211.
37. J. P. Jones, E. W. Roberts, *Surf. Sci.*, **1977**, *62*, 415.
38. L. C. A. Stoop, *Thin Solid Films*, **1979**, *62*, 115.
39. J. G. Dean, F. L. Bosqui, K. H. Lanouette, *Environ. Sci. Technol.*, **1972**, *6*, 518.
40. Y. Otani, C. Kanaoki, C. Usui, S. Matsui, H. Emi, *Environ. Sci. Technol.*, **1986**, *20*, 735.
41. Y. Otani, C. Kanaoki, H. Emi, I. Uchijima, H. Nishino, *Environ. Sci. Technol.*, **1988**, *22*, 708.
42. S. V. Krishnan, B. K. Gullett, W. Jozewicz, *Environ. Sci. Technol.*, **1994**, *28*,

1506.

43. T. Y. Yan, *Ind. Eng. Chem. Res.*, **1996**, *35*, 3697.
44. H. -C. Hsi, S. Chen, M. Rostam-Abadi, M. J. Rood, C. F. Richardson, T. R. Carey, R. Chang, *Energy Fuels*, **1998**, *12*, 1061.
45. K. Kadirvelu, C. Faur-Brasquet, P. Le Cloirec, *Langmuir*, **2000**, *16*, 8404.
46. A. Brennsteiner, J. W. Zondlo, A. H. Stiller, P. G. Stansberry, D. Tian, Y. Xu, *Energy & Fuels*, **1997**, *11*, 348.
47. D. Li, J. Lin, W. Li, S. Lee, G. Vidali, P. A. Dowben, *Surf. Sci.*, **1993**, *280*, 71.
48. Y. Tamishiro, M. Fukuyama, K. Yagi, *Mater. Res. Soc. Symp. Proc., Evolution of Surface and Thin Film Microstructure*, **1993**, *280*, 109.
49. J. -Y, Veuillen, J. M. Gomez-Rodriguez, A. M. Baro, R. C. Cinti, *Surf. Sci.*, **1997**, *377-379*, 847.
50. D. L. Sparks, A. M. Scheidegger, D. G. Strawn, K. G. Scheckel, *ACS Sympos. Ser. No.* 715, **1998**, 108.
51. K. C. Swallow, D. N. Hume, F. M. M. Morel, *Environ. Sci. Technol.*, **1980**, *14*, 1326.
52. M. F. Schultz, M. M. Benjamin, J. F. Ferguson, *Environ. Sci. Technol.*, **1987**, *21*, 863.
53. H. Abdel-Samad, P. R. Watson, *Appl. Surf. Sci.*, **1998**, *136*, 46.
54. C. Namasivayam, K. Ranganathan, *Ind. Eng. Chem. Res.*, **1995**, *34*, 869.
55. M. G. Burnett, C. Faharty, C. Hardacre, J. M. Mallon, G. C. Saunders, R. M. Ormerod, *JCS Chem. Comm.*, **1998**, 2525.
56. D. W. Brown, *ACS Sympos. Ser.*, **1979**, *93*, 237.
57. N. Jaffrezic-Renault, H. Poirier-Andrade, D. H. Trang, *J. Chromatogr.*, **1980**, *201*, 187.
58. S. Kaneko, S. Okuda, M. Nakamura, Y. Kubo, *Chem. Lett.*, **1980**, 1621.
59. N. Mikami, M. Sasaki, K. Hachiya, T. Yasunaya, *J. Phys. Chem.*, **1983**, *87*, 5478.
60. A. -M. M. Jacobsson, R. S. Rundberg, *Mater. Res. Soc. Sympos. Proc.*, **1997**, *465*, 797.
61. S. Cavallaro, N. Bertuccio, P. Antonucci, N. Giordano, J. C. J. Bart, *J. Catal.*, **1982**, *73*, 337.
62. A. W. Struempler, *Anal. Chem.*, **1973**, *45*, 2251.
63. W. Haerdi, E. Gorgia, N. Lakhova, *Helv. Chim. Acta*, **1971**, *54*, 1497.
64. R. Celis, M. C. Hermosin, J. Cornejo, *Environ. Sci. Technol.*, **2000**, *34*, 4593.
65. A. M. Liu, K. Hidajak, S. Kawi, D. Y. Zhao, *JCS, Chem. Comm.*, **2000**, 1145.
66. S. T. Beatty, R. J. Fischer, D. L. Hagers, E. Rosenberg, *Ind. Eng. Chem. Res.*, **1999**, *38*, 4402.
67. P. Meyers, *EPA Metals Adsorption Workshop*, Cincinnati, OH, May 1998, pp.29-32.
68. Y. Yin, H. E. Allen, C. P. Huang, D. L. Sparks, P. F. Sanders, *Environ. Sci. Technol.*, **1997**, *31*, 496
69. R. W. Klusman, C. P. Matoske, *Environ. Sci. Technol.*, **1983**, *17*, 251.
70. R. A. Griffin, N. F. Shimp, *Environ. Sci. Technol.*, **1976**, *10*, 1256.
71. G. R. Lumpkin, T. E. Payne, B. R. Fenton, T. D. Waite, *Mater Res. Soc. Sympos.*, **1999**, *556*, 1067.
72. R. T. Pabalan, J. D. Prikryl, P. M. Muller, T. B. Dietrich, *Mater. Res. Soc.*

*Sympos.*, **1993**, *294*, 777.
73. K. Idemitsu, K. Obata, H. Furuya, Y. Inagaki, *Mater. Res. Soc. Sympos.*, **1995**, *353*, 981.
74. C. H. Ho, D. C. Doern, *Canad. J. Chem.*, **1985**, *63*, 1100.
75. J. R. Bargar, R. Reitmeyer, J. A. Davis, *Environ. Sci. Technol.*, **1999**, *33*, 2481.
76. U. Neubauer, B. Nowack, G. Furrer, R. Schulin, *Environ. Sci. Technol.*, **2000**, *34*, 2749.
77. M. P. Papini, Y. D. Kahie, *Environ. Sci. Technol.*, **1999**, *33*, 4457.
78. G. N. Freeland, R. M. Hoskinson, R. J. Mayfield, *Environ. Sci. Technol.*, **1974**, *8*, 943.
79. K. Hayakawa, H. Yamkita, *J. Appl. Polym., Sci.*, **1977**, *21*, 665.
80. S. R. Shukla, V. D. Sakhardande, *J. Appl. Polym. Sci.*, **1991**. *42, 829.*
81. D. Karatza, A. Lancia, D. Musmarra, *Environ. Sci. Technol.*, **1998**, *32*, 3999.
82. P. Ricou, I. Lecuyer, P. Le Cloirec, *Water Sci. Technol.*, **1999**, *39,* 239.
83. S. Nakamura, M. Amano, Y. Saegusa, T. Sato, *J. Appl. Polym. Sci.*, **1992**, *45*, 265.
84. S. M. Ritchie, L. G. Bachas, T. Olin, S. K. Sikdar, D. Bhattacharyya, *Langmuir*, **1999**, *15*, 6346.
85. A. Denizli, D. Tanyolac, B. Salih, A. Ozdural, *J. Chromatogr, A.*, **1998**, *793*, 47.
86. M. I. Khalil, S. Farag, *J. Appl. Polym. Sci.*, **1998**, *69*, 45.
87. M. H. Ansari, A. M. Deshkar, P. S. Kelkar, D. M. Dharmadhikari, M. Z. Hasan, R. Paramasivam, *Water. Sci. Technol.*, **1999**, *40*, 109.
88. T. Hirotsu, S. Katoh, K. Sugasaka, N. Takai, M. Seno, T. Itagaki, *Ind. Eng. Chem. Res.*, **1987**, *26*, 1970.
89. M. Nakayama, K. Uemura, T. Nonaka, H. Egawa, *J. Appl. Polym. Sci.*, **1988**, *36*, 1617.
90. T. Hirotsu, S. Katoh, K. Sugasaka, N. Takai, M. Seno, T. Itagaki, *J. Appl. Polym. Sci.*, **1988**, *36*, 1741.
91. B. L. Rivas, H. A. Maturana, X. Ocampo, I. M. Peric, *J. Appl. Polym. Sci.*, **1995**, *58*, 2201.
92. J. Reichert, J. G. P. Binner, *J. Mater. Sci.*, **1996**, *31*, 1231.
93. A. Denizli, B. Salih, E. Piskin, *J. Chromatogr. A*, **1997**, *773*, 169.
94. C. Peniche-Covas, L. W. Alvarez, W. Arguelles-Monal, *J. Appl. Polym. Sci.* **1992**, *46*, 1147.
95. Y. Bara, H. Kirakawa, *Chem. Lett.*, **1992**, 1905.
96. Y. Kawamura, M. Mitsuhashi, H. Tanibe, H. Yoshida, *Ind. Eng. Chem. Res.*, **1993**, *32*, 386.
97. S. Tan, Y. Wang, C. Peng, Y. Tang, *J. Appl. Polym. Sci.*, **1999**, *71*, 2069.
98. Y. Yang, J. Shao, *J. Appl. Polym. Sci.*, **2000**, *77*, 151.
99. H. F. Koch, D. M. Roundhill, *Sep. Sci. Technol.*, **2001**, *36*, 137.

# 3

# SOIL WASHING AND *IN SITU* STABILIZATION METHODS

## 1. INTRODUCTION

The contamination of soils by chemically toxic or radioactive metals presents a major environmental problem. This problem is a combination of several factors. One problem is that metals in an aqueous environment are usually in a cationic form. Since soils are generally zeolites, which are negatively charged structures, metals are strongly bound to them. This charge association between metals and soils is especially problematic when the metal ion has a high positive charge such as +3 and +4. In these cases the charge pairing interaction is high, and the metal ion cannot be easily extracted by ion exchange with singly charged cations such as protons or sodium(I). Solvent washing is also inefficient in such cases because the solvation energies are usually not sufficiently strong for the metals to be eluted with small volumes of fluid extractants. Another problem with contaminated soils is that they are fixed in place, which contrasts with waters that can often be transported to another more convenient site for subsequent cleansing. With soils, however, excavation and transportation to an alternative location is at best very expensive. In many cases of soil remediation, numerous other problems must also be addressed. One problem is that excavation and removal may not be viable because of the presence of buildings or trees on the site that are not scheduled for removal. In these cases, electrokinetic extraction (Chapter 4) of metals is a viable alternative, especially for the areas around and beneath these immovable objects. Another problem with the removal of metals by soil washing is that it is usually desirable for the soil to be reclaimed to its original state before the contamination occurred. Since natural soils contain a range of desirable cationic components, these ions must be subsequently replaced if a consequence of soil washing is that they are removed along with the toxic and/or radioactive metals. Soils also have particular properties that result from the particle size distribution within them, and the cleansing process must not destroy these individual characteristics that makes the soils useful for cultivation applications. In addition to soils, another fixed environment that presents particular problems for the cleansing of metal contaminants is contaminated buildings. In some cases the building can be demolished with the materials transported for off-site remediation, but in other cases it is only a portion of the building that is contaminated, and it is desirable to retain the entire structure.

For soils, an alternative to washing is *in situ* stabilization. In some cases the metal contamination is not creating an immediate problem in the location, but it remains of concern because if the metals are not removed from the site they may be slowly leached out of the soil by rain or other sources of water into locations where they are more problematic. In these cases, *in situ* stabilization to ensure that the toxic metals do not migrate is a viable alternative to washing them out with water or other fluids. To achieve *in situ* stabilization of toxic metals in soils it is necessary to convert them into a highly insoluble form that is not transferred into water. This presents a significant challenge, however, since such stabilization needs to be effective for long periods of time. One method of achieving *in situ* stabilization is to convert the metal into the chemical form that it was present in its original mineral composition. In many cases this involves converting the metal into its oxide or sulfide form. Another *in situ* stabilization method involves cementation, especially if the chemical cementation process can be induced to incorporate the metal ions into its structure. Since cementation is an inorganic cross-linking process, the solid material that results is likely to be very insoluble in water, and highly resistant to having its components leached from it.

## 2. SOIL WASHING AND FLOODING

Much of the prior engineering work on *in situ* soil reclamation has focused on the removal of organics using technologies that were developed for the recovery of oil through polymer and/or surfactant flooding of porous media.[1-3] In some cases these methods can be directly applied to metals, but in other cases chemical modifications are necessary in order to achieve selectivity for particular metal ions

## 2.1. Soil Washing

The use of flooding methods for the removal of toxic metals from soils by washing with highly alkaline or acid water has been investigated, but to a lesser extent than for organic. Chromium has been effectively removed from soil by using a washing solution at pH 10.4.[4] An *in situ* application of this technique was recently carried out in Europe, where an acidic solution was used to leach metals from more than 30,000 cubic meters of soil.[5]

In some studies the flooding mixture contains chelating agents such as EDTA to enhance metal species extraction.[6] Aqueous solutions of EDTA and NTA (nitrilotriacetic acid) have been used to extract lead from soils contaminated from the manufacturing of automobile batteries.[7,8] EDTA, however, is a non-selective chelating agent and its use results in the removal of both toxic and non-toxic metals from the soil.

Since it is important that the reclamation of heavy metals from soils does not leave a residue of organic complexants in the soil it is important that it can be recovered for re-use. This concept of recycling the complexant is also an important consideration because it factors into the cost effectiveness of the entire extraction process. One of the more common complexants that has been used for the extraction of heavy metal ions from soils is EDTA. Although this complexation has low selectivity, it has the advantages of being effective and inexpensive. Nevertheless, it still needs to be recycled. One method of recovering the metal from its complex and recycling the free EDTA is to use an electrochemical process.[9] The current for this electodeposition process must be kept low in order to improve the current efficiency. Nevertheless, water electrolysis is still an important problem since during the electrodeposition process the pH in the vicinity of the cathode increases from an intial value of 2.0 to 12.0. The results of such an electrodeposition study with lead, cadmium and mercury are shown in Table 1. In the absence of EDTA mercury is not efficiently electroplated from solution because it forms a precipitate of mercury(II) hydroxide in the vicinity of the cathode. For the other metal ions the current efficiencies remain constant until about 70% of the metal has been deposited, after which they decrease. In the presence of EDTA the method can be effectively used for lead(II) and mercury(II). High recoveries of both the metal and EDTA are observed, and the current efficiency is high. For chromium(III) and

**Table 1. Electrodeposition of Metals from Solutions of Free Ions and of their EDTA Complexes**

| Metal Species | % Metal Recovery | % EDTA Recovery | Current Efficiency |
|---|---|---|---|
| Lead(II) | 99.3 | — | 0.79 |
| Lead (II)-ETDA | 99.7 | 99.4 | 0.62 |
| Cadmium(II) | 70.0 | — | 0.27 |
| Cadmium(II)-EDTA | 0.0 | 0.0 | 0.00 |
| Chromium(III) | 67.9 | — | 0.18 |
| Chromium(III)-EDTA | 22.9 | 24.1 | 0.18 |
| Mercury(II) | 0.0 | — | 0.00 |
| Mercury(II)-ETDA | 92.0 | 88.0 | 0.60 |

EDTA the recovery percentages and current efficiency is lower. For cadmium(II) the method fails because the electrode potential required to reduce the cadmium(II)-EDTA complex to cadmium metal is greater than that required to electrolyzed water.[10] A flow-through system has been devised whereby a continuous metal deposition-EDTA recycling sequence can be employed.

A water soluble chelating polymer, Metaset-Z, available from Microset Corporation, has been used for the removal of lead from soils. This chelating polymer is prepared by functionalizing the BASF Polymin Water Free Polyethyleneimine (PEI; molecular weight 600-750,000) with acetate groups. The polymer chelates lead(II) in aqueous solution within seconds. In a surrogate soil the lead is effectively removed in a single batch extraction. In real soil samples two removal rates for lead were observed, one for the loosely bound lead, and the second for the more tightly bound. Metaset-Z is more selective for calcium than EDTA, and has comparable selectivity for lead over calcium as NTA. The removal efficiency for lead from surrogate soils is greater than that observed for EDTA.[11]

Soils also contain humic acid materials, therefore for a complexant to be effective as an extractant it must be competitive with the metal ion binding sites that are available in these materials. Although this topic is covered in more detail in Chapter 12, some mention of it is made in this chapter because it must be considered when choosing extractants for treating soils. The binding sites for several humic acid samples from different locations has been investigated by a range of different techniques. The chosen metals are copper(II), iron(III) and manganese(II), which have the respective coordination numbers of 4, 6 and 6 in the humic materials. The X-ray absorption near-edge structures (XANES) and extended X- ray absorption fine structure (EXAFS) also indicate several binding sites, with one having principally carboxylate groups. Binding free energies and electron paramagnetic resonance spectroscopy data suggest that the free metal ions rapidly transfer between specific humic acid binding sites, which affects the rates of metal ion release and transfer to other matrices.[12] The metal can be removed by washing with 0.1M HCl.

The weathering of soils also presents a challenge to extracting metals from them. In many cases the metal ion intially binds to the surface of the soil particles, but over an extended time period the metal imigrates further inside the soil matrix. This phenomena has been observed for nickel sorbed into clay mineral, and the sample maintained for 1 year at a pH of 7.5. Over this time period the nickel becomes increasingly resistant to extraction with EDTA. Intially the nickel(II) nitrate is adsorbed as a nickel-aluminum layered double hydroxide, but over time EXAFS shows that the anionic species in the interlayer space of the layers changes from nitrate to silicate. The nickel in this silicate matrix is more resistant to extraction by EDTA.[13]

## 2.2. *In Situ* Stabilization

Another approach to dealing with metal contamination in soils is *in situ* stabilization. The *in situ* stabilization can involve converting the metal to an insoluble form such as an insoluble sulfide, thereby converting the metal into the same chemical composition as the mineral from which it was originally extracted. Another alternative involves solidification and stabilization into Portland cement prior to burial and disposal. An agricultural-based alternative is also viable. This method uses green plants and soil amendments to achieve *in situ* stabilization.[14]

This stabilization technique depends on changing the contaminant chemistry by adding and incorporating soil amendments. The soil amendments must be able to effectively change the trace element chemistry while having a neutral to positive effect on plant growth. The ability to speciate trace elements in complex matrices such as soil is limited because they are found in low concentrations and in a variety of physicochemical forms. A possible solution, however, is to use electrokinetics to preconcentrate the metal ions.[15]

Although *in situ* stabilization can be achieved by converting the metals into their sulfides, other options are also available. Among these options are oxides (hydroxides) and phosphates. In addition to the simple binary oxides and hydroxides, other oxide systems have been found to be useful in other specific cases. An example of this latter situation is found with lead(II), where managanese(IV) oxide is particularly strong adsorbent for it. A method of studying the effectiveness of a material for *in situ* stabilization is to investigate its subsequent bioavailability. Such a study has been carried out for lead stabilized in either a manganese (IV) oxide or phosphate host matrix. In each case the bioavailability of lead wad decreased, with a further reduction when a combination both adsorbents was present. The reductions observed with phosphate alone ranges from 15 to 41%, whereas the reductions observed with the combination ranges from 23 to 67% as compared to the control. From X-ray diffraction data it is apparent that more pyromorphite-like minerals are formed in the presence of manganese(II) oxide and phosphate.[16]

Removal of these metals from soils presents another challenge. The adsorption and/or movement of both organic chemicals and inorganic pollutants in the soil matrix of waste sites presents a formidable problem to scientists and engineers who are attempting to devise efficient and cost-effective remediation strategies. Heavy metals in particular form stable complexes in soils with inorganic and organic ligands and their presence in even small amounts can pose a serious health risk. Since metal pollutants migrating into the ground water supplies from unengineered facilities can threaten the nation's health, low cost and efficient means of treating large volumes of soil in a timely fashion would be immensely useful.

Probably the greatest challenge for extracting heavy metal ions is their *in situ* removal from soil and earth samples. For contaminated fluids it is plausible to use precipitation or evaporation techniques to obtain concentrated solids which can be disposed of in a safe manner. For soil and earth samples, the metals are likely to be complexed into the zeolite structures of the soil and, therefore tightly bound. Although the retention of metals by soil, clay, and other surface materials has been the focus of much research in recent years,[17] there is no consensus among experts in the field as to the best method for removing heavy metals from the soil matrix and concentrating the resulting waste. Contaminated soil may be excavated and transported to a cleaning facility, cleaned on-site after excavation, or remediated *in-situ*. It is generally recognized that *in situ* heavy metal treatment methods are preferable due to their inherent simplicity and low cost. Unfortunately, there have been very few studies in the literature documenting the use of *in situ* soil remediation strategies and little is known regarding the fundamental principles which govern the process.

## 3. ANALYTICAL TECHNIQUES

Before strategies can be developed, and that metals can be removed from soil and water, it is necessary that they can be detected and quantitatively measured. Commonly used techniques are atomic absorption (AAS) spectroscopy or inductively coupled plasma-atomic emission spectroscopy (ICP-AAS) that are monochromated for the absorption and emission wavelengths of these elements. Other techniques are, however, available. One of these is square wave anodic stripping voltammetry (SWASV) with carbon disk ultramicroelectrodes. [18] Laser-excited atomic fluorescence has been used to detect lead in ultratrace amounts that has been electrothermally atomized in a graphite furnace. A detection limit of 10 fg/ml was achieved.[19] A similar technique, with optical fibers being used to transport the laser pulse to and from the sample, has been used to determine lead concentrations.[20] The technique has improved sensitivity over the widely used x-ray fluorescence technique.

Subsurface heavy metal detection has been carried out using a laser-induced breakdown spectroscopy penetrometer system (LIBS).[21] The LIBS probe includes a laser as part of the optical LIBS sensor section for metals analysis, and a core and sleeve soil classification module as its tip. This *in situ* method eliminates the necessity to obtain core samples, since the probe is inserted directly into the ground. Soil classification data are acquired during the push, while LIBS data are recorded during probe retraction. A block diagram of the equipment is shown in Figure 1.

Because some metals accumulate in biological organisms, attempts have been made to monitor heavy metal toxicity by using aquatic plants and animals as biomonitors,[22,23] or by using immunoassay methods. Such methods have been used for cadmium. A monoclonal antibody (2A81G5)

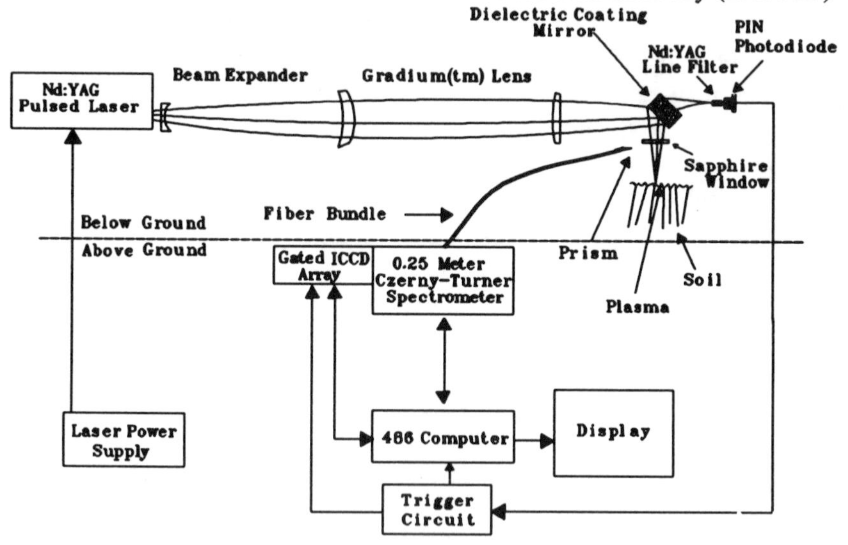


**Figure 1. Diagram of the Laser-induced breakdown spectroscopy penetrometer system(LIBS).**

that recognizes cadmium-EDTA complexes has been produced by the injection of BALB/c mice with a metal chelate complex covalently coupled to a carrier protein. A negative logarithm plot of the equilibrium dissociation constant for metal-EDTA complexes plotted against the difference between the atomic volume of cadmium(II) minus that of M(II) shows a roughly linear relationship. An immunoassay for measuring cadmium(II) in aqueons samples used this same antibody.[24] An inhibition immunoassay format was used where the antibody binds tightly to the cadmium-EDTA complex, but not to EDTA itself. The assay measures cadmium(II) in the presence of excess iron(III), magnesium(II) and lead(II). The presence of mercury(II) in concentrations above 1μM causes interference. The affinity of metal-EDTA complex for the 2A81G5 antibody correlates with its inhibitory effect in competitive immunoassay(Table 2).

The removal of these metals from mammals, soils and waters usually involves the use of a chelating ligand. An alternative strategy with soils

**Table 2. Affinity of Metal-EDTA Complex for the 2A81G5 Antibody Correlates with Its Inhibitory Effect in Competitive Immunoassay**

| metal-EDTA complex | equilibrium dissociation contant (M) | inhibitory concentration in immunoassay (M) |
|---|---|---|
| Magnesium(II) | $2.2 \times 10^{-4}$ | $> 1 \times 10^{-3}$ |
| Lead(II) | $7.4 \times 10^{-5}$ | $> 1 \times 10^{-3}$ |
| Iron(III) | $5.4 \times 10^{-5}$ | $> 1 \times 10^{-3}$ |
| Zinc(II) | $2.5 \times 10^{-6}$ | $1 \times 10^{-3}$ |
| Nickel(II) | $2.1 \times 10^{-6}$ | $1 \times 10^{-3}$ |
| Manganese(II) | $4.1 \times 10^{-7}$ | $1 \times 10^{-4}$ |
| Indium(II) | $6.2 \times 10^{-7}$ | $1 \times 10^{-4}$ |
| Mercury(II) | $2.6 \times 10^{-8}$ | $1 \times 10^{-5}$ |

involves stabilizing them in place by conversion into an insoluble form that does not migrate into aquifers and ground waters. In devising chelate strategies it is useful to consider size and charge. Lead and cadmium exist as a 2+ cation. For a monomeric 2+ cation, the relative sizes are cadmium < lead. Although the relative size difference is quite small, it may be the only property that can lead to selectivity.

## 4. EXTRACTION FROM SOILS

One technique, which can be used for all metals, is soil washing. Another is electrokinetic extraction. Techniques other than soil washing and electrokinetic extraction have been used to remove metals from soils. One of these is Chelate Assisted Pressurized Liquid Extraction (CAPLE) which uses subcritical water and chelating agents for the extraction of metals from soils and sediments. Good extraction is observed with aqueous EDTA at 150 to 200 atm presssure and 100°C to 200°C temperature.[25] Composted sludge and other biomass have a high affinity for heavy metals, with the uptake being particularly high for lead.[26]

Another approach to heavy metal remediation involves microwave energy. One application is vitrification, where the microwave radiation fixes the lead within the soil matrix such that it cannot be subsequently extracted with nitric acid.[27] Alternatively, a chelate assisted microwave extraction (CAME) method can be used for metals in soils.[28]

## 5. COMPLEXANT DESIGN FOR SOIL EXTRACTION

A class of heavy metal complexing agents involves chelators consisting of hydrophobic molecular platforms with convergent hydrophilic groups which are carboxylic acids and carboxylic acid derivatives. These compounds offer a number of distinct advantages: they lack significant toxicity, they may be tailorable to specific metals using molecular modeling design techniques, and the metal-chelator complexation reaction can be made reversible, thus allowing the chelating agent to be reused. Four such compounds, NTA, EDTA, EGTA and DCyTA have been used to study the desportion/complexation/dissolution charactersitics of cadmium from kaolin clay. Both DCyTA and EGTA complexed strongly with cadmium (100% cadmium dissolution) over a wide pH range (2.5-12.0). The zeta potential of kaolin did not change, and no cadmium readsorption is found, after addition of EGTA and DCyTA. The capacity of the four chelators for removing cadmium from kaolin is in the order DCyTA > EGTA > EDTA > NTA.[29] In terms of the selectivities for the removal of metals from kaolin clay, the metal ordering for chelation and dissolution is Cd > Pb in each case. The chelator EGTA is the most effective in selectively removing cadmium in the presence of adsorbed alkaline earth metals, with approximately 90% of the adsorbed calcium and magnesium being retained on the koalin until over 80% of the cadmium is desorbed by the EGTA.[30]

Chelators can also be used to remove heavy metals from soils. The ability to thoroughly yet selectively remove toxic heavy metals from soils at superfund sites while leaving intact non-toxic metal species (e.g., sodium(I), calcium(II), and magnesium(II)) is a goal of these studies. In this regard, a primary objective of such studies is to synthesize and test two new classes of chelating agents which complex with heavy metals. One group of compounds (dithiocarbamates, xanthates, and dithiophosphates) form chelate complexes via binding to the metal through their sulfur atoms. These compounds should be particularly effective in removing metals from soils because they are known to form chelate complexes with zinc, cadmium, lead, and mercury.[31] The resulting sulfur-heavy metal linkages will be considerably stronger than those formed using more traditional chelating agents, such as EDTA. Thus, these compounds may be highly effective in removing heavy metals which are tightly bound to the soil matrix. These chelating agents will preferentially bind to heavy metals, as opposed to non-toxic metallic species, which will further enhance the efficiency of extraction. In addition, this class of chelators can be readily synthesized from inexpensive precursors and their alkyl chain length can be easily varied, which will alter their surfactant properties.

One goal is to discover chelating agents with surfactant properties which will selectively extract toxic heavy metals and also solubilize the organic contaminants. The flooding mixture that emerges out of the washed soil, is a hazardous emulsion/suspension containing the heavy metals, the hazardous organic phase and the used chelating agents. A great deal of work has been done for the improvement of waste water treatment processes involving hazardous emulsions/suspensions. This includes the use of optimal flocculation schemes,[32-34] membranes,[35] flotation,[36,37] microbial biomass for complex petroleum emulsions,[38] metal precipitation followed by activated carbon treatment and ion exchange.[39]

# REFERENCES

1. A. S. Abdul, T. L. Gibson, *Environ. Sci. Technol.,* **1991**, *25*, 665.
2. A. N. Clark, P. J. Plumb, T. K. Subramanyam, D. *J.* Wilson, *Sep. Sci. Technol.,* **1991**, *26*, 301.
3. O. K. Gannon, P. Bibring, K. Raney, J. A. Ward, D. J. Wilson, J. L. Underwood, K. A. Debelak, *Sep. Sci. Technol.,* **1989**, *24*, 1073.
4. H. N. Hsieh, D. Raghu, J. Liskowitz, *Hazard. Ind. Wastes, 22nd. Conf.,* **1990**, 459.
5. T. H. Pfeifer, T. J. Nunno, J. S. Walters, *Environ. Prog.,* **1990**, *9*, 79.
6. P. M. Esposito, B. B. Locke, J. Greber, R. P. Traver, *EPA Document 600/9-88/021.* 1988.
7. H. A. Elliot, G. A. Brown, *Water, Air, Soil Pollut.,* **1989**, *45*, 361.
8. H. A. Elliot, J. H. Linn, G. A. Shields, *Hazard. Waste Hazard. Mater.,* **1989**, *6*, 223.
9. H. E. Allen, P.-H. Chen, *Environ. Prog.,* **1993**, *12*, 284
10. S. B. Martin, Jr., H. E. Allen, *Chemtech,* **1996**, 23.
11 C. G. Rampley, K. L. Ogden, *Environ. Sci. Technol.,* **1998**, *32*, 987.

12. G. Davies, A. Fataftah, A. Cherkasskiy, E. A. Ghabbour, A. Radwan, S. A. Jansen, S. Kolla, M. D. Paciolla, L. T. Sein, Jr., W. Buermann, M. Balasubramanian, J. Budnick, B. Xing, *JCS, Dalton Trans.*, **1997**, 4047.
13. R. G. Ford, A. C. Scheinost, K. G. Scheckel, D. L. Sparks, *Environ. Sci. Technol.*, **1999**, *33*, 3140.
14. W. R. Berti, S. D. Cunningham, *Environ. Sci. Technol.*, **1997**, *31*, 1359.
15. D. L. Lake, P. Kirk, J. N. Lester, *J. Environ. Qual.*,**1984**, *13*, 175.
16. G. M. Hettiarachchi, G. M. Pierzynski, M. D. Ransom, *Environ. Sci. Technol.*, **2000**, *34*, 4614.
17. L. J. Evans, *Eviron. Sci. Technol.*,**1989**, *23*, 1046.
18. B. J. Feldman, J. D. Osterioh, B. H. Hata, A. D. D'Alessandro, *Anal. Chem.*, **1994**, *66*, 1983.
19. E. P. Wagner, II, B. W. Smith, J. D. Winefordner, *Anal. Chem.*, **1996**, *68*, 3199.
20. B. J. Marquardt, S. R. Goode, S. M. Angel, *Anal. Chem.*, **1996**, *68*, 977.
21. B. Miles, J. Cortes, *Field Anal. Chem. and Technol.*, **1998**, *2*, 75.
22. E. Baatrop, *Comp. Biochem. Physical.*, **1991**, *100C*, 253.
23. P. Chakrabarti, F. M. Hatcher, R. C. Blake, II, P. A; Ladd, D. A. Blake, *Anal. Biochem.*, **1994**, *217*, 70.
24. M. Khosraviani, A. R. Parlov, G. C. Flowers, D. A. Blake, *Environ. Sci. Technol.*, **1998**, *32*, 137.
25. K. O'Leary, G. L. Long, *Abstr. 50th S. E. Regional ACS Meeting*, Research Triangle Park, N. C., Nov. 1998, Abstr. 223.
26. I. T. Urasa, N. Mwebi, *Abstr. 50th S. E. Regional ACS Meeting*, Research Triangle Park, N. C., Nov. 1998, Abstr. 226.
27. R. A. Abramovitch, D. A. Abramovitch, E. Hicks, J. M. Sinard, *Abstr. 50th S. E. Regional ACS Meeting*, Research Triangle Park, N. C., Nov. 1998, Abstr. 224.
28. U. Chatreewongsin, G. L. Long, H. M. McNair, *Abstr. 50th S. E. Regional ACS Meeting*, Research Triangle Park, N. C., Nov. 1998, Abstr. 222.
29. J. Hong, P. N. Pintauro, *Water, Air and Soil Pollut.*, **1996**, *86*, 35.
30. J. Hong, P. N. Pintauro, *Water, Air and Soil Pollut.*, **1996**, *87*, 73.
31. J. A. Cras, J. Willemse, *Comprehensive Coordination Chemistry,* Wilkinson, G. Ed., Pergamon, Oxford, 1987, Vol. 2, Ch. 16.4.
32. K. D. Papadopoulos, H. Y. Cheh, *Plating and Surf. Finishing,* **1982**, *69*, 122.
33. P. C. Y. Huang, Y. C. Wu, K. C. Ou, Bornholm, J. K. *Proc. 39th. Ind. Waste Water Conf., Purdue Univ.* May, 1984.
34. E. Gomolka, B. Gomolka, *Acta Hydrochim. Hydrobiol,*. **1985**, *13*, 91.
35. G. K. Anderson, C. B. Saw, *Environ. Tech. Letts.*, **1987**, 8, 121.
36. I. N. Myasnikov, L. N. Butseva, L. V. Gandurina, *Environ. Protect. Eng.*, **1985**, *11*, 119.
37. V. C. Gopalratnam, G. F. Bennett, R. W. Peters, *Environ. Progr.*, **1988**, *7*, 84.
38. N. Kosaric, Z. Duvnjak, W. L. Cairns, *Environ. Progr.*, **1987**, *6*, 33.
39. M. H. Bates, J. N. Veenstra, J. Barber, R. Bernard, J. Karleskint, P. Khan, R. Pakanti, M. Tate, *Environ. Systems,* **1990**, *19*, 237.

# 4

# ELECTROKINETIC EXTRACTION OF METALS FROM SOILS

## 1. INTRODUCTION

Removal of metals from soils presents a different challenge than their removal from waters. In particular, in certain locations it is advantageous to carry out the *in situ* remediation of soils without having to excavate and remove them for off-site cleansing. As discussed in chapter 3, this situation is particularly important if the soil is covered by buildings or trees that need to be retained in an undamaged state. The analytical challenges can also be problematic, yet before a strategy for metal removal is decided upon, it is important that the amounts and distributions of the metals be precisely known. Because the metals are strongly occluded into the zeolite soil structure they are difficult to release for accurate assays, therefore an *in situ* method must be used. If the metals are present in moderate to high quantities an electrokinetic approach may be the method of choice.

## 2. ELECTROKINETIC EXTRACTION

An alternative to soil washing or injection flooding is the use of *in situ* electric fields to dissolve and then transport heavy metal ions in soils and clays. The efficiency of this electrokinetic method depends on the type and water content of the soil. For dry and sandy soils the method requires the addition of water in order for there be a mobile phase through which the metals can migrate. This electrochemical (electrophoretic) processing of soils by the use of an *in situ* DC current results in the development of electrical, hydraulic, and chemical gradients within the soil matrix. These gradients produce a net movement in both the water (electroosmosis) and ionic species (electromigrations). Because the technique involves the application of an electrical potential there is a comcomitant electrolysis of water into hydrogen and oxygen. This electrolysis of water is a significant cost deterrent to the use of electrokinetic extraction since the major cost is the electrical power, and a portion of this is consumed in the water electrolysis, a process of no commercial value in the remediation. The primary electrochemical reactions at the anode and cathode produce protons ($H^+$) and hydroxyl anions ($OH^-$), which are also transported through the soil, with other ions such as $M^+$.

$$\text{Anode:} \quad 2H_2O - 4e^- \rightarrow O_2 + 4H^+$$
$$\text{Cathode:} \quad 4H_2O + 4e^- \rightarrow 2H_2 + 4OH^-$$

The generation of an acid front within the soil which moves from the anode to the cathode results in both the solubilization of basic metal hydroxides, carbonates, or adsorbed species, and the protonation of any basic organic functional groups in the soil humus, giving these components more cationic character. An ion exchange process occurs within the soil, with the solubilized ionic species being carried by both the electric field and the electroosmotic movement of water to the cathode. At the cathode, reduction of the metal ions to the metal can occur as in an electrolytic purification process. More commonly, however, the electrokinetic extraction process is not carried out to this degree of completeness, but is instead used to achieve concentration of the metal ions in the vicinity of the cathode. This concentration of heavy metals or radioactive metals into a chosen region of the soil site is particularly valuable, especially if this concentrates the metals into a location where a relatively small amount of soil can be easily excavated and cleansed off-site by soil washing or other techniques. Mixtures of contaminants and the presence of organic material in the soil do not necessarily cause problems.[1] However, a problem with the method is selectivity, since the applied voltage migrates all cations in the direction of the cathode.

The selectivity problem can be addressed by the addition of chelating agents to the contaminated soil. These chelating agents are chosen as ones that selectively bind to the metals that are targeted for removal. These agents should discriminate against other metals such as sodium, calcium, magnesium and iron, which are normal components of soils, and need to be left in place. The use of chelating agents will improve the efficiency of the metal contaminant removal, especially if the resulting metal complexes preferentially migrate. The use of complexants or chelates presents different challenges. One of these is that the complexant must be non-toxic because it is unacceptable to leave residues of toxic compounds in the soil after the metal contaminants have been removed. An alternative, and usually preferable strategy, is to employ complexants or chelates that are completely biodegradable under the ambient conditions prevalent in the soil. Such a situation eliminates all questions as to whether the chosen method for metal removal has left the site fully remediated for all future uses. Another challenge in choosing a complexant for the electrokinetic extraction process is that the complex formed with the metal to be removed must have an overall charge in order that it can be transported to an electrode. Furthermore, since the complex must compete with other ions in the transport process, it needs to have low molecular weight to ensure a high electrokinetic mobility. The challenge therefore is to obtain a complex that has a high charge:molecular weight ratio. An interesting challenge is to develop a system that is catalytic in the complexant, while meeting all of the other criteria outlined in this paragraph. Such a strategy has the advantage of requiring the use of only very small quantities of complexant, which is beneficial in terms of both cost and final removal of the complexant from the remediated soil. This catalytic condition can be achieved if the complexant and its complex with the metal have opposite overall charges. For a negatively charged complexant and a positively charged metal complex, the two species will move in opposite directions under the influence of a DC potential. If, therefore, after migration of the positively charged complex to the cathode the complexant dissociates from the metal, it will migrate back toward the anode until it binds to a metal ion migrating in the opposite direction, and reverses back to migration to the cathode. A limited number of laboratory scale studies and simplified mathematical analyses have shown that an *in situ* electrophoretic acid treatment of contaminated soil is a potentially useful and cost effective remediation technique[1-3] Nevertheless, in some case the cost of the electrical power causes the method to be rejected. The energy consumption depends on the type of soil, the water content of the soil, and the tightness of binding of the metal ions into the soil structure. For soils having a high water content, the electrolysis of this water into hydrogen and oxygen may be the most wasteful energy step. Lower applied to the electrodes voltages can assist in reducing this electrolytic reaction, as can adjustments to the soil acidity.

## 2.1. Electrophoretic Transport Model

The soil (clay) matrix can be viewed as an array of parallel cylindrical pores, with a uniform distribution of ion exchange sites on the pore wall. Such a description has been used to accurately model transport in the complex pore structure of polymeric membranes.[4] When the radius of the membrane pores is of the same order of magnitude as the Debye screening length (which will be the case in porous soils due to the low concentration of dissolved salts in the interstitial water), electroneutrality within the pores is no longer valid. The existence of a non-zero space charge results in coupling between electrical forces in the pore, and mass transfer and fluid flow, which manifests itself in two ways: (1) the interaction of fluid flow and the electric field in the pore generates a body force term in the momentum balance equation and (2) the large magnitude of the electrostatic potential in the space charge region of the pore causes counterion (cation and proton) enrichment and co-ion (anion) exclusion near the surface of the soil particles. The interaction between the space charge region and the transport driving forces directed parallel to the walls of the pores results in such electrokinetic phenomena as streaming potential and electroosmosis.

The following mathematical analysis represents an extension of existing electrophoretic transport models and includes such effects as ion size and hydration and a variable solvent dielectric constant. In addition, the relevant hydrodynamic, mass transfer, and electrokinetic equations are solved in a two-dimensional coordinate system in order to accommodate the possibility of modeling a nonuniform diameter pore structure. The model predicts ion and water movement of the case of a single heavy metal contaminant. The electrophoretic transport model consists of the following equations:

1. Two velocity equations (1, 2) for fluid (water) motion in the radial and axial pore directions that involve **u**, the fluid velocity, $\eta$ the solvent viscosity, **P** the pressure, $c_i$ the concentration of species i, and $z_i$ the charge number of species i, and $\phi$ is the electric potential.

$$-\nabla P + \eta \nabla^2 u - \left( \sum_i z_i c_i \right) \nabla \phi = 0 \qquad (1,2)$$

2. The equation (3) of continuity for an incompressible fluid.

$$\nabla \bullet \mathrm{u} = 0 \tag{3}$$

3. Poisson's equation (4) with a radially dependent solvent dielectric constant which describes the electric potential field generated by a non-zero space charge distribution in the pore involving $\epsilon$ the solvent dielectric constant, and **F** the Faraday's constant.

$$\nabla \bullet [\epsilon(r)\nabla\phi] = -F \sum_{i} z_i \, c_i \tag{4}$$

4. A modified Boltzmann equation (5) for the radial pore distribution of each anion and cation species, with the second exponential term on the right side of this equation being a Born model ion hydration correction factor.[5]

$$c_i = (c^0)_i \exp\left[\frac{-z_i F(\phi - \phi^0)}{\mathrm{RT}} + \frac{A_i}{\mathrm{RT}}\left(\frac{1}{\epsilon(r)} - \frac{1}{\epsilon^o}\right)\right] \tag{5}$$

5. The Nernst Planck flux equations (6, 7) for diffusion, migration, and convection of each anionic and cationic species in the radial and axial pore directions and the continuity equation for the ionic species involving the mobility and diffusion coefficient of species i, respectively. A term in the equation represents the increase in metal ion or charged metal/complex concentration due to dissolution at the soil particle surface. This term is dependent on the local solution pH and/or chelate concentration. The model assumes that the dissolution and complexation reactions are fast.

$$N_i = -\mathrm{D}_i \nabla c_i - z_i \lambda_i F c_i \nabla\phi + c_i u \tag{6}$$

$$\frac{\partial c_i}{\partial t} = -\nabla \bullet N_i \pm R \tag{7}$$

6. An equation derived by Booth (8) which describes the dependence of the solvent dielectric constant on the electric field strength.[6] This equation

$$\epsilon = n^2 + (\epsilon^0 - n^2)\left[\frac{3}{\beta \nabla\phi}\right]\left[\coth(\beta \nabla\phi) - \frac{1}{\beta \nabla\phi}\right] \tag{8}$$

involves the refractive index of the solvent and a constant that is a known function of temperature and the solvent dipole moment.

These equations are solved simultaneously for the pressure (**P**), electric potential ($\phi$), cation and anion species concentrations ($C_i$), solvent dielectric constant ($\epsilon$), radial and axial velocities ($\mathbf{u}_r$ and $\mathbf{u}_x$), and the anion and cation fluxes ($\mathbf{N}_i$) in the axial and radial pore directions for any time (t) after current application. The initial condition for the model is given by the concentration of $H^+$ (pH), free metal ions, and chelator (if present) prior to current application. The boundary conditions at the two soil/electrode interfaces include zero axial flow constraints and modified Ohm's Law relationships which relate potential, current, ion concentration gradients, and fluid flow.[7] A number of symmetry conditions at the pore centerline (involving dielectric constant, pressure, velocity, potential and mass flux) are also specified. At the pore wall, the fluxes and fluid velocity are zero and Gauss's Law is applied to relate the potential gradient and fixed cation exchange capacity of the soil (clay) particles. The model also contains a number of parameters which must be specified, including: (1) ion diffusion coefficients, which will be determined from infinite dilution equivalent ionic conductance data in the literature, (2) ionic mobilities, which are computed from diffusion coefficients.[8]

## 2.2. Experimental Method

The eletrokinetic method involves inserting electrodes in the soil and applying a potential across them. The electrodes are usually inserted into a well that can be kept in an aqueous environment in order to ensure a good electrical contact. An important aspect of this method is the presence of water in the soil that can facilitate ion migration. For soils that have a high water content, the method is generally applicable. For dry sandy soils, however, water must be added to the area under treatment. By moving the location of the electrodes periodically the remediated metals can be migrated across the site in a manner that resembles zone refining. In this technique the application of an *in situ* direct current produces an electroosmotic water flow, and $H^+$ and $OH^-$ fronts which move through the soil in opposite directions.[3, 9-12] The acid front that moves from the anode to the cathode dissolves metals that are absorbed in the soil (Figure 1). The highly alkaline interstitial water near the cathode then causes heavy metals to precipitate,

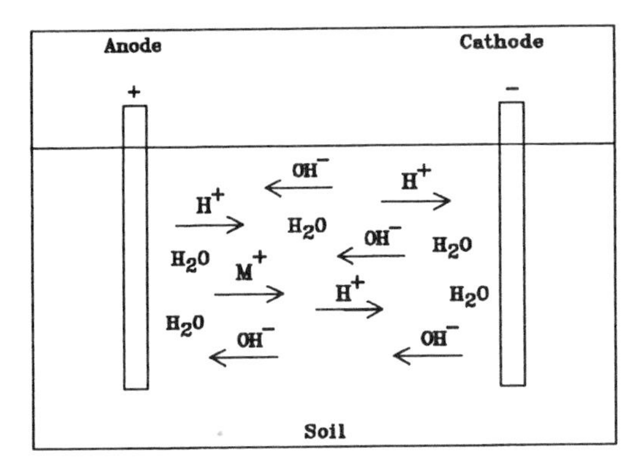


**Figure 1. Electroosmotic Flow**

limiting the extent of metal ion migration and the effectiveness of the reclamation process. This precipitation of metal hydroxides can be avoided if the metal is strongly bound to a complexant. Nevertheless, the electrokinetic extraction of metals from soils does still have its limitations. One limitation is that the consumption of electrical power can be high because of the concurrent electrolysis of water to hydrogen and oxygen. Another is that the metal may be bound in the soil in a form that resists electromigration. This feature is particularly problematic for lead, cadmium and mercury, because they are metals that form highly insoluble sulfides. Electrokinetic extraction has also been used for uranium and chromium, along with other radioactive metals that have become bound to soils and building materials.

An important aspect of soil decontamination is that the application of an *in situ* electric field under field conditions can modify the characteristics of the soil. Soil characteristics are highly critical, and important properties include the permeability, adsorption capacity, buffering capacity, and degree of saturation. Spatial variation in permeability can nullify the decontamination process if electroosmotic flow in the fine soils induces a pressure gradient that causes a return flow through the coarser region. Although soil properties generally cannot be altered, the selection of operating conditions and the introduction of reagents by electroosmotic purging can lead to successful metal removal. In electromigration the energy cost per mole of contaminant removed will be proportional to the resistivity and current. In electroosmosis, however, it will be proportional to the strength of the electric field and inversely proportional to the contaminant concentration in the pore solution. Thus, the cost to move the pore fluid is proportional to the electroosmotic flux, and independent of the contaminant concentration.[13] Pourbaix diagrams provide a means for identifying pH and redox conditions suitable for mobilizing metals, and optimum operating conditions can then be determined using a mathematical model that incorporates the appropriate metal speciation chemistry.[14]

## 2.3. Electrokinetic Extraction of Copper, Lead, Cadmium and Mercury

As was noted in the previous section, these metals are problematic because they may be bound to the soil in a form such as highly insoluble sulfides that resist electromigration. Nevertheless, electrokinetics has been used to remove lead(II) from kaolinite. A test bed is generated by loading kaolinite specimens with 118 μg to 145 μg of lead(II) per gram of dry soil, which is below the cation exchange capacity of the mineral. Specimens are placed in a horizontal configuration, and the current applied by a colostat that has a maximum voltage output of 150V. A small constant external hydraulic gradient is maintained at the anode side, and the influent-effluent pH, conductivity of effluent, the outflow, and the current, are continuously monitored. After a selected period of time the specimen is removed, and sliced into 10 sections. Each section is analyzed for water content, pH, conductivity, and lead content. The lead content in the different sections shows progressively decreasing amounts for sections that are closer to the anode. Across the entire test bed there is 75-95% removal. The removal of lead(II) is due to flushing of the acid generated at the anode across the specimen, resulting in desorption of lead(II) together with its migration and advection by electroosmotic flow. In short duration time tests where the acid front does not completely flush across the specimen, lead(II) is removed at the acidified anode section while it accumulates at the cathode section.

Pore fluid measurements show that a higher fraction of the lead(II) is in the pore fluid in the anode section than in the cathode section, where little lead(II) is present. This low lead(II) level in the pore fluid can be explained if it is precipitated as lead hydroxide in the higher-pH environment, or adsorbed on the clay surface. The increase in the pore fluid pH and the decrease in conductivity in the cathode section are in agreement with these explanations. Possibly the decrease in conductivity in the cathode section is due to both an ion depletion by precipitation of lead(II), and a migration of other ions, as well as the formation of water by the combination of protons and hydroxyl ions.

During electroosmosis there is a decrease in the electroosmotic flow rate and an increase in the electrical potential gradient across the test bed. The increase in the electrolyte concentration due to the diffusion and advection of the protons into the specimen at the anode and the saturation of the clay surface by protons is the main reason for the decrease in flow rate with time. The increase in the electrical potential across the test bed results from the decrease in electrolyte concentrations due to the increase in pH, and the removal of anions by migration and cations by precipitation at the cathode section.[15]

Electrokinetics can be used for the removal of a range of different metals from a series of different soils. Data obtained for laboratory samples of agrillaceous sand and river mud are collected in Tables 1 and 2 respectively. For the agrillaceous sand samples the energy demand is 30 kWh/ton, and for the river mud it is 100 kWh/ton.

**Table 1. Electrokinetic Extraction from Agrillaceous Sand**

| Metal | Decrease(%) |
|---|---|
| Cadmium | 99 |
| Chromium | 91 |
| Nickel | 85 |
| Lead | 64 |
| Mercury | 67 |
| Copper | 91 |
| Zinc | 81 |

**Table 2. Electrokinetic Extraction from River Mud**

| Metal | Decrease(%) |
|---|---|
| Cadmium | 59 |

| | |
|---|---|
| Chromium | 64 |
| Nickel | 91 |
| Lead | 54 |
| Mercury | 60 |
| Copper | 71 |
| Zinc | 94 |

A difficulty that arises with the electrokinetic extraction of mercury from soils, however, is that it can be reduced from the normal mercury(II) to the mercury(I) state by naturally occurring reducing agents such as humic acids that are already present in soils. Once formed, the mercury(I) state can disproportionate into a mixture of mercury(II) and elemental mercury. Since elemental mercury is uncharged, it cannot be removed by electrokinetic methods. Furthermore, since elemental mercury is volatile, the application of electric currents to soils containing mercury(II) in an immobilized form may result in its reduction to elemental mercury, with the subsequent creation of a greater environmental problem than was originally present at the site.

Field tests on a site contaminated with lead and copper have also been conducted. The site was engineered with the cathode installed horizontally, and the anodes placed vertically, in the sand about two meters apart. The experiment was conducted for 430 hours. For 26 sampling locations the average decrease in lead and copper was 74%.[10]

Various enhancement techniques have been used to remove or avoid precipitates being formed in the cathode compartment. Any scheme needs to have the following characteristics:

- the precipitate should be solubilized or precipitation should be avoided.
- ionic conductivity across the specimen should not increase excessively in a short period of time both to avoid a premature decrease in the electroosmotic transport and to allow the transference of species of interest.
- the cathode reaction should possibly be depolarized to avoid generation of hydroxide and its transport into the specimen.
- if a complexant is used its metal complexes should be soluble across

the entire pH range of the electrokinetic cell.

Two complexants that have been used in the electrokinetic removal of lead are hydrochloric acid and acetic acid. A concern with the former, however, is its possible oxidation to chlorine at the anode. For acetic acid, lead is removed from the region around the anode, but precipitation occurs in the middle section of the soil sample, resulting in clogged soil pores and a cessation of metal transport.[16]

## 2.4. Electrokinetic Extraction of Uranium and the Other Radionuclides

Electrokinetics is a useful method for the removal of radioactive metals from soils. The technique can be used to concentrate the metals to a zone from which they can be readily removed, or they can be migrated into a region of the soil where they are less hazardous. This latter situation prevails for a situation such as the Chernobyl nuclear reaction disaster where large quantities of radioactive material, especially cesium-137, were spread over a wide area of the ground surface. Although complete removal is the preferred solution to the problem, if electrokinetics could be used to move these radioactive metals to lower surface layers before they are dispersed by winds to other locations, it would be a useful interim solution.[17] The ELECTROSORB™ cell developed by Isotron Corporation has been applied to these surface soils to accomplish this goal (Figure 2). Electrokinetics is also a viable method for the removal of actinides from soils. Test bed experiments have been carried out to evaluate the effectiveness of the method for the remediation of uranium(IV). For actinide removal the test cell can be fabricated of teflon. The use of complexants such as the citrate ion to aid the removal of uranium by electrokinetics can also be advantageous because in its common +4 oxidation state the metal is electrostatically tightly bound to the negatively charged zeolite soil structure, and therefore difficult to mobilize. A negatively charged complexant such as citrate reduces this electrostatic attraction, and thereby allows for greater ease of mobilization of uranium(IV) or plutonium(IV).

Another application of electrokinetics is in the removal of radionuclides from solid materials such as steel and concrete. In the era of weapons development, much radioactive material was processed, and as

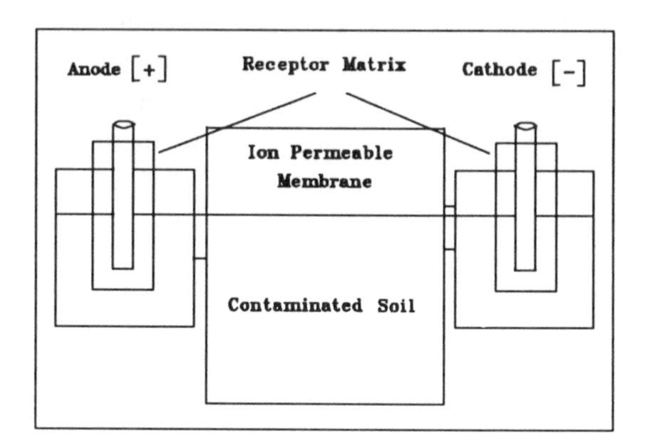


**Figure 2. ELECTROSORB™ cell**

a result many containment areas have become contaminated by radioactive material. The steel and concrete in these areas that has become contaminated by these long-lived radionuclides must therefore be cleansed before these building materials can be disposed of in a conventional manner. Electrokinetics is one of the few methods that can be safely used for the *in situ* remediation of such materials. The need for such techniques will grow as nuclear reactors become decommissioned, especially since there are still close to two dozen Chernobyl-type reactors still in operation.

Depolarizing agents such as ammonium sulfate have been used to enhance the electrokinetic remediation of radionuclides. Ammonium hydroxide is injected at the anode compartment and sulfuric acid at the cathode compartment. The protons generated by electrolysis at the anode are neutralized by the hydroxide ions in ammonium hydroxide, and the hydroxide ions generated by electrolysis at the cathode are neutralized by the protons of sulfuric acid. Ammonium sulfate is then transported across the cell under the electrokinetic conditions.[18, 19]

## 2.5. Electrokinetic Extraction of Chromium

The oxidation-reduction chemistry of chromium is an important factor in determining remediation strategies for soils. Certain forms of chromium(III) can be oxidized to chromium(VI), and chromium(VI) can be reduced to chromium(III) under different soil conditions. Because of the high toxicity of chromium(VI) it is important that its amount can be accurately estimated. It has been proposed that a Potential Chromium Oxidation Score (PCDOS) be used as the basis for setting acceptable maximum limits for chromium(III)-containing wastes in particular soil environments to maintain chromium(VI) levels at or below health-based limits. This PCDOS is based on solubility and site potential of chromium(III), levels of manganese(III) and manganese(IV) since they are species that can act as oxidants of chromium, soil potential for chromium(VI) reduction, and soil pH.[20] Much of the prior engineering work on *in situ* soil reclamation has focused on the removal of organics using technologies developed for the recovery of oil through polymer and/or surfactant flooding of porous media, although many of these strategies can also be used for metals such as chromium.[21-23] Alternatively it is possible to use the method for the simultaneous removal of both metal and organic contaminants, which is a very attractive feature of the technique for sites that are contaminated with both chromium and toxic organics.

Electrokinetic remediation has been used to treat chromium contaminated soils(Figure 3). Complete cleansing of a sample of 50-100 mesh sand containing 10% water and 100 ppm chromium(VI) is obtained

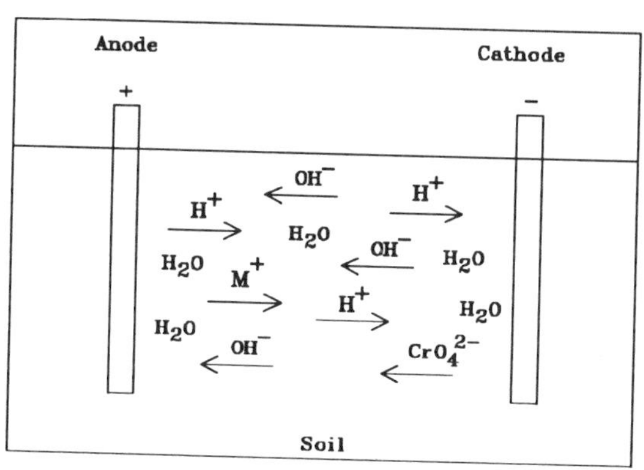


**Figure 3. Chromium(VI) Migration to the Anode**

after 22 hours of treatment.The chromium(VI) that collects near the anode migrates at a rate of 0.40 cm/hr, with a pure water current density of 2.26 mA/cm$^2$.[24] The results of such a homogeneous run with chromate contaminated sand are shown in Figure 4. During the run the current remains constant at 10.3 mA, while the voltage rises from the initial value of 90V to 150V after the 22 hours. No chromium is detected in the sample, except adjacent to the anode. These studies indicate that the electromigration rates normalized by pure water current densities in homogeneous studies can

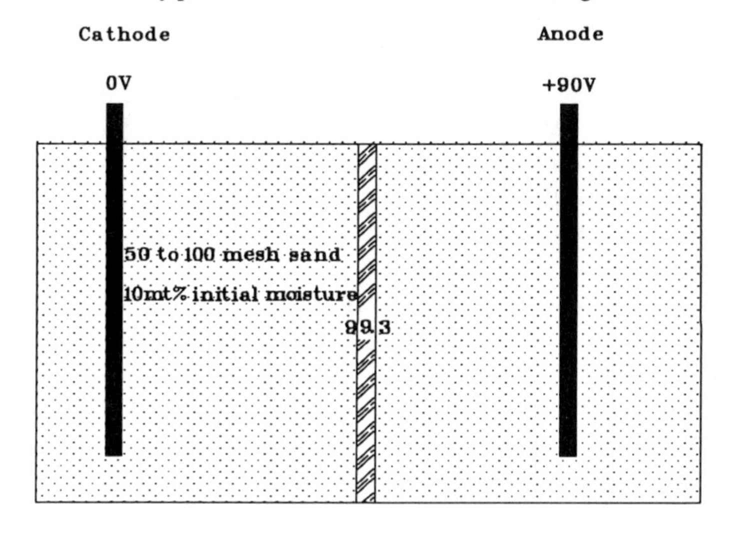


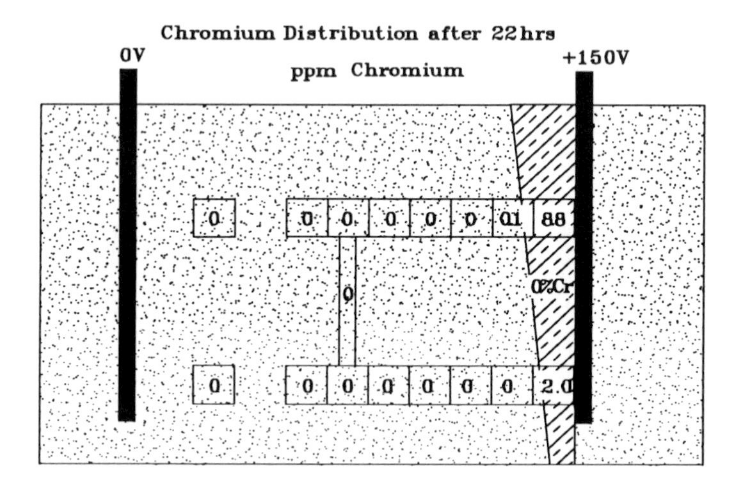


**Figure 4. Electrokinetic Cleansing of Chromium(VI)**

also be applied to heterogeneous soil systems. The *in situ* electrokinetic remediation of chromium(VI) is dependent on the type of clay soils onto which it is adsorbed. In a study of both the clay kaolin and a glacial till, differences were found between the two materials. In kaolin, a sharp pH gradient ranging from ~2 near the anode to 12 near the cathode was developed. However, in the glacial till, alkaline conditions (pH > 8) exist throughout the soil because of its high carbonate-buffering capacity. The migration of anionic chromium(VI) toward the anode is therefore more efficient in the glacial till, and also near the cathode region of kaolin, because of low adsorption of chromium(VI) in high pH environments. Adsorption of chromium(VI) in the low pH regions near the anode in kaolin results in low chromium(VI) removal.[25]

# REFERENCES


1. R. Langeman,"Demonstration of Remedial Action Technologies for Contaminated Land and Ground Water" *NATO/CCMS Pilot Study*, Copenhagen, Denmark, May, 1989.
2. Y. B. Acar, R. J. Gale, G. A. Putnum, J. Hamed, R. Wong, *J. Environ. Sci. Health*, **1990**, *25*, 687.
3. D. D. Runnells, J. L. Larson, *Ground Water Monitoring Rev.*, **1986**, *85*.
4. A. Guzman-Garcia, P. N. Pintauro, M. W. Verbrugge, R. Hill, *AIChE Journal*, **1990**, *36*, 1061.
5. Y. Gur, I. Raina, A. J. Babchin, *J. Colloid Interf. Sci.*, **1978**, *64*, 333.
6. G. J. Booth, *Chem. Phys.*, **1951**, *19*, 391.
7. J. Newman, *Electrochemical Systems*, Prentice Hall, Englewood Cliffs, N. J. 1973.
8. R. M. Noyes, *J. Am. Chem. Soc.*, **1962**, *84*, 513.
9. D. D. Runnells, J. L. Larson, *Ground Water Monitoring Rev.*, **1986**, 81.
10. R. Lageman, W. Pool, G. Seffinga, *Chem. Ind.*, **1989**, 585.
11. A. P. Shapiro, P. Renauld, R. Probstein, *Physicochemical Hydrodynamics*, **1989**, *11*, 785.
12. R. R. Turner, Proc. Electrokinetics Workshop, U.S., D.O.E, Atlanta, GA, Jan. 1992.
13. R. F. Probstein, R. E. Hicks, *Science*, **1993**, *260*, 498.
14. R. E. Hicks, S. Tondorf, *Env. Sci. Technol.*, **1994**, *28*, 2203.
15. J. Hamed, Y. B. Acar, R. J. Gale, *J. Geotech. Eng.*, **1991**, *117*, 241.
16. Y. B. Acar, A. N. Alshawabkeh, *Env. Sci. Technol.*, **1993**, *27*, 2638.
17. D. J. Kelsh, Proc. Electrokinetics Workshop, U.S., D.O.E, Atlanta, GA, Jan. 1992.
18. Y. B. Acar, E. E. Ozsu, A. N. Alshawabkeh, M. F. Rabbi, R. J. Gale, *Chemtech* **1996**, 40.
19. Y. B. Acar, M. F. Rabbi, E. E. Ozsu, *J. Geotech. Geoenviron. Eng.*, **1997**, 239.
20. B. R. James, J. C. Petura, R. J. Vitale, G. R. Mussoline, *J. Soil Contamin.*, **1997**, *6*, 569.
21. A. S. Abdul, T. L. Gibson, *Environ. Sci. Technol.*, **1991**, *25*, 603.
22. A. N. Clarke, P. J. Plumb, T. K. Subramanyam, D. J. Wilson, *Sep. Sci. Technol.*, **1991**, *26*, 301.

23. O. K. Gannon, P. Bibring, K. Raney, J. A. Ward, D. J. Wilson, J. L. Underwood, K. A. Debelak, *Sep. Sci. Technol.*, **1989**, *24*, 1073.
24. E. R. Lindgren, E. D. Mattson, M. W. Kozak, Proc. Electrokinetics Workshop, U.S., D.O.E., Atlanta, GA, Jan. 1992.
25. K. R. Reddy, V. S. Parupudi,. *J. Soil Contam.*, **1997**, *6*, 391.

# 5

# SELECTIVE EXTRACTION WITH CHELATES, MACROCYCLES AND CALIXARENES

## 1. INTRODUCTION

When choosing a compound for use as an extractant for metal ions several factors must be considered. One is how strongly it binds to the metal, and although this stability constant of the metal complex is not the only factor, it is still an important consideration. Since it is likely that the *in situ* environment contains complexants such as chloride, nitrate or sulfate ions, possibly at high concentrations, it is necessary that the added extractant is a competitve complexant under such conditions. As a result, extractants are usually not simple monodentate complexants. Multidentate complexants such as chelates have higher metal ion stability constants than do monodentates, mainly because of entropic effects. They are therefore often chosen as extractants.

A chelating agent by definition has at least two donor atoms that are coordinated to the same metal.[1] Such atoms are usually oxygen, nitrogen and sulfur, but selenium, tellurium, phosphorus, arsenic and others may also be the donors. In designing extractants for positively charged metal ions an additional advantage can be gained if the chelating agent has acidic hydrogens that can be substituted by the metal ions during complex formation. Such a strategy can lead to the formation of an uncharged complex, which can be expected to have a higher compatibility with a hydrophobic organic solvent. The complex to be extracted, even if uncharged, should be as hydrophobic as possible, especially if it is desired to employ a non-polar solvent for the phase transfer extraction. One of the earliest reports of chelate extraction was that of Cazeneuve, who in 1900 described how chromium could be extracted into a benzene solution containing 1,5-diphenylcarbohydrazide.[2] Now there are many examples of organic chelates being employed as extractants for heavy and precious metal ions. The following sections in this chapter summarize the most commonly used chelate extractant agents, along with the heavy metal and precious metal ions for which they have been used. In later chapters the application of these multidentates and macrocycles to the extraction of individual groups of metal ions will be described. In this particular chapter, however, the sequence order is based on complexant type rather than on the metals. The multidentate[3] and macrocyclic[4] agents in the following sections in this chapter are divided into four major groups depending on their structure and donor atom types. Because a large part of this book is focused on heavy metals that preferentially coordinate to soft donor ligands, the topic sequence in this chapter reflects this priority.

Macrocycles are also widely used as extractants, especially for alkali and alkaline earth metal ions. These complexants can be synthesized with the same donor atoms that are used in the design of chelates, but macrocycles have two additional advantages. One of these is that there is a macrocyclic effect that gives such complexes a higher stability over their open chain analogs. A second one is that selectivity of metal ion binding can be achieved by matching the size of the cavity of the macrocycle with that of the metal ion. The complexation of both multidentate and macrocyclic ligands have been investigated by molecular mechanics calculations, and patterns observed that can aid in choosing selective complexants.[5]

Recently calixarenes, cyclic condensation products derived from a phenol and formaldehyde, are being used as extractants for metal ions. These are conformationally mobile oligomers that can be chemically modified to function as preorganized chelate-type complexants. An advantage of calixarenes is that in addition to having a cavity for occlusion of molecules, they can also be designed to offer an array of functional groups that can coordinate with the metal ion. The current state of the art in extractant design is the incorporation of a macrocycle into the calixarene framework. This has allowed for the development of new extractants that have extremely high selectivities for metal ions such as cesium(I).

Other strategies have been used to occlude metal ions, and these can also be potentially applied to the liquid-liquid phase extraction of metals. One of these involves the use of concave hydrocarbons as host molecules. These compounds represent three-dimensionally clamped analogs of $\pi$-

prismands, and they show exceptionally high selctivity for metal ions.[6] Another approach involves the use of molecular replication to generate molecules that have selective metal ion recognition. These systems are often assembled using hydrogen bonding and molecular self-assembly to generate the compounds.[7,8] Metallomacrocycles are yet another class of compound that may become important in the development of new extractants, especially for situations where very high selectivity is required.[9]

Linear free energy relationships and pairwise interactions in supramolecular chemistry have been reviewed. This review has relevance for both organic molecules and metal ions as guests in supramolecular systems.[10]

## 2. MULTIDENTATE LIGANDS

This section on multidentate ligands covers all types of complexants that can coordinate to a single metal ion *via* two or more ligating atoms. This categorization therefore includes chelate ligands that coordinate through two donor atoms. All ring sizes are included, but the fact that a complexant has the potential to form a chelate complex does not necessarily mean that it must bind to a metal ion in this manner. This situation is particularly relevant for potential chelates such as carboxylates and dithiocarbamates where coordination can be either *via* a single oxygen or sulfur donor to give a monodentate complex, or through a pair of these donor atoms to give four-membered ring chelate complexes. In general, the most stable complexes are found with five-membered rings, but four-, six- and other ring sizes are also observed.[3,11,12]

Multidentate ligands can be prepared that are either neutral, anionic, or cationic. Anionic derivatives are often prepared by attaching carboxylate functionalities to terminal sites, and cationic structures are frequently obtained by the inclusion of alkylammonium groups within the structure. Because carboxylic acids and alkylamines are respectively weak acids and bases, by changing the solution pH it is possible to control the charge on the ligand and metal complex, and to use it to induce a charge-switched metal binding-release equilibrium. When choosing multidentate ligands for extraction purposes it is useful to be aware that for many complexes the preferred coordination number is six, and that polarizable soft and non-polarizable hard metal cations and ligand bases have a preference for each other.

Preorganization is also an important aspect of multidentate complexation.[13] Although binding to multiple coordination positions leads to a favorable translational entropy contribution, the non-bonding interactions that result from the ligand adopting the necessary conformation to achieve multidentate coordination can reduce this entropic advantage. These unfavorable non-bonded interactions can be reduced by having a geometric arrangement in its uncomplexed *apo* form that is close to that adopted in its metal complexed form.

## 2.1. Dithio- and Diselenocarbamates

Dithiocarbamates, especially as their ammonium or sodium salts, are widely used as liquid-liquid extractants for transferring heavy metal ions from an aqueous into an organic phase. In general, an excess of dithiocarbamate is added to the aqueous solution containing the metal ions to be extracted, and the complexed neutral species are then extracted into the organic phase. The extraction properties of sodium diethyldithiocarbamate (Na-DDC),[14, 15] zinc dibenzyldithiocarbamate (Zn-DBDTC),[16] hexamethyleneammonium hexamethylenedithiocarbamate (HMA-HMDC),[17-19] ammonium pyrrolidine dithiocarbamate (APDC),[20] diethylammonium diethyldithiocarbamate (DDDC),[20] ammonium tetramethylene dithiocarbamate (ATMDTC),[21] ethylene *bis*-dithiocarbamate, propylene *bis*-dithiocarbamate, butylene *bis*-dithiocarbamate, pentamethylene *bis*-dithiocarbamate and hexamethylene *bis*-dithiocarbamate have all been used for heavy metal ions.[22] The structure of the dialkyldithiocarbamate anion **1** shows the presence of both thiolate and thiocarbonyl sulfurs.

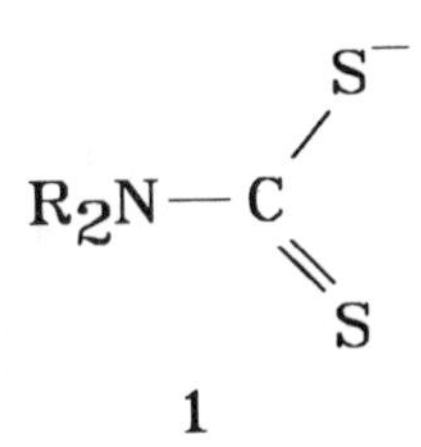


1

Sodium dibutyldiselenocarbamate is the only selenocarbamate that has been widely used for the extraction of heavy metal ions.[23] Both dithio- and diselenocarbamates show high a pH dependent extractability but overall poor selectivity. These ligands form covalent species with most transition and heavy metals, with the latter being almost always extracted together into the organic phase. Nevertheless, despite this lack of selectivity, there are several methods available for using these ligands to concentrate heavy metals into an organic phase until they reach limits detectable by atomic absorption spectroscopy (AAS) and inductively coupled plasma-atomic emission spectroscopy (ICP-AES). For example, dithiocarbamate complexes of heavy metals have different retention times, and thus they can be used in the gas chromatographic (GC) and high pressure liquid chromatographic (HPLC) determination of heavy metals present in aqueous solution.[14, 18, 19] A second example involves *bis*-(carboxylmethyl)dithiocarbamate (CMDTC) that has been pre-treated with stannous chloride and hydrochloric acid. This product can be incorporated onto the polystyrene-divinylbenzene resin (XAD-4), and the modified resin used for the preconcentration of palladium, platinum and rhodium.[24] These resins can be very beneficial in selectively transferring heavy metals from aqueous solutions onto the resin, with subsequent elution resulting in their concentration in the eluate.

Long chain alkyl and choline substituted xanthates have also been used as extractants for cadmium(II), although dithiocarbamates are generally preferred because of the ready loss of carbon disulfide from the xanthate ligand.[25] Other sulfur derivatives are also potential extractants. Among these are *bis*- dithiocarboxylates **2** that have different spacers between the dithiocarboxylate groups.

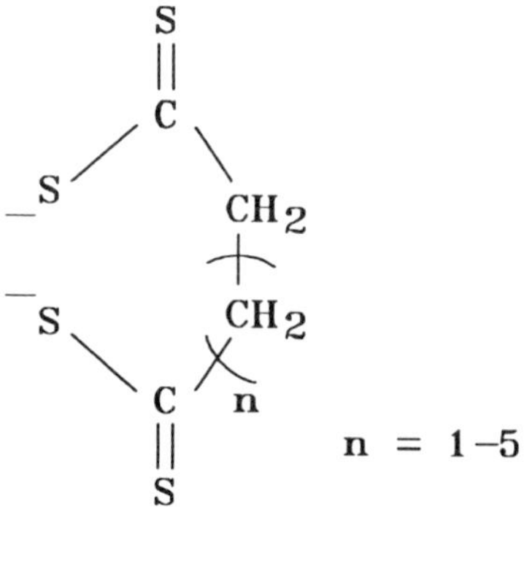

$n = 1\text{–}5$

**2**

## 2.2. Aminopolycarboxylic acids, amides and aminimides.

Aminopolycarboxylic acids are a class of water soluble chelating agents that are widely used for the extraction of heavy metal ions from soils and other solid phases into water or, when attached to polymeric support, for heavy metal removal from aqueous solution. Iminodiacetic acid, for example, when attached to a cellulose sorbent, leads to a resin **3** that is used for the pH dependent removal of chromium(III), copper(II), lead(II) and nickel(II) from water.[26]

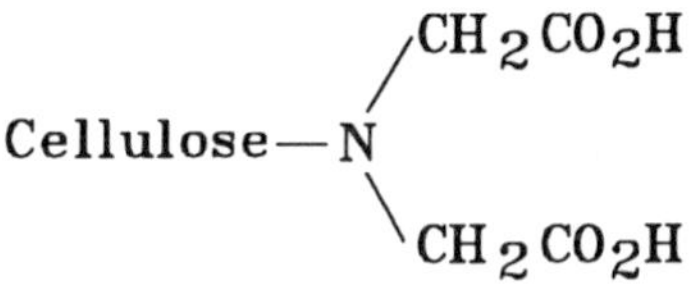

Cellulose—N($CH_2CO_2H$)$_2$

3

Metal recovery from the material can be accomplished by elution with 3M nitric acid. The derivative *N*-2-acetamidoiminodiacetic acid (ADA, **4**) selectively extracts Zn(II) from spiked soils in the presence of iron(III) and calcium(II),[27] and both **4** and *S*-carboxymethyl-*L*-cysteine (SCMC, **5**) are

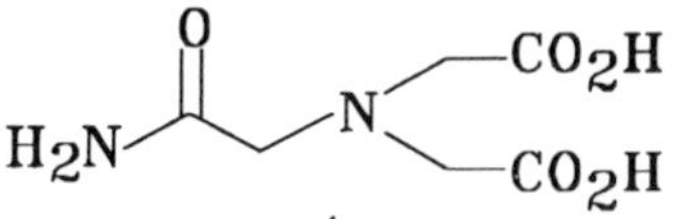

$H_2N$–C(=O)–$CH_2$–N($CH_2CO_2H$)$_2$

4

$H_2N$–CH($CO_2H$)–$CH_2$–S–$CH_2$–$CO_2H$

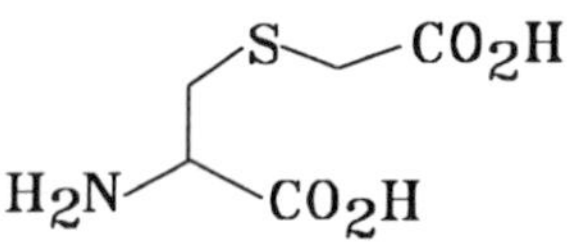

5

used for the extraction of both lead(II) and copper(II) from contaminated soil.[28,29] One of the best known chelates, the hexadentate ethylenediaminetetraacetic acid (EDTA),[30] has been used for the removal of cadmium, chromium, copper, iron, lead, nickel and zinc from sludge. The high toxicity of EDTA, however, precludes its more widespread use. The closely analogous compounds nitrilotriacetic acid (NTA) and ethyleneglycol *bis*-(2-aminoethylether) tetraacetic acid (EGTA)[31] have been similarly used. Diethylenetriaminepentaacetic acid (DTPA) has been used for the extraction of lead, nickel, copper and zinc.[32] Another water soluble agent used for the modification of silica gel is Methylthymol Blue, which is the 3,3'-bis[*N,N*-di-(carboxymethyl)aminomethyl]thymol sulfonephtalein sodium salt **6**.[33] The

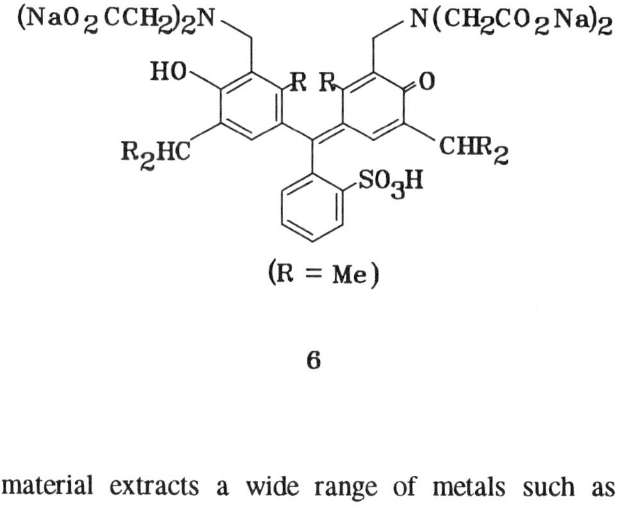

$(NaO_2CCH_2)_2N$ $N(CH_2CO_2Na)_2$
HO R R O
$R_2HC$ $CHR_2$
$SO_3H$
(R = Me)

**6**

modified material extracts a wide range of metals such as calcium, magnesium, aluminium, copper, iron, cobalt, nickel, zinc, cadmium, lead, chromium and bismuth.

Extractants can also be used to mobilize heavy metals from a soil matrix. Thus the concentration of soluble nickel, lead and zinc increases in soil-sludge mixtures upon addition of the complexant nitrilotriacetic acid (NTA). A lesser effect is found for copper, and only minimal effect is observed with this extractant for manganese.[34] After the chelating extractant APDC has been used to preconcentrate cadmium, copper, lead and zinc from seawater, nitric acid can then be used to recover these metals back from the organic phase.[35] Sandy soils contaminated with cadmium, zinc, copper and lead can be cleansed by sequential washing first with $Na_4P_2O_7$, then with the chelate extractant EDTA.[36]

Commercial extractants have found application for heavy metal removal. A microporous hydrophobic hollow fiber membrane-based solvent extraction device, along with the chelate *anti*-2-hydroxy-5-nonylacetophenone oxime (LIX84), can be used to extract copper(II).[37] Silica gel impregnated with a mixture of Aliquat 336 (methyltricaprylammonium chloride) and Calcon **7** can be used for the preconcentration of magnesium, calcium, copper, zinc, aluminum, chromium and iron. The reagent is an effective extractant for even trace concentrations

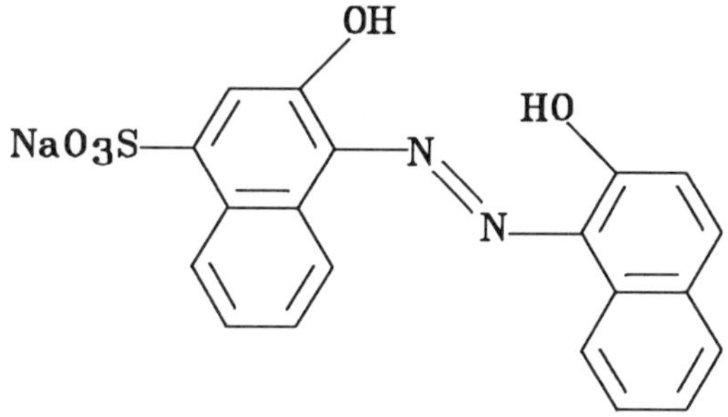

**7**

of these metals. The entire study covers thirty three metals over pH range of 1-9. The material can be used to separate trace amounts of copper(II) and zinc(II) from higher concentrations of calcium(II), magnesium(II), sodium(I), and potassium(I). This separation, along with the adsorption of the other heavy metals, is most effective when the solution containing the metal ion has a pH in the range of 2-4. The retained copper and zinc can be subsequently eluted from the column with 0.5M perchloric acid, although for concentrations of up to 5M perchloric acid or 10M hydrochloric acid, no Calcon is removed from the support. Since Aliquat 336 has an ion-exchangeable chloride ion, the material can also be used for the adsorption of the oxyanions of chromium(VI), molybdenum(VI) and selenium(IV). Silica impregnated with a mixture of Aliquat 336 and Titan Yellow can be used as a extractant for the metals calcium, magnesium, aluminum, copper, iron, nickel, cobalt, cadmium, zinc, lead, mercury and chromium. Elution from the resin can be effected with a dilute solution of either hydrochloric acid or perchloric acid.[39]

Amides and aminimides bearing lipophilic groups are water insoluble complexants that are useful for the extraction of metal ions from water into an organic phase. Disubstituted pyridines **8-10** selectively extract

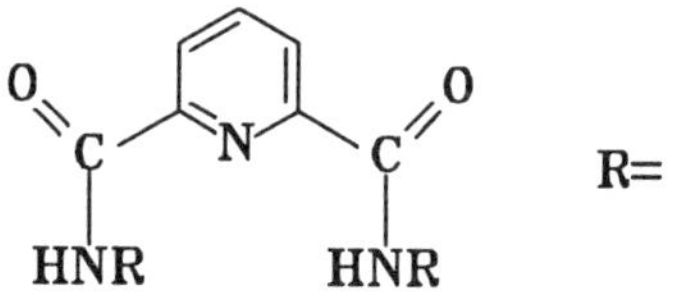

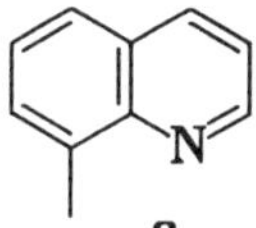

R=

8

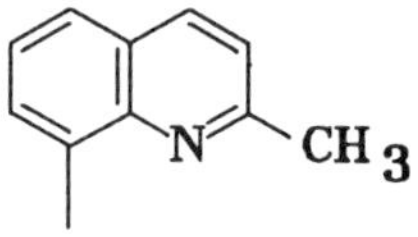

9

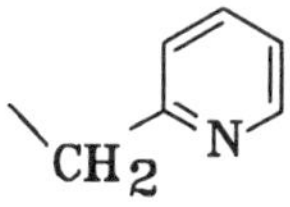

10

copper(II) in the presence of nickel(II), cobalt(II) and zinc(II).[40] *N*-Acylated ureas, thioureas and selenoureas **11** extract almost all heavy metal and

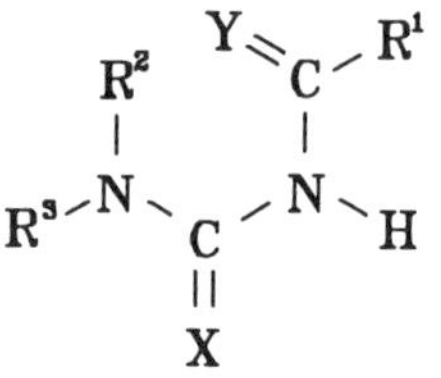

X = O, S, Se
Y = O, S
$R^1$, $R^2$, $R^3$ = alkyl

11

transition metal ions.[41] The aminimide **12**, however, selectively extracts copper(II), nickel(II) and silver(I).[42]

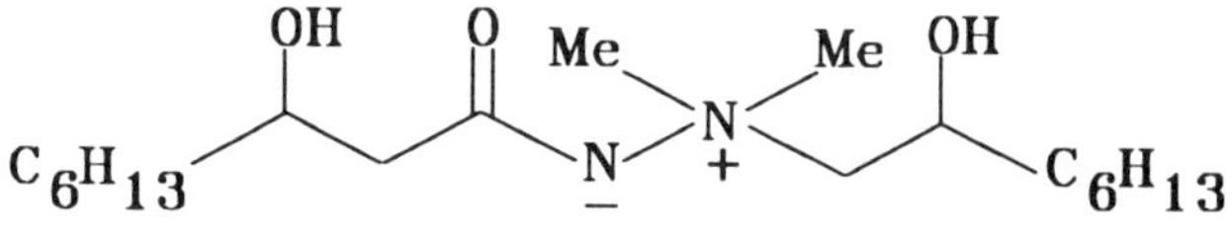

12

The series of azoles pyrazole, imidazole, 1,2,4-triazole and tetrazole, when immobilized onto either poly(glicidyl methacrylate) **13** or its sulfur analog **14,** are highly selective extractants for copper(II) in the presence of

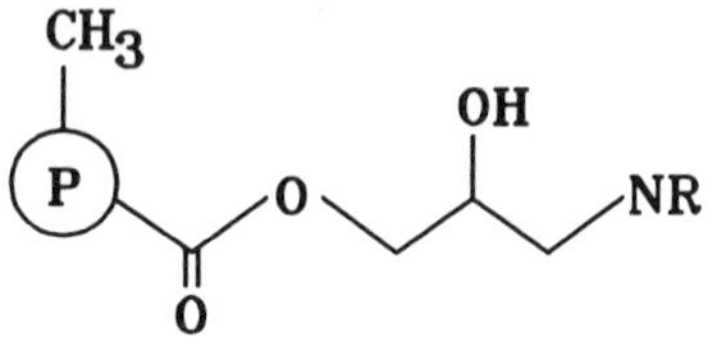

$CH_3$ OH P O NR O

13

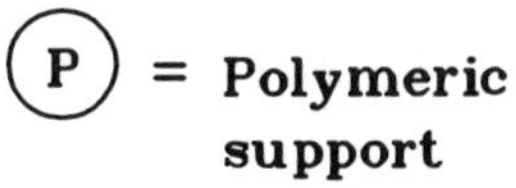

P = Polymeric support

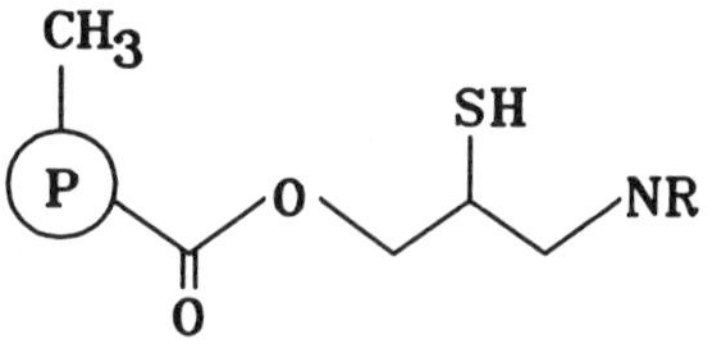

$CH_3$ SH P O NR O

14

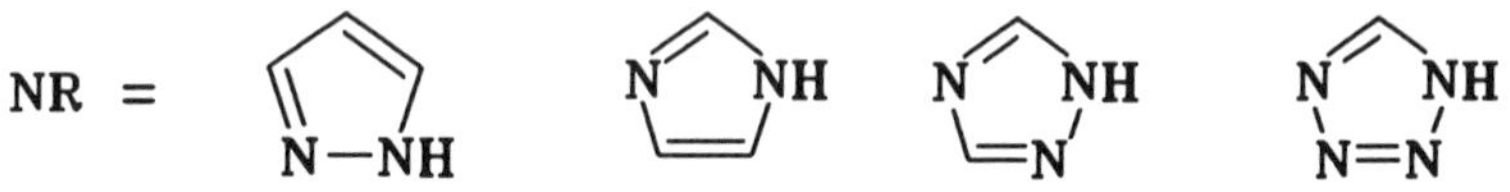

NR = N—NH N NH N NH N NH N N=N

nickel(II), cobalt(II), zinc(II), and cadmium(II). There is no metal ion uptake onto these modified resins below pH 2.5, and the resins can be regenerated by the addition of a 1M sulfuric acid solution.[43]

## 2.3. Oximes and hydroxyaromatic derivatives.

Water soluble oximes and hydroxyaromatic derivatives form stable chelates with many metal ions that are extractable into organic solvents. The compound 1-(2-hydroxy-5-β-hydroxyethylsulfonyl-phenylazo)-2-naphthol (Hyphan I, **15**) forms stable extractable 1:1 complexes with copper(II)

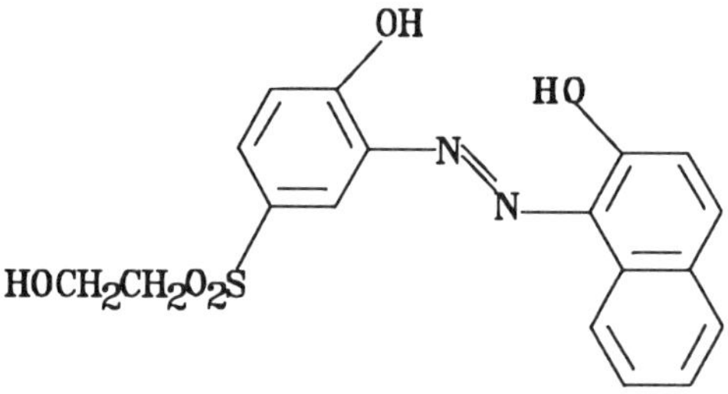

**15**

and uranium(VI) in weakly acidic solutions, while at pH > 8 this complexant can be used to quantitatively extract beryllium(II), bismuth(III), chromium(III), manganese(II), lead(II) and zinc(II).[44] A similar derivative, 1-(2-pyridylazo)-2-naphthol (PAN, **16**) can be used in conjunction with

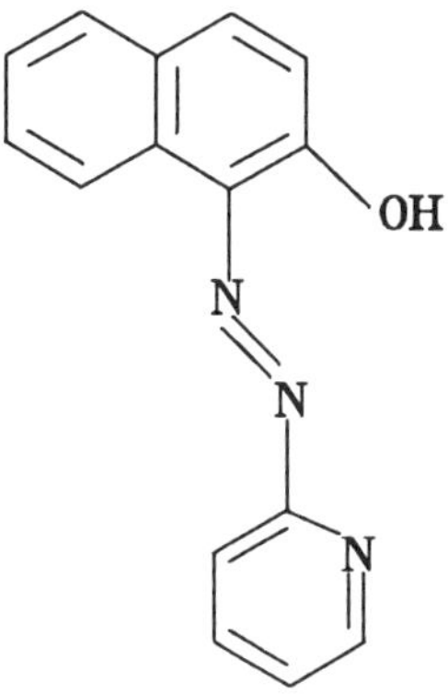

**16**

liquid chromatography for the separation of the platinum group metals palladium(II), platinum(II) and rhodium(III) from nickel(II), copper(II), cobalt(III) and other heavy metals.[45] The rapid separation of palladium(II) from other metals has been achieved by extraction with the acetophenone hydroxylamine complexant HINAP, **17**.[46] The complexants HINAP, 5,7-dichloro-8-quinolinol, the di-2-pyridylimine complexant salicyloylhydrazone

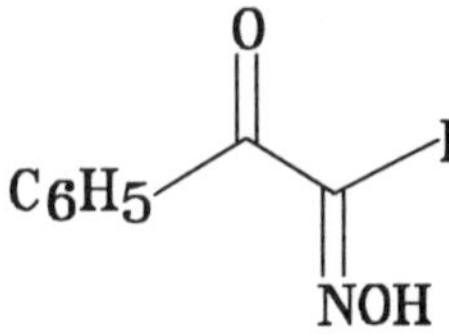

**17**

(DPKSH, **18**), and its benzoylhydrazone derivative (DPKBH, **19**) are used for the preconcentration and subsequent determination of traces of heavy metals.[47, 48] The compound 5,7-dichloro-8-quinolinol has also been used

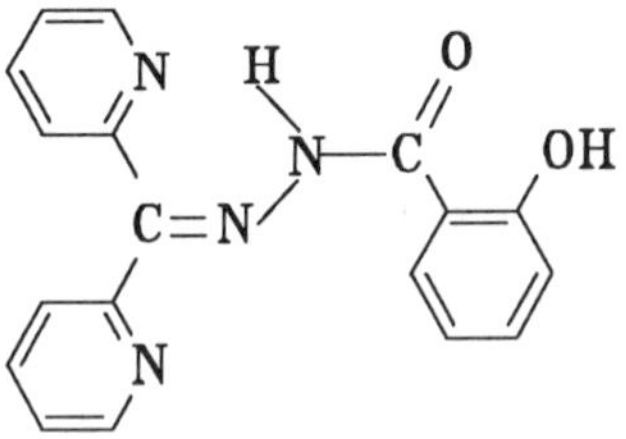

**18** 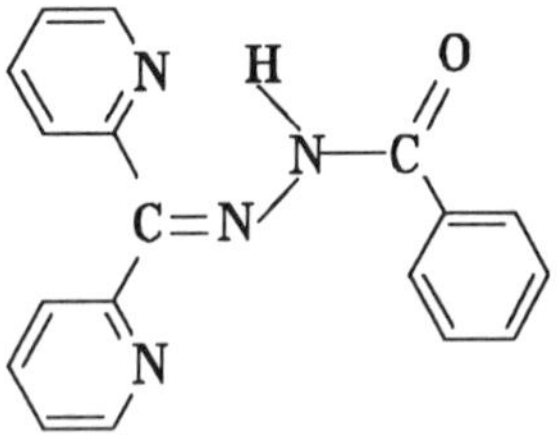 **19**

the preconcentration of for cadmium, lead, zinc, iron, copper, nickel, molybdenum, and vanadium, and the compounds DPKSH and DPKBH have also been used for iron, cobalt, nickel, copper and zinc. Another chelating agent from this group that has been used is the 3-[(dioctylamino)methyl] alizarin **20**. Compound **20** has been used as an extractant for the metals

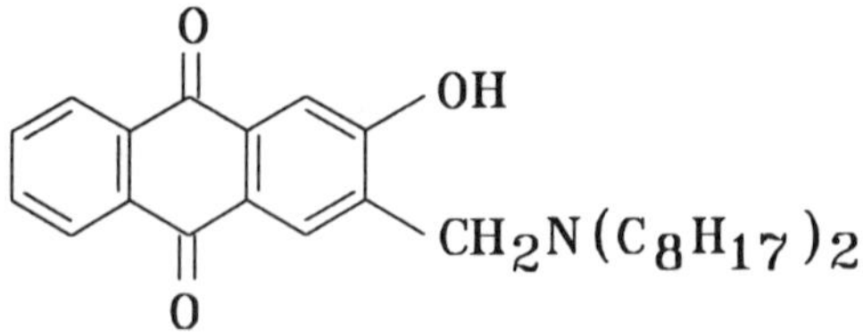

**20**

copper, zinc, cadmium, cobalt, manganese, nickel and iron.[49] Other similar type extractants that have been used are 2-nitroso-1-naphthol, **21** 1-nitroso-

2-naphthol, **22**[50] and 4-(2-pyridylazo) resorcinol (PAR, **23**).[51] Alkyl-8-hydroxyquinolines show high extractablilities for copper(II), nickel(II),

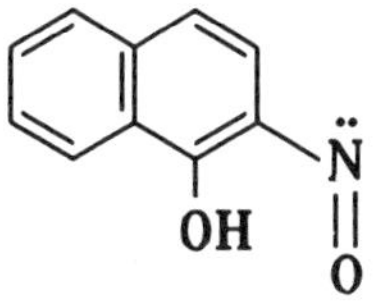

zinc(II), cadmium(II) and mercury(II). The addition of decanol to the organic phase causes a significant increase in metal extraction because of the alcohol solvation of the extracted complex.[52] Of these complexants that have been investigated, the 2-nonyl-8-hydroxyquinoline derivative (*N,O*) is the best extractant. Molecular modeling calculations for the extraction of cadmium(II) support the formation of a hexacoordinate cadmium complex of the $Cd(NO)_2(decanol)_2^{2+}$ or $Cd(NO)_2(H_2O)(decanol)^{2+}$ type.[53] Salicylic acid **24** in the presence of triisoamyl phosphate (TAP) has

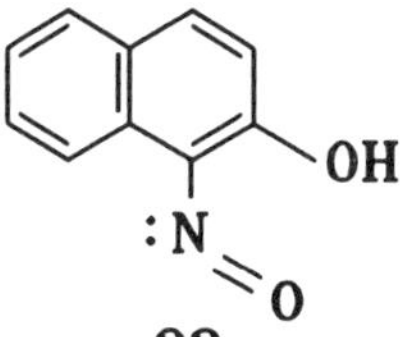

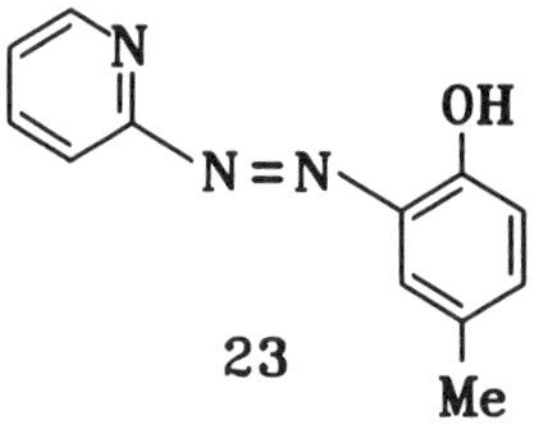

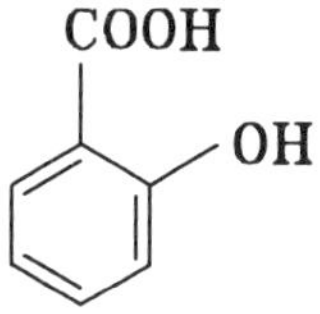

been used for the extraction and separation of copper(II) and iron(III). This mixture has also been used for the partial extraction and separation of a mixture of chromium(III), manganese(II), nickel(II), zinc(II), lead(II) and cadmium(II).[54]

## 2.4. Others.

Numerous other complexants have been used for the extraction of metal ions. These are included in this section in an uncharacterized listing. The majority of them bind to the metal ion *via* nitrogen or oxygen donor atoms, but other extractant types are included in this section.

A series of chelating agents that have been used for the extraction of heavy metal ions are lipophilic and oligomeric polyethers **25** and **26** that

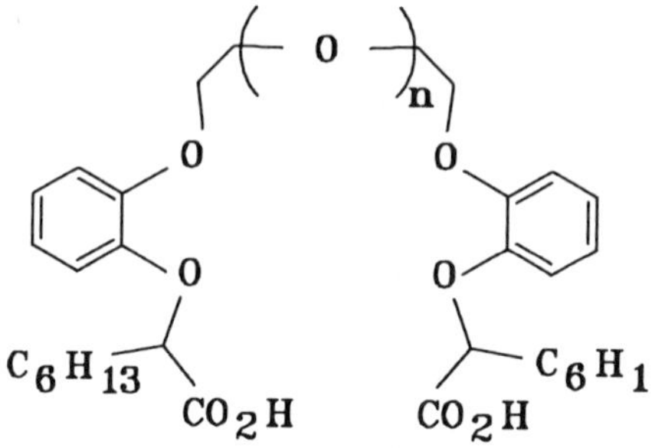

$n = 0, 1$

25

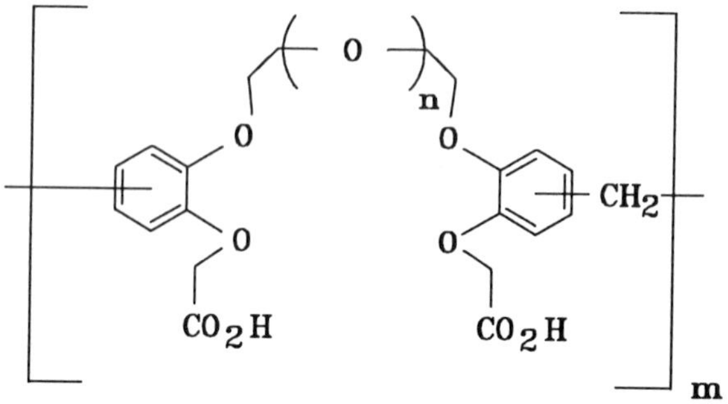

$m = 0, 1$

26

exhibit high selectivity toward lead(II) over copper(II).[55] This selectivity for lead(II) may be due to its preference for oxygen donor ligands; by contrast, copper(II) forms strong complexes with nitrogen donor ligands. Another class of compounds are the bipyridyl-substituted polyethers **27** that selectively extract zinc(II) and copper(I) over iron(III), cobalt(II), nickel(II), manganese(II) and chromium(III).[56] Multidentate tripodal ligands such as

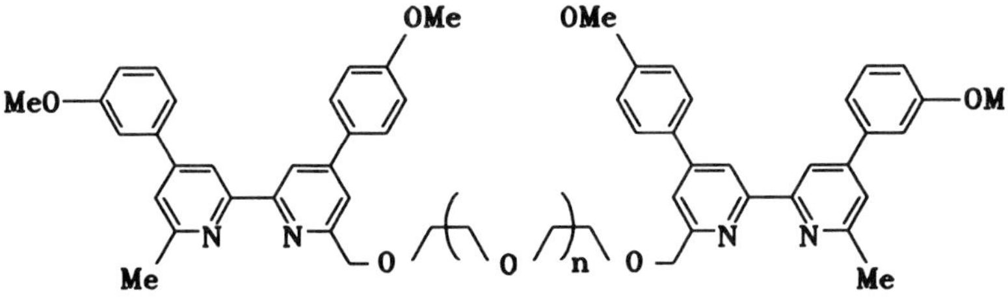

n = 3, 4

27

those with oxygen, nitrogen and sulfur donor functionalities **28** (A-E = O, N, S; R = Ph. benzyl, $C_6$-$C_{12}$ alkyl groups)can be good metal ion extractants. Two such *N*, *S* compounds **29** and **30** exhibit high selectivity for silver(I)

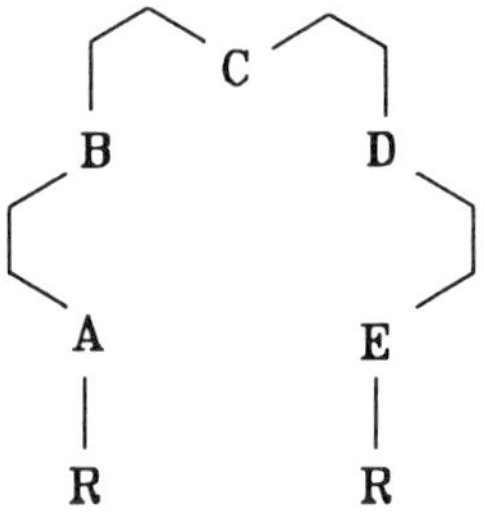

28

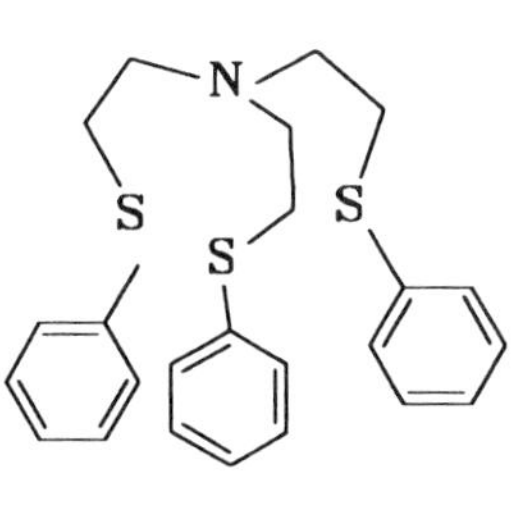

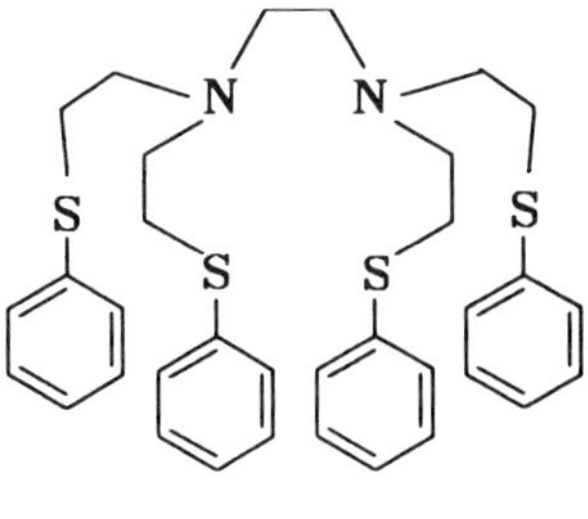

29 30

over lead(II).[57] A series of thiophosphonyl derivatives **31-33** that can coordinate *via S*, *O* or *S*, *S* combinations efficiently extract palladium(II) and

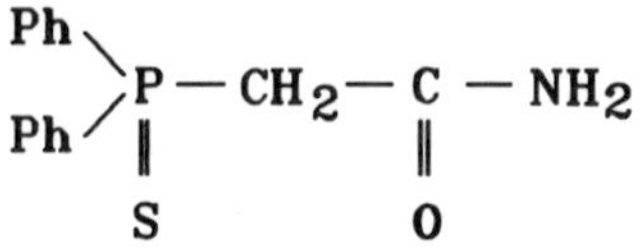

Ph
P — $CH_2$ — C — $NH_2$
Ph
S O

31

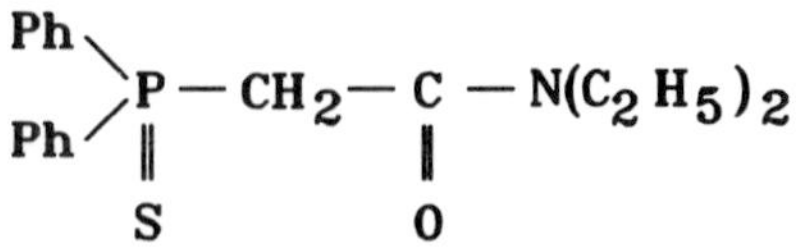

Ph
P — $CH_2$ — C — $N(C_2H_5)_2$
Ph
S O

32

Ph
P — $CH_2$ — C — $N(C_2H_5)_2$
Ph
S S

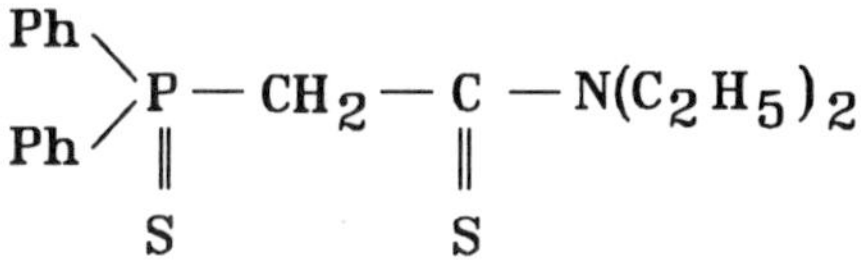

33

silver(I).[58] A series of preorganized multidentate *N,O*-chelates **34-36** have been used for the extraction of lead(II) over cadmium(II), zinc(II) and copper(II).[59]

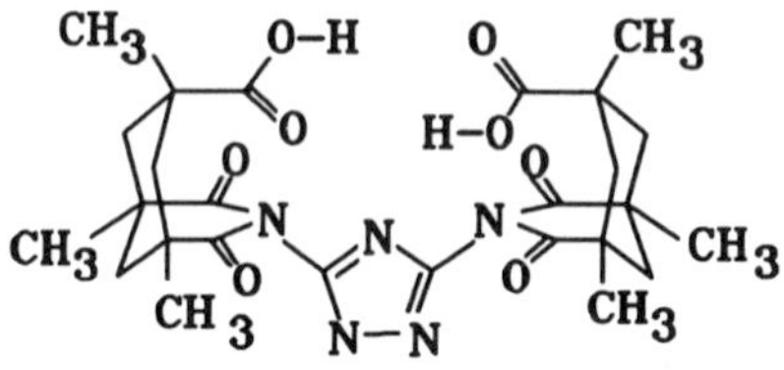

$CH_3$ O–H O $CH_3$
O H–O
O O
$CH_3$ N N N $CH_3$
O O
$CH_3$ N–N $CH_3$

34

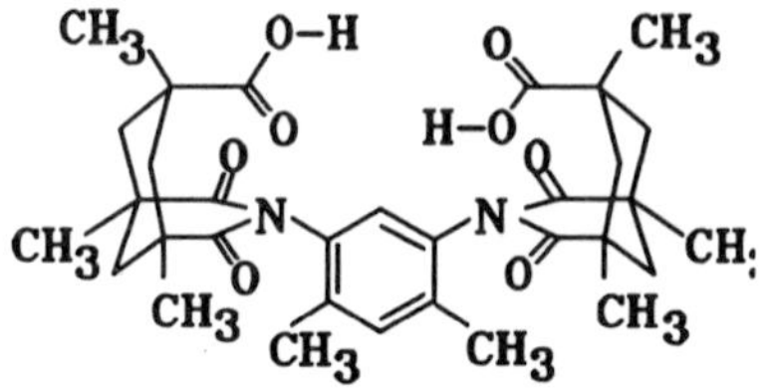

$CH_3$ O–H O $CH_3$
O H–O
O O
$CH_3$ N N $CH_3$
O O
$CH_3$ $CH_3$ $CH_3$ $CH_3$

35

$CH_3$ O–H O $CH_3$
O H–O
O O
$CH_3$ N N N $CH_3$
O O
$CH_3$ $CH_3$

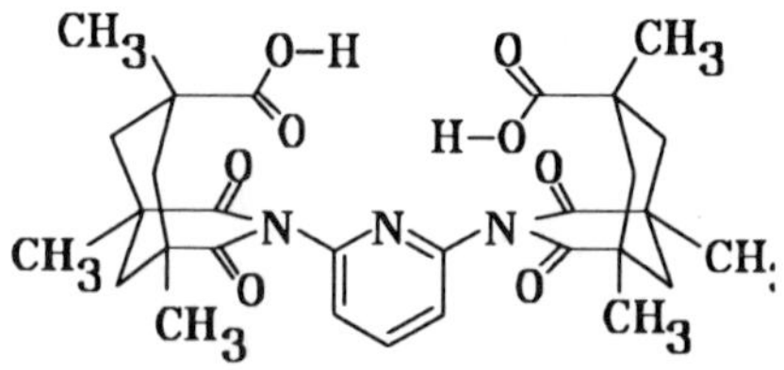

36

Bis(phenoxyalkyl) sulfane podands have been prepared that can be structurally modified to be selective extractants for silver(I). The highest extractability is observed with the *bis*-[(5-hydroxy-3-oxapentyloxy)-3-

phenoxy-prop-2-yl] sulfane.[60] Comparative extractabilities of mercury(II), zinc(II), cadmium(II) and copper(II) by Aliquat 336, di(2-ethylhexyl) phosphoric acid, acyl thiourea and oxathiaazaalkanes have been examined. The advantages and disadvantages of each system are discussed, which results in the finding that Aliquat 336 is effective for mercury(II) chloride solutions, and also for the extraction of cadmium and zinc from copper. Metal recovery can be achieved using ethylenediamine.[61] A systematic study of the extraction of mercury(II) chloride by a series of thiaoxaazaalkanes has shown that the extractability decreases in the order NN > NS > NO > SS > SO. The highest extractability is observed where five-membered chelate ring formation can occur.[62] A series of 1,2-dithiolate ligands **37-39** (X,Y

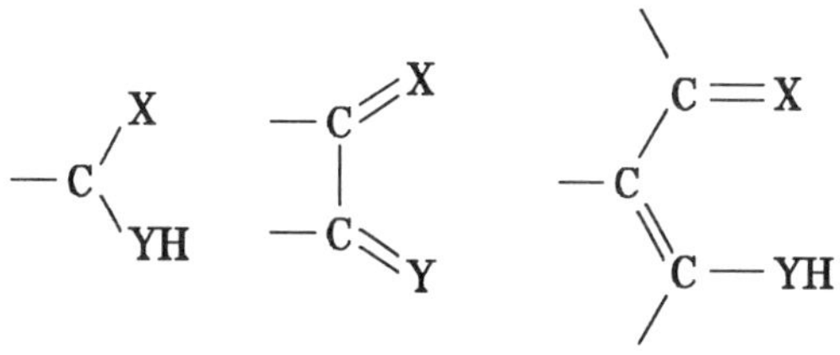

37 38 39

= O, S, Se) have been used as extractants for a broad range of transition metal ions.[63] Selectivity, however, is quite low.

The compound pyridine-2-aldehyde-2-quinolylaldehyde (PAQH,**40**) is an effective extractant for transferring nickel(II), copper(II), cobalt(III)

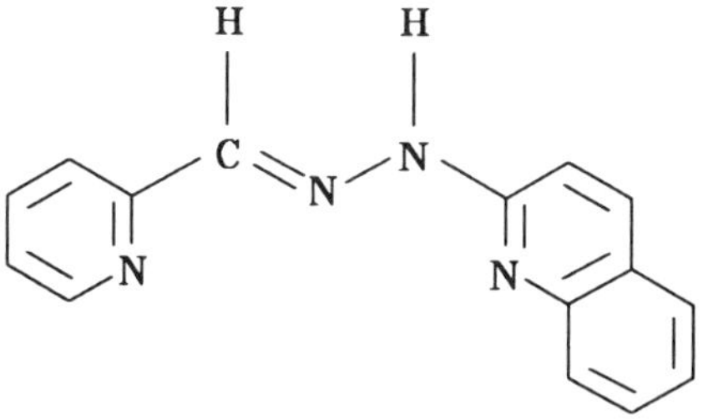

40

iron(II), zinc(II) and cadmium(II) into benzene, isoamyl alcohol and methyl isobutyl ketone.[64] A 30% *N,N*-methylene-*bis*-(acrylamide) cross-linked poly[*N*-((acryloylamino)methyl)mercaptoacetamide] resin **41** has been

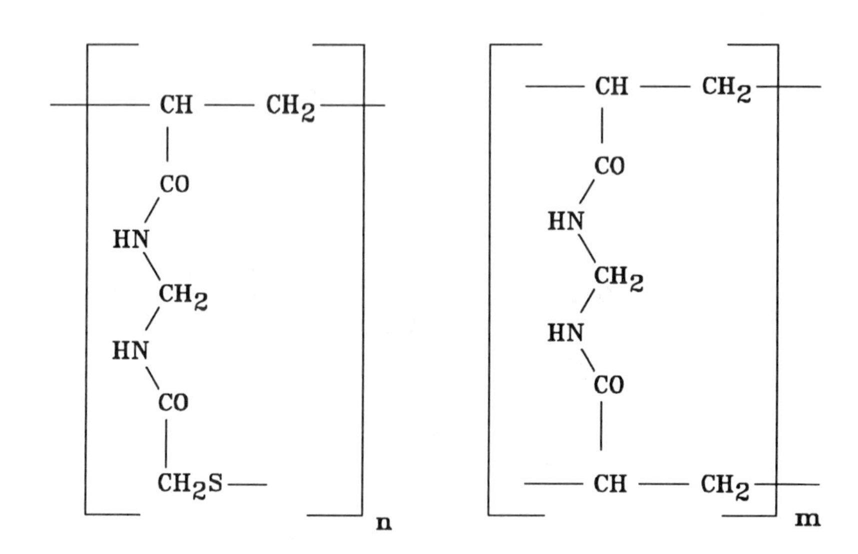


**41**

used for the extration of manganese(II), iron(II), cobalt(II), nickel(II), copper(II), zinc(II), silver(I), cadmium(II), mercury(II), lead(II), and uranium(VI). The resin activity at pH 5.5 follows the metal ion sequence zinc(II) < cadmium(II) < lead(II) < copper(II) < mercury(II).[7]

## 2.5. Extractants for Supercritical Carbon Dioxide

Supercritical carbon dioxide, along with chelates that are soluble in liquid carbon dioxide, have been used for the extraction of mercury, arsenic and lead from soils. Chelate solubility in carbon dioxide has been achieved by the incorporation of long fluorinated or silicone-based tails. Krytox ($CF_3[CF_2CF(CF_3)O]_{14}CF_2CF_2$-) has been used as the fluorinated tail, and the functionality -$O(CH_2)_3[SiO]_3Si(CH_2)_3Si[OSi(CH_3)_3]_3$ as the silicone tail.[65] The compound 3-(2-aminoethylamino)propyltrimethoxysilane (Dow Corning Z-6020) has been attached to silica gel. This reagent, which structurally has a substituted ethylenediamine moiety present, can be used to extract copper(II), zinc(II), chromium(III), manganese(II), lead(II) and mercury(II).[66] Synergistic extraction of cadmium(II), cobalt(II), copper(II), nickel(II), lead(II) and zinc(II) with dithizone and tributylphosphine oxide has also been achieved. The pH limit for each metal ion has been calculated.[67] A group of fluorinated chelating agents **42-45** that possess high solubilities in liquid

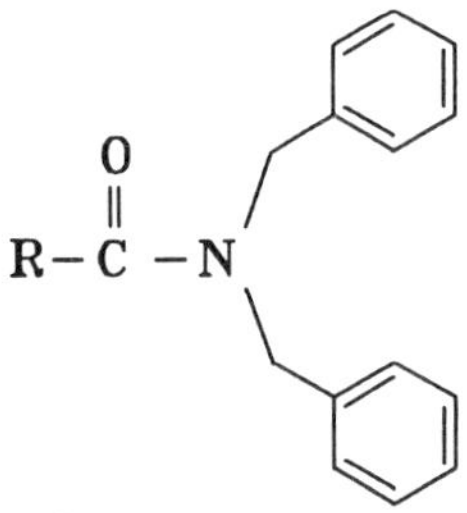

R–C(=O)–N($CH_2C_6H_5$)$_2$

42

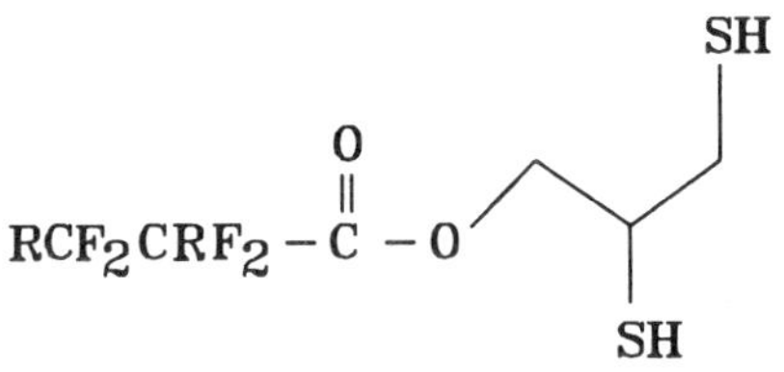

$RCF_2CRF_2$–C(=O)–O–$CH_2$–CH(SH)–$CH_2$SH

43

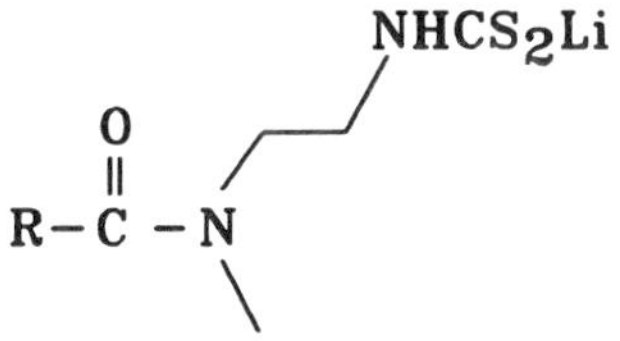

R–C(=O)–N($CH_3$)–$CH_2CH_2$–$NHCS_2Li$

44

$$R = CF_3 \left[ CF_2-\underset{\displaystyle CF_3}{\underset{|}{CF}}-O \right]_{14} CF_2CF_2$$

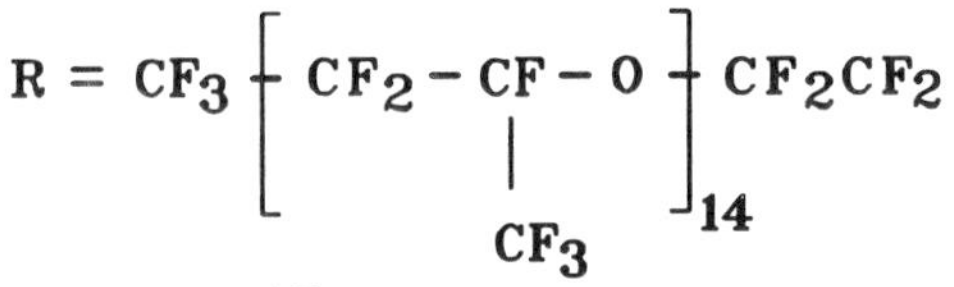

45

carbon dioxide can be used for the supercritical extraction and recovery of lead(II), mercury(II) and arsenic(III) from soil matrices.[68]

## 3. MACROCYCLIC LIGANDS

Recently there has been a considerable amount of work carried out on the interaction of heavy and precious metals with macrocyclic compounds, including crown and lariat ethers, cryptands and calixarenes.[69-71] Within this work there have been many studies of macrocycles binding with transition metal ions in a single solvent.[72,73] Nevertheless only a few extraction studies of transition metals salts with macrocyclic ligands have been reported.[73,74] In general oxygen macrocycles such as crown ethers are effective for the extraction of alkali and alkaline earth metal salts, but not for transition metal salts. By contrast, the nitrogen analogue macrocycles are effective for transition metal salts but not for alkali metal salts.[75-87-] The following sections summarize the available macrocyclic extractant agents and the metal ions that they extract.

## 3.1. Oxygen macrocycles

While macrocyclic compound bearing only oxygen donor atoms are poor extractants for heavy and precious metal ions, they do, however, appear to be good extractants for oxophilic ions such as silver(I), thallium(I), and barium(II). A series of ring-contracted and ring-enlarged crown ethers **46-58** have been reported, some of which selectively extract silver(I), thallium(I) and barium(II) in the presence of other alkali and alkaline earth metals.[88-91] For compound **46**, n = 1-4; for compound **47**, m = 1, n = 2; m = 2, n = 2; m = 2, n = 3; m = 2, n = 4; m = 3, n = 3; m = 2, n = 6; m = 4, n = 4; and for compound **57**, m = 1, n = 2; m = 2, n = 1; m = 2, n = 2. Interestingly, the

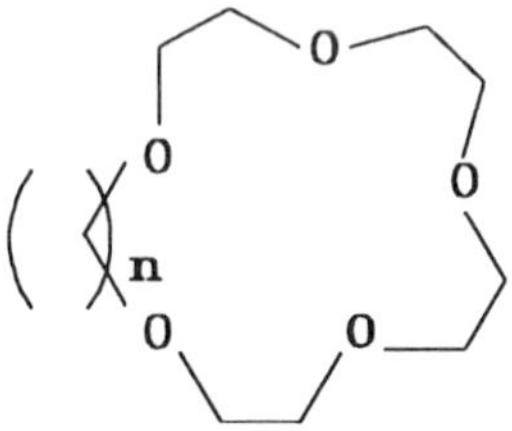

46

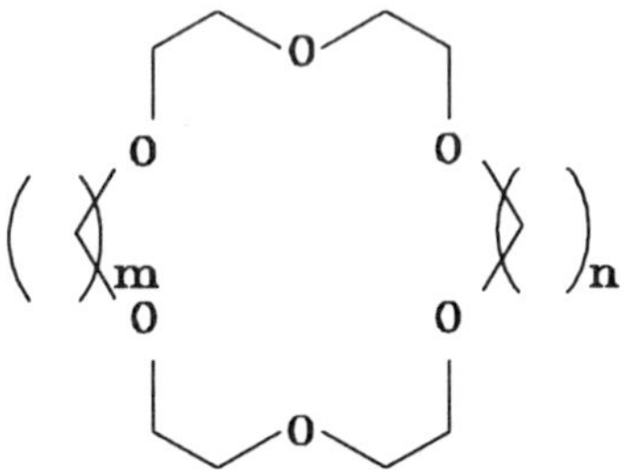

47

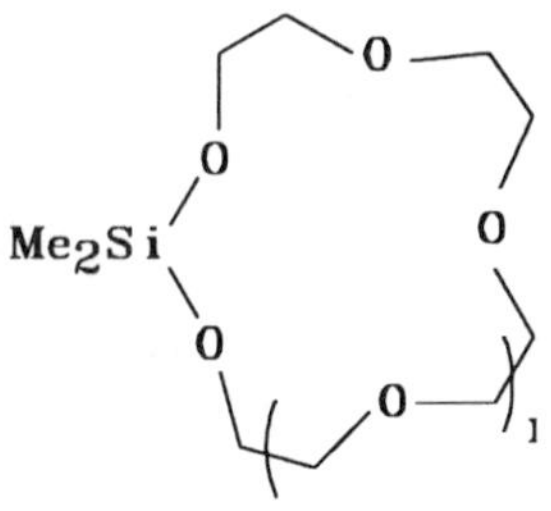

48

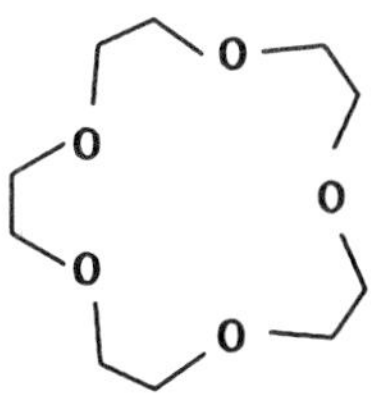

49 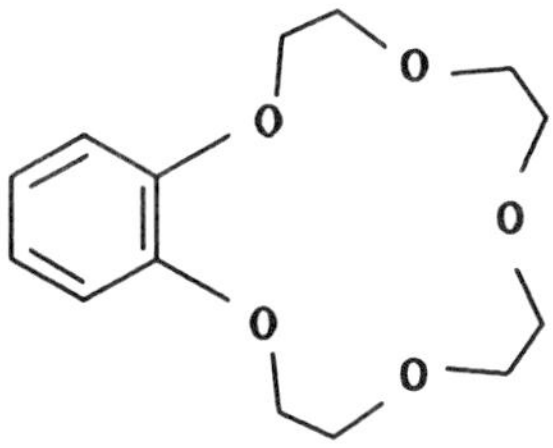 50 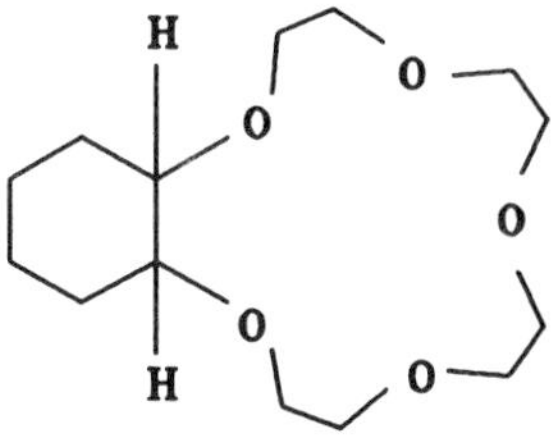 51

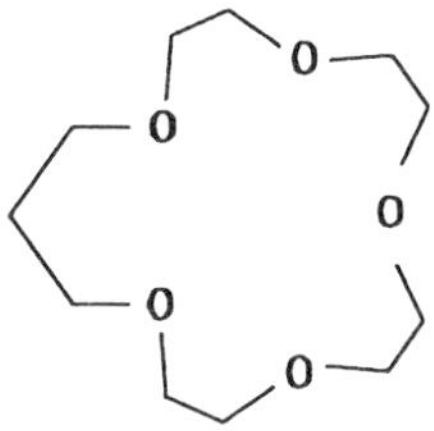

52 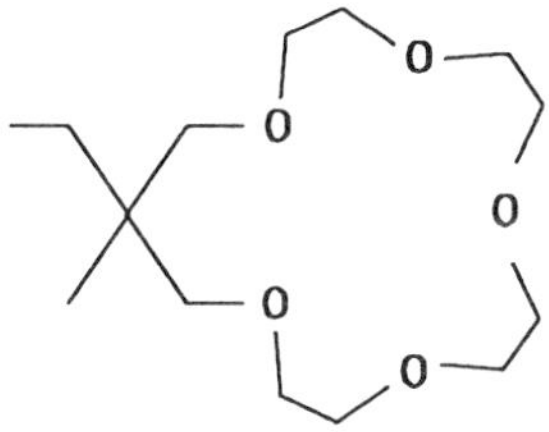 53

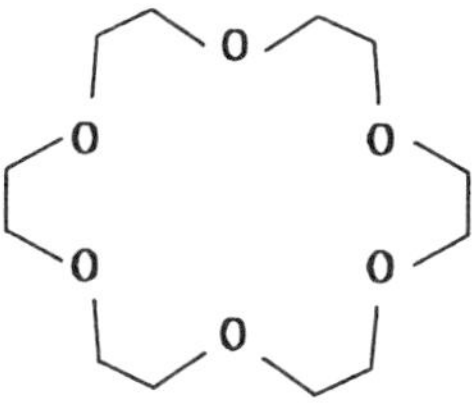

54 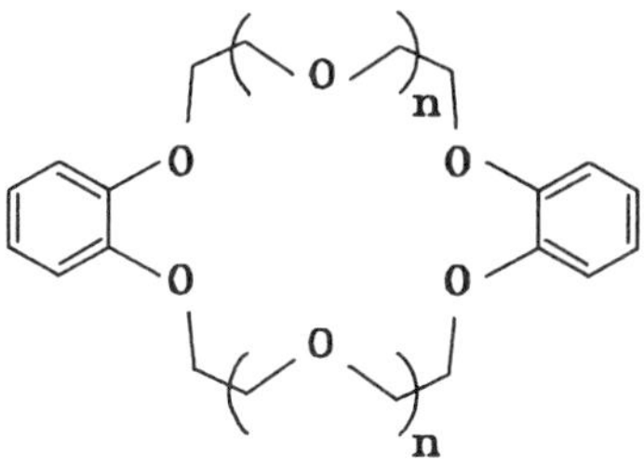 55

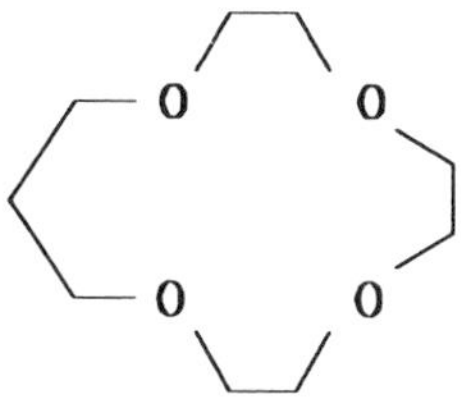

56 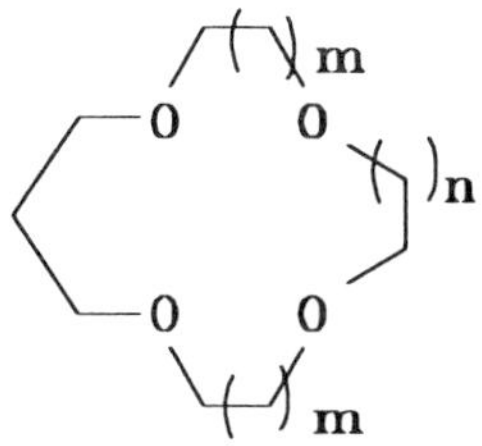 57

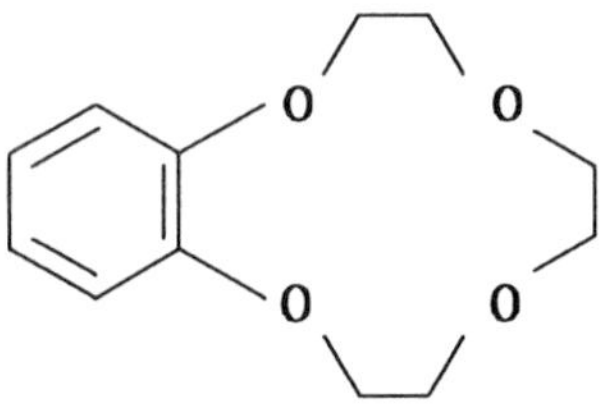

58

extractabilities of these macrocycles do not decrease monotonously with ring expansion.

## 3.2. Azamacrocycles and Thiamacrocycles

One of the earliest literature reports concerning the use of a macrocyclic agent for the extraction of heavy and precious metals are photoizomerizable azobenzene-bridged crown ethers **59-63**. These compounds have been used in the liquid-liquid extraction of mercury(II) and

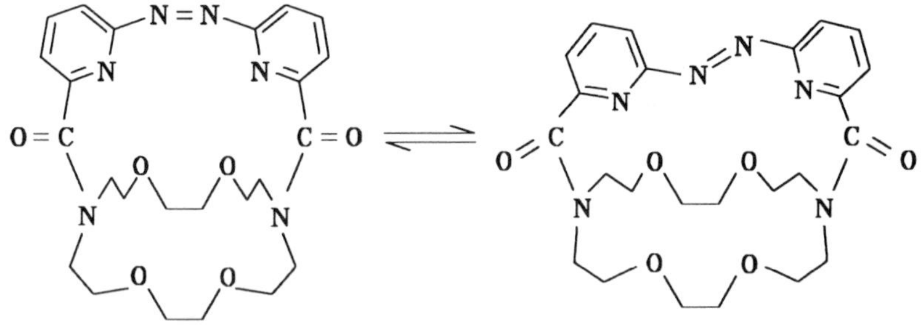

59

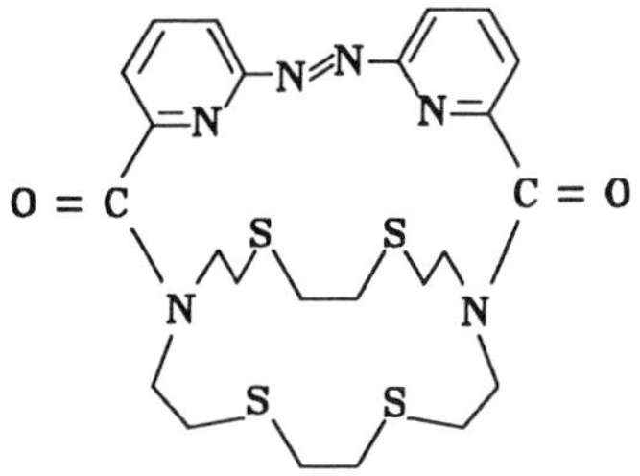

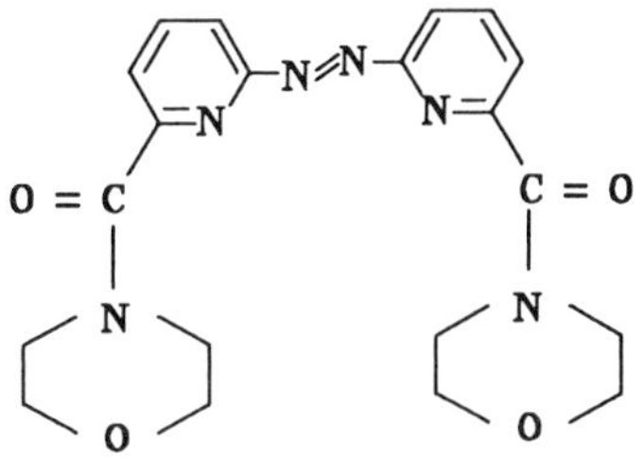

N=N
N N
O = C C = O
S S
N N
S S

N=N
N N
O = C C = O
N N
O O

60 61

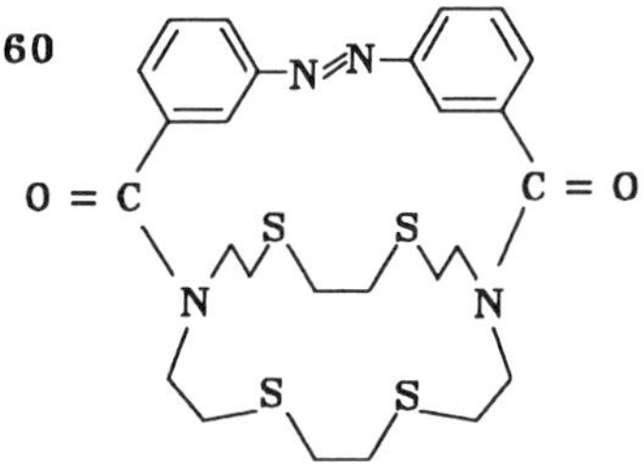

N=N
O = C C = O
S S
N N
S S

62

S S
O N — X — N=N — X — N O
S S

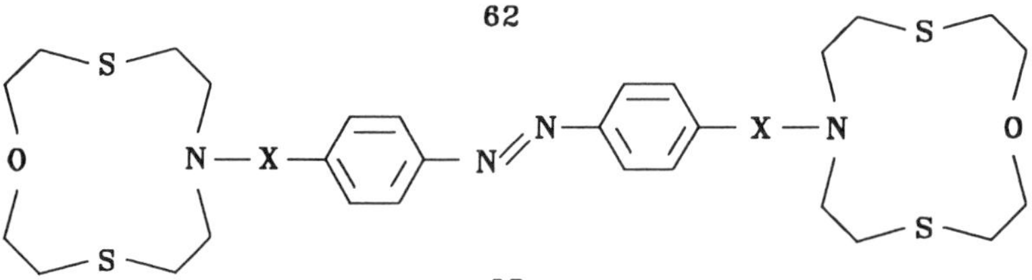

63

copper(II). These photoirradiated systems have been compared in their metal ion extraction properties with similar macrocycles **64-66**.[92-94] The *trans*

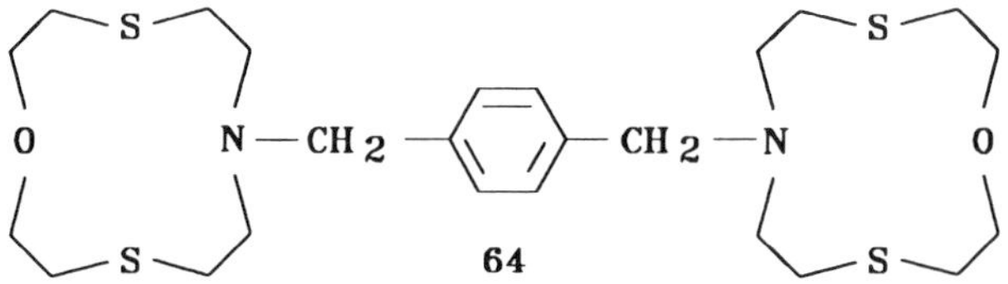

S S
O N — CH 2 — CH 2 — N O
S 64 S

S
O2 N — CH 2 — N O
65 S

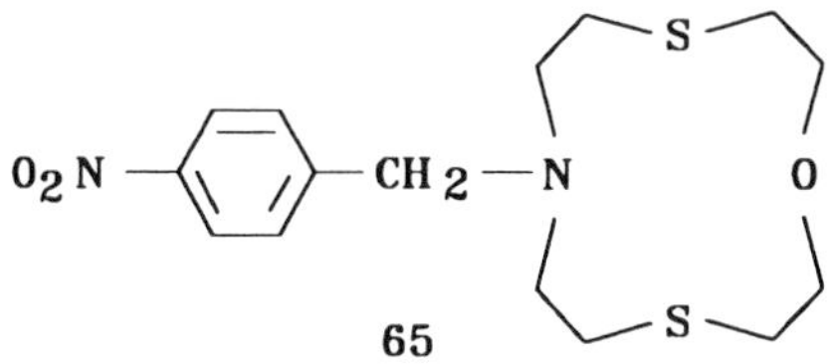

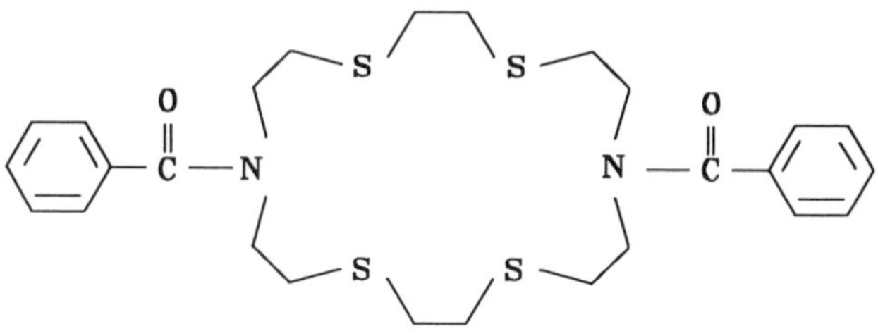

66

isomer extracts metal ions to a much greater extent than does the *cis* isomer. This is the first example of a photoresponsive cryptand for the solution extraction of heavy metal cations. A similar set of thiacrown ethers extract copper(II), mercury(II), cobalt(II), nickel(II) and lead(II). Compound **64** has the extraction sequence copper(II) > mercury(II) = lead(II) >> nickel(II) > cobalt(II), and **65** has the sequence mercury(II) >> lead(II) > nickel(II) > cobalt(II) > copper(II).[95] The liquid-liquid extraction of cobalt(II), nickel(II), copper(II), zinc(II), cadmium(II), lead(II), and silver(I) with the tetraazamacrocycles **67-69** shows that the extraction efficiency is sensitive

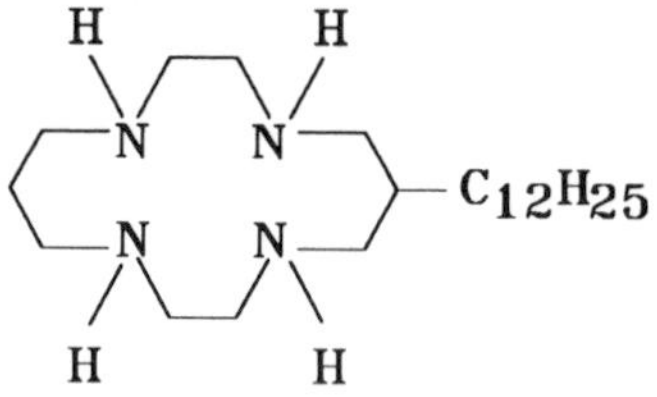

67

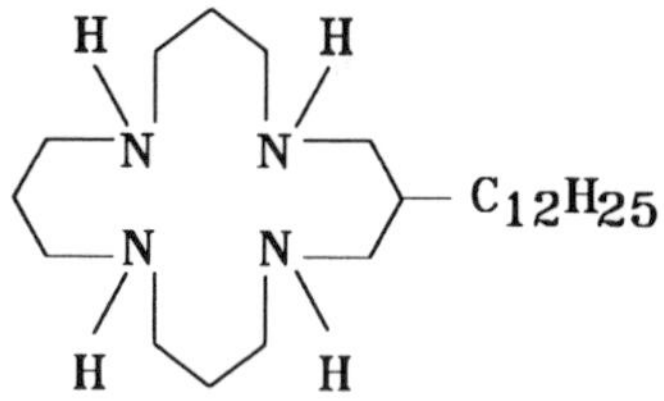

68

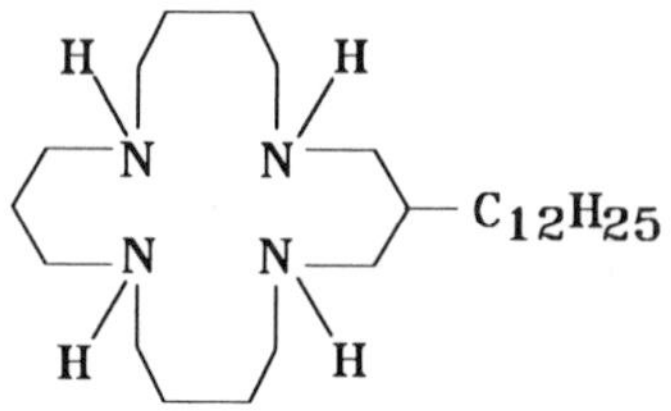

69

to the extent to which the macrocycle cavity matches the metal ion size.[96] Thus, while **68** is the best extractant for cobalt(II), nickel(II), copper(II), zinc(II), cadmium(II), lead(II) and silver(I), **67** is a better extractant for zinc(II) and cadmium(II) than is **69**. The reverse is observed for lead(II). Another set of lipophilic hexamide and hexamine derivatives of azacrown[18]-N6 (**70**) have been found to selectively extract copper(II),

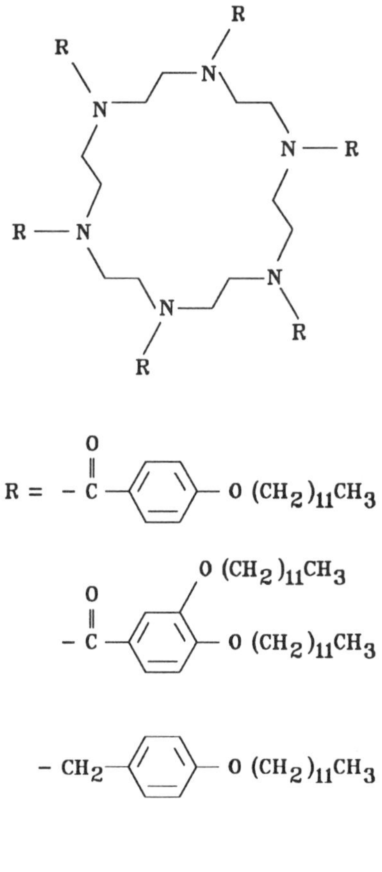

R = $-C(=O)-C_6H_4-O(CH_2)_{11}CH_3$

$-C(=O)-C_6H_3-(O(CH_2)_{11}CH_3)_2$

$-CH_2-C_6H_4-O(CH_2)_{11}CH_3$

**70**

silver(I), and mercury(II) but not cadmium(II) and some base metals.[97] A series of dihydroxycrownophanes **71** show a high affinity toward lead(II) even under acidic conditions, and excellent selectivity toward silver(I).[98]

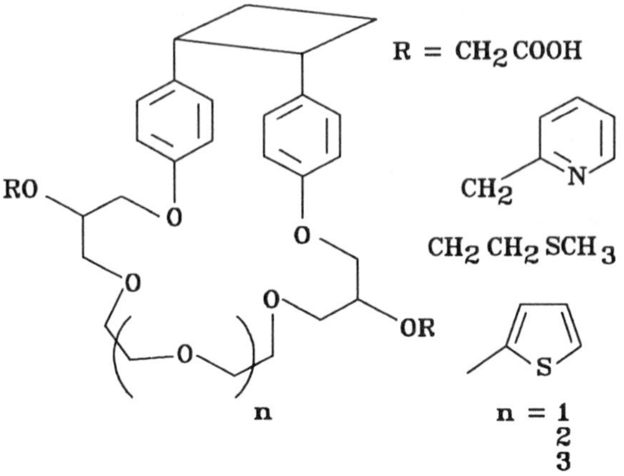

$R = CH_2COOH$

$CH_2CH_2SCH_3$

n = 1, 2, 3

71

Recently two chromogenic and fluorogenic crown ether compounds **72** and**73** have been prepared and used for the selective extraction and determination of heavy metal ions.[99] Each complexant exhibits a selectivity for mercury(II)

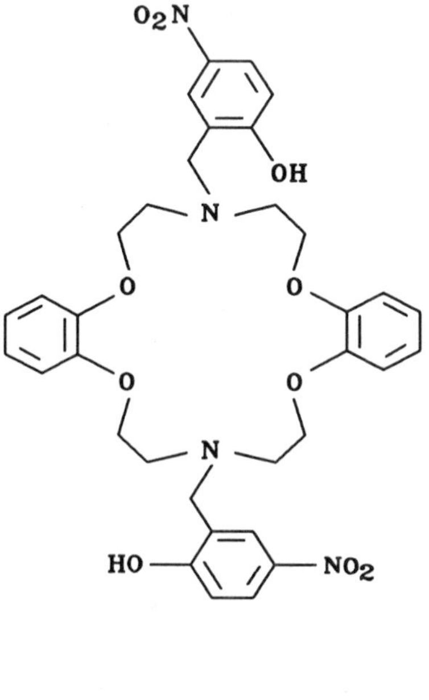

72

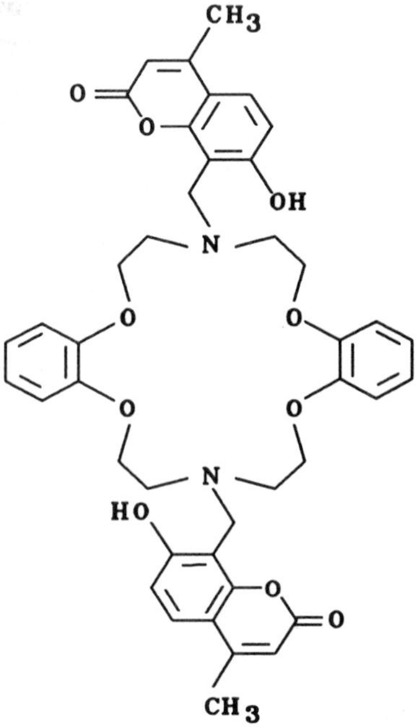

73

that is greater than $10^6$ over the next best extracted cation.

There is a literature example of selenacrown ethers being used as extractants for heavy and precious metal ions. These compounds **74** are good extractants for copper(I), palladium(II), mercury(I), and methyl

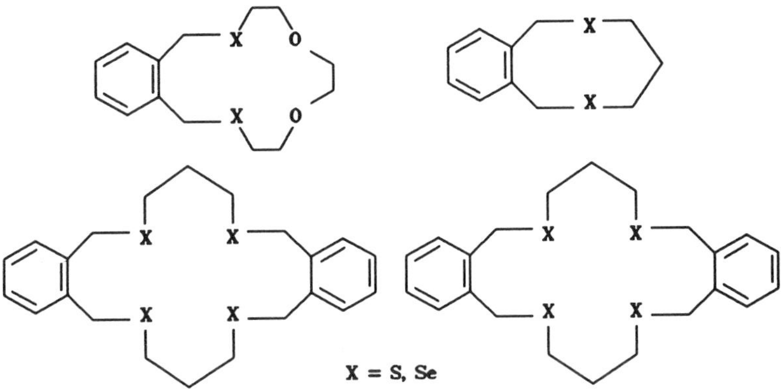

X = S, Se

**74**

mercury(II) in the presence of cobalt(II), nickel(II) and copper(I).[100] Interestingly, the authors report that the extractabilities of these selenacrown ethers toward methyl mercury(II) are higher than those of the structurally corresponding thiacrown ethers. A study describes the use of thia and azacrown ethers **75-79** for the separation of heavy metal ions in a neutral

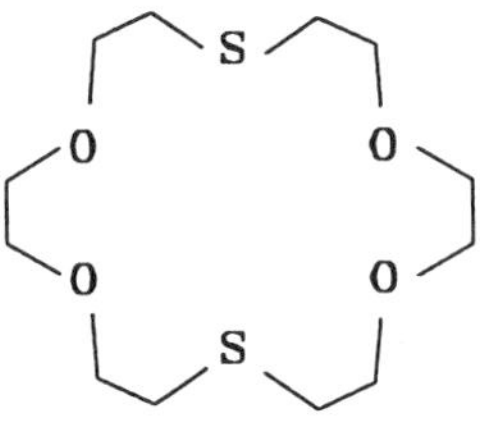

75

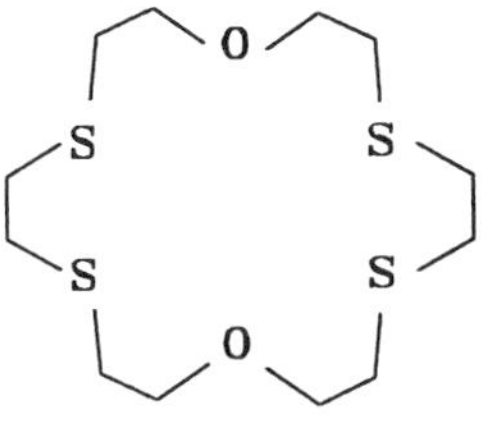

76

77 78

79

macrocycle-mediated emulsion liquid membrane system.[101] It is reported that copper(II) is transported at higher rates than the other ions (manganese(II), cobalt(II), nickel(II), zinc(II), strontium(II), cadmium(II), and lead(II)) in the solution mixture. Others have chemically bonded the macrocycles **80** and **81**

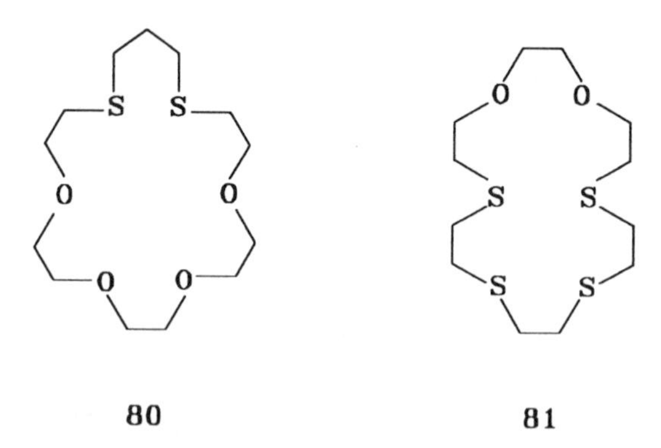


80 81

to silica gel and shown that they bind copper(II), iron(III) and other base metals less strongly than they do palladium(II), gold(III), silver(I), and mercury(II).[102] Crown ethers containing a pendant ether or thioether group **82-86** selectively extract silver(I) and mercury(II) in the presence of manganese(II),cobalt(II), nickel(II), copper(II), zinc(II), cadmium(II), and

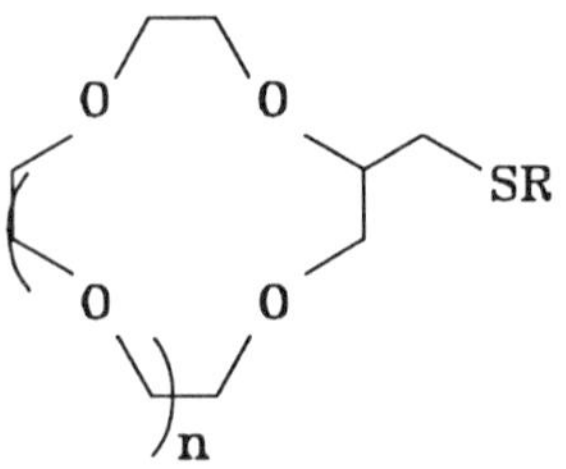

82

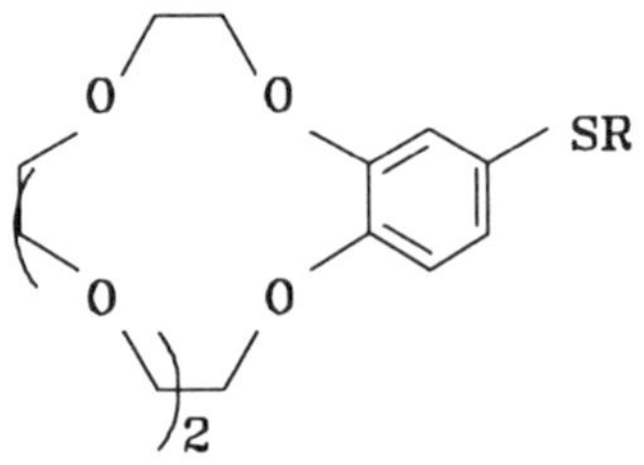

83

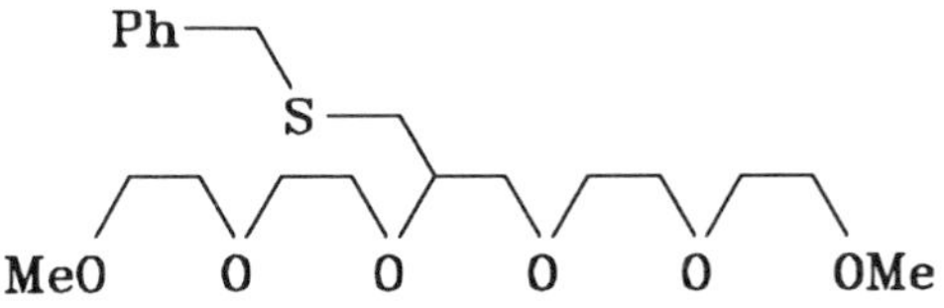

84

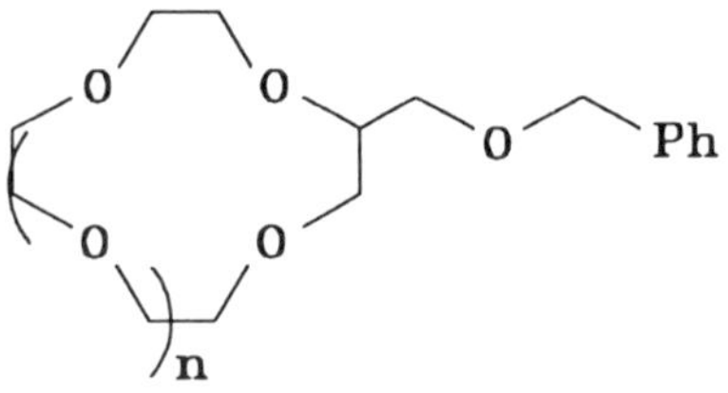

85

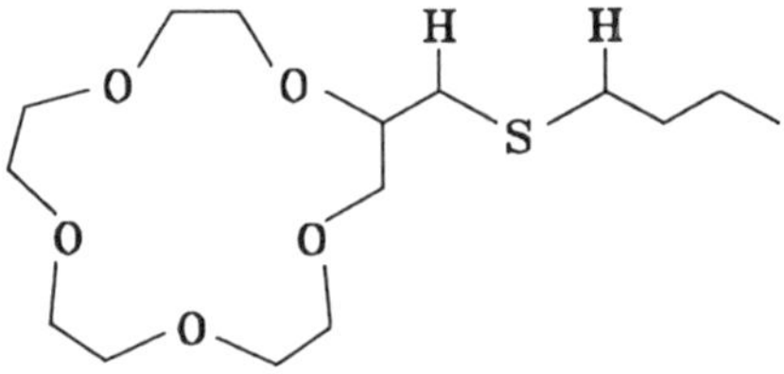

86

lead(II).[103] For compound **82**, R = benzyl, n = 1, 2; R = butyl, n = 1, 2; R = dodecyl, n = 2; for **83**, R = benzyl or butyl; and for **85**, n = 1, 2. Other thiacrown ethers that exhibit selectivity for mercury(II) and silver(I) are **87-89**,[104] and the series of pendant (thiolariat) ethers **90-92** (for compound **90**, n = 1, m = 1, X = S-benzyl, S-butyl, butoxy; n = 2, m = 1, X = S-benzyl, S-butyl, dodecyl, $SCH_2CH_2OMe$, $SCH_2CH_2SBu$, S(O)benzyl, $S(O_2)$benzyl, butoxy; n = 3, m = 1, X = S-benzyl, S-dodecyl, butoxy; n = 2, m = 3, X = S-benzyl; for compound **92**, R = benzyl or butyl) that exhibit a remarkable selectivity for silver(I) and lead(II) in the presence of other alkali and base metals.[105] Such compounds are useful because crown ethers usually coordinate to alkali metal ions. A series of coronands containing 1,3,5-

triazine thiols as redox switchable subunits can be used as extractants for

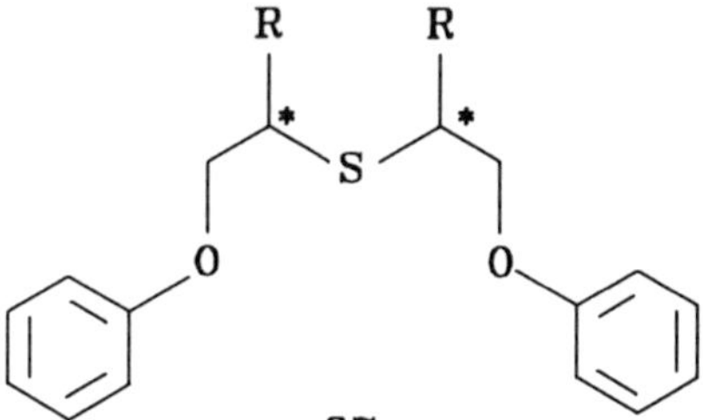

R R S O O S S S

87 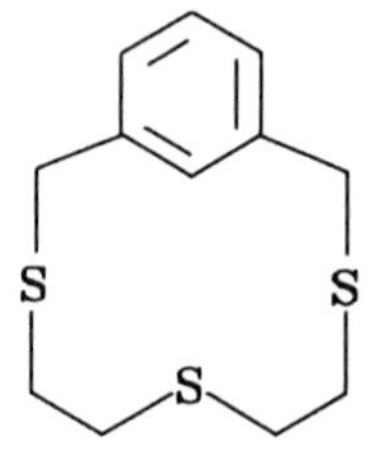 88

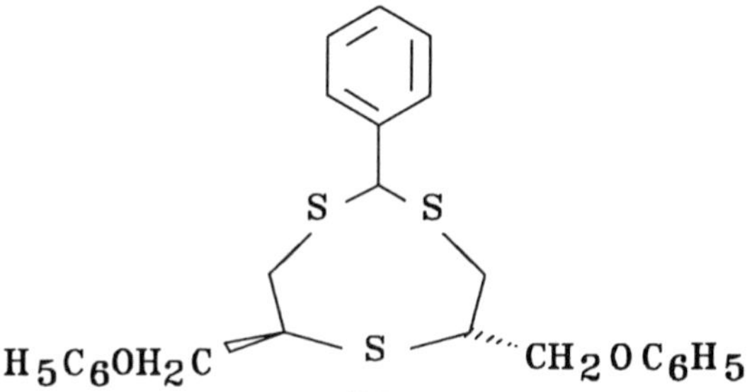

S S S $H_5C_6OH_2C$ $CH_2OC_6H_5$

89

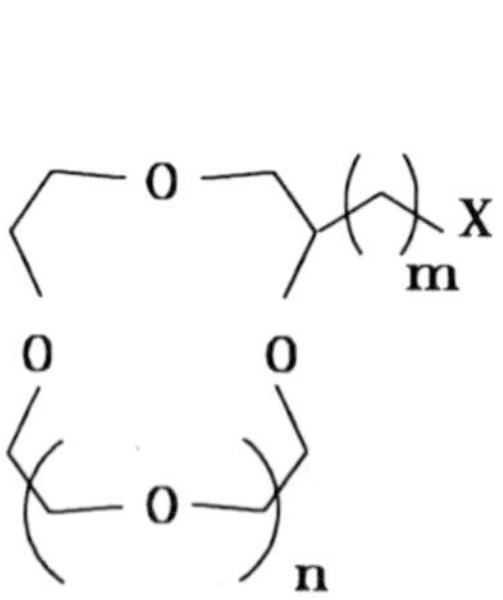

O X m O O O n O O O SBu O SBu O O n

90 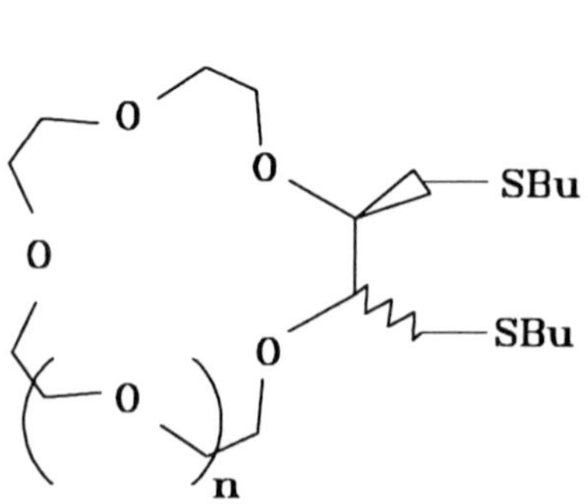 91

thallium(I), silver(I) and mercury(II), but the corresponding disulfides are poorer extractants for all three metal ions.[106]

Molecular modeling studies on open-chain and cyclic thia compounds, along with their silver(I) and mercury(II) complexes, show that the structures of the two sets of complexes are very similar with coordination occurring *via* three S donor atoms and the pyridine nitrogen. The higher extraction of silver(I) is a consequence of the shorter bonding distances.[107] A series of tri- to hexa-dentate sulfur containing macrocycles **93-104** incorporating aromatic and heteroaromatic subunits are extractants for silver(I), mercury(II), gold(III) and palladium(II). A trithiacrown with a benzo subunit preferentially extracts silver(I) over mercury(II), and selectivity for gold(III) and mercury(II) over palladium(II) with the pyridine substituted macrocycles **96, 97, 102–104** is observed.[108] The extraction

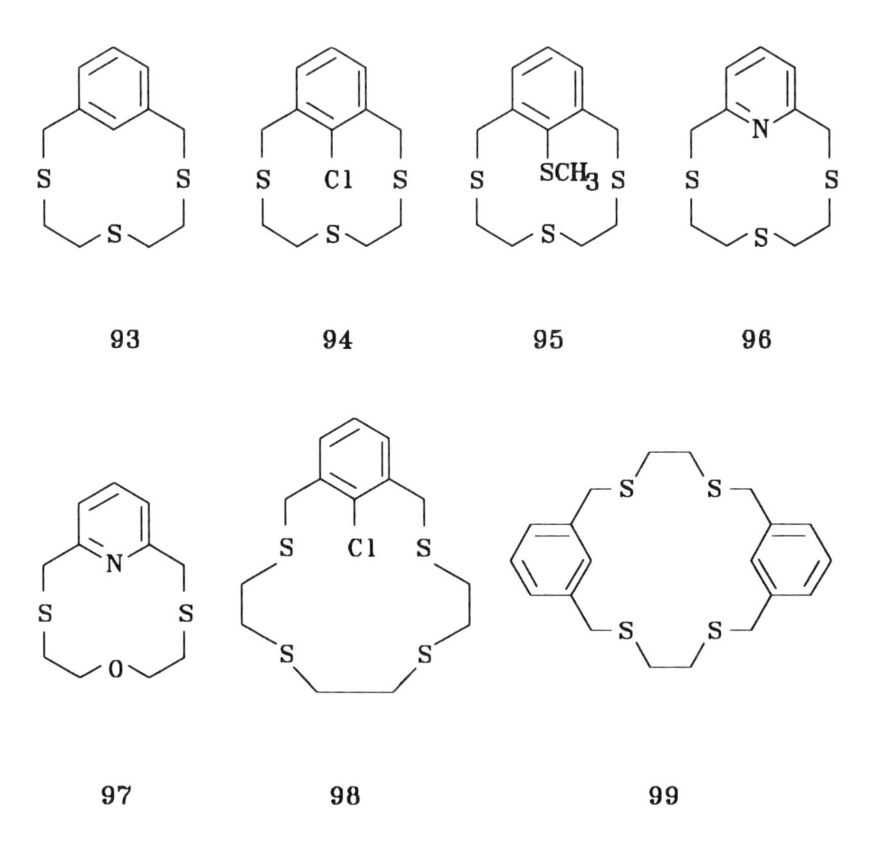

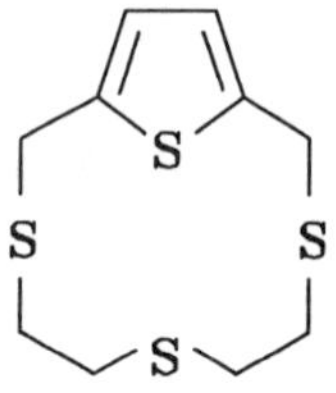

100 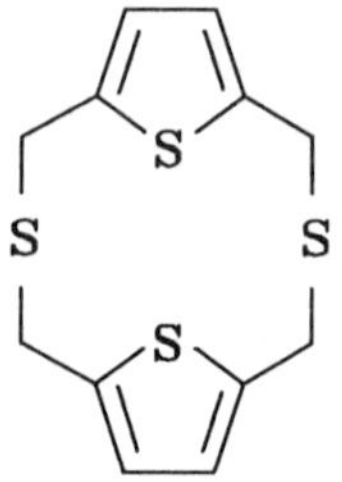 101 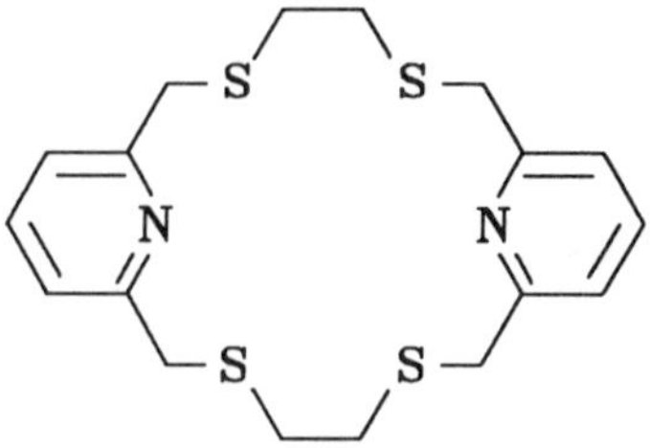 102

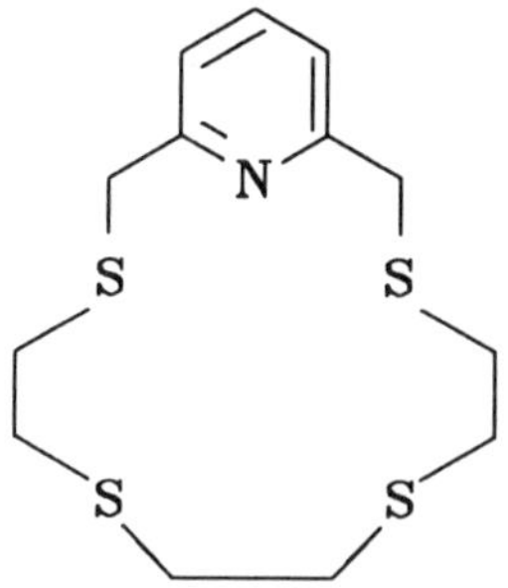

103 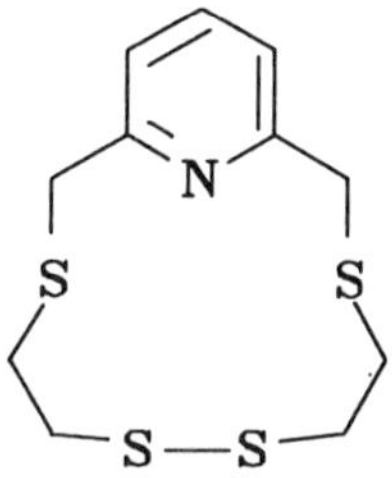 104

behavior of crown and related open-chain compounds with mixed O, S donor sets toward silver(I) shows that a SOOS donor set is the most favorable for complex formation.[109] Bis-crown ethers having benzo-15-crown-5 units linked to 1,1-positions of iron or ruthenium metallocenes **105** have high

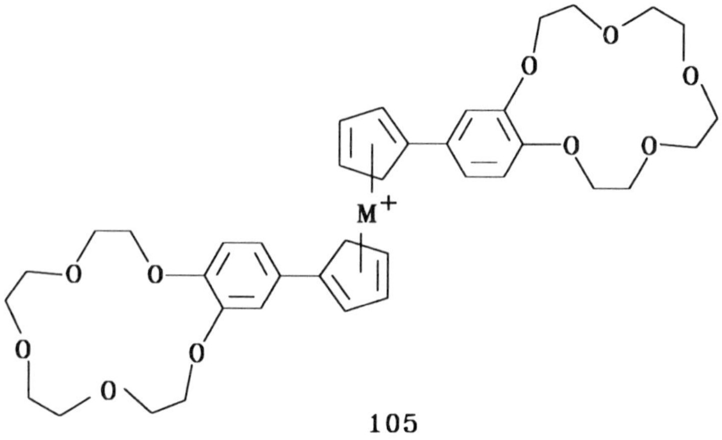

105

affinities for lithium(I), sodium(I), potassium(I) and rubidium(I), silver(I), and thallium(I). The extractibility of metals using these bis-crown compounds is larger than that for observed for the corresponding mono-crowns.[110] For the lipophilic monothia-15-crown-5, dithia-15-crown-5, and dithia-18-crown-6 compounds, and their oxygen analogs **106-110** with R =

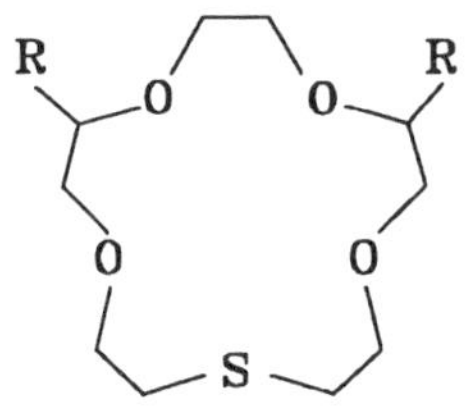

**106** 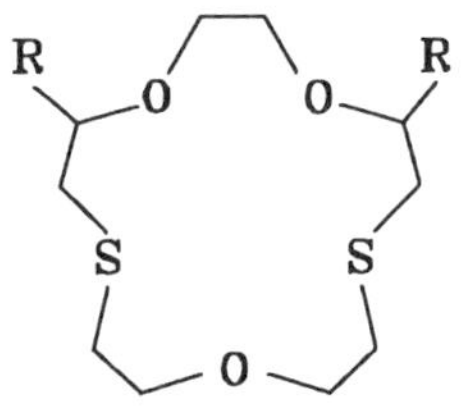 **107**

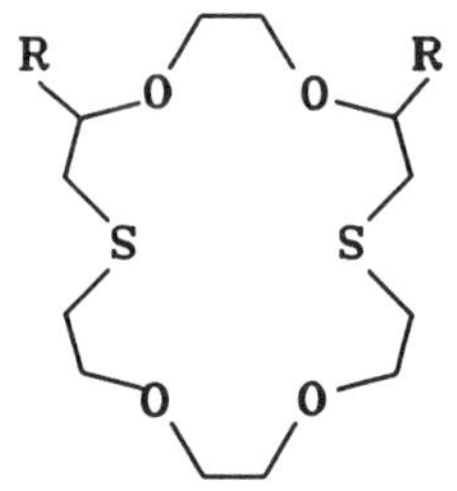

**108** 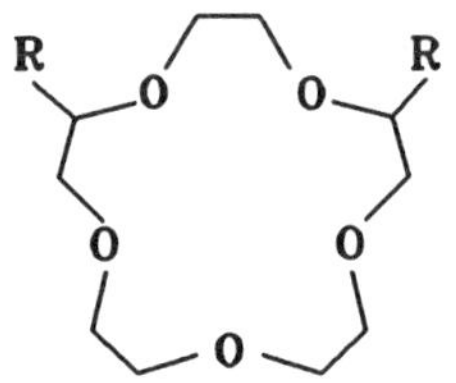 **109** 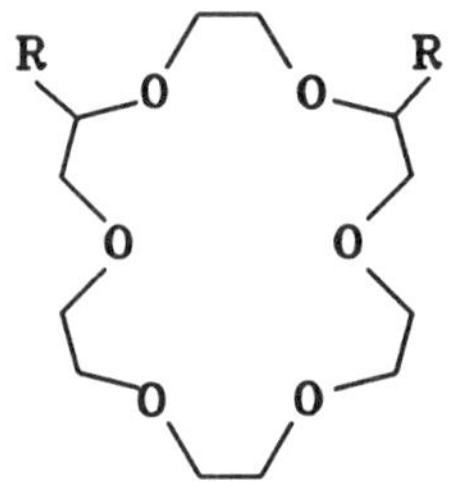 **110**

$C_{12}H_{24}$, it has been found that the thia derivatives have a strong preference for silver(I) extraction over alkali metal cations or thallium(I), but for the oxygen analogs the extraction of silver(I) is low.[111]

## 4. CALIXARENES

A recent development in the complexant field is the recognition that calixarenes may be useful extractants for metal ions. Since calixarenes were originally developed because of their potentially useful phase transfer properties, this makes them likely candidates for functioning as metal extraction agents. To date this optimism is justified, and it appears that calixarenes can possibly make a contribution toward alleviating some of the problems associated with the application of the traditional crown ethers and cryptands for extraction purposes. This optimism results from calixarenes, like crowns, being preorganized complexants, yet, unlike crowns, they can be readily synthesized in large quantities. Calixarenes are cyclic oligomers obtained by condensation reactions between *p*-tert-butyl phenol and formaldehyde. By judicious choice of base, reaction temperature, and time, calixarenes having different ring sizes can be prepared in high yield. The structures and abbreviations used for designating the 4-membered ring calix[4]arenes are shown in **111**. Other ring sizes are available, but few

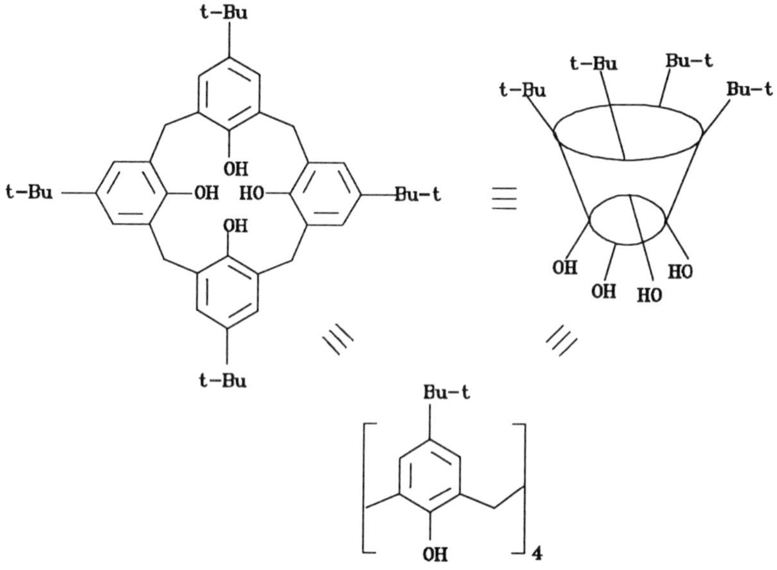

**111**

metal ion extractions have been carried out with them. The calix[4]arenes, for which the majority of the extraction studies have been carried out, have both a wide (upper) and a narrow (lower) rim that can be chemically modified to produce extractants that are selective for particular metal ions. Because of the conical geometry of the calix[4]arene, the cavity size of the wide rim is larger than that of the narrow rim. In the cone conformation, which is often the one having the highest stability, the preorganization puts all of the functional groups in positions where they can coordinate with a metal ion. As a result, functionalized calix[4]arenes can be designed to have a large number of donor atoms that can bind to a metal center. This feature is particularly important for the extraction of metals such as lanthanides which commonly have high coordination numbers such as 8, 9 or 10. Calix[4]arene complexants offering such high coordination numbers can be prepared from the parent calix[4]arene in single-step reactions. This is a particularly useful feature of calixarenes because the design and synthesis of high coordination number complexants usually requires a multistep approach which leads to lower yields and higher costs.

At present there are only a few examples in the literature of calixarenes being used as extractants and ligands for heavy and precious metals. One study reports that a series of Scihff base *p*-tert-butylcalix[4]arenes **112** are good extractants for copper(II) and lead(II).[112]

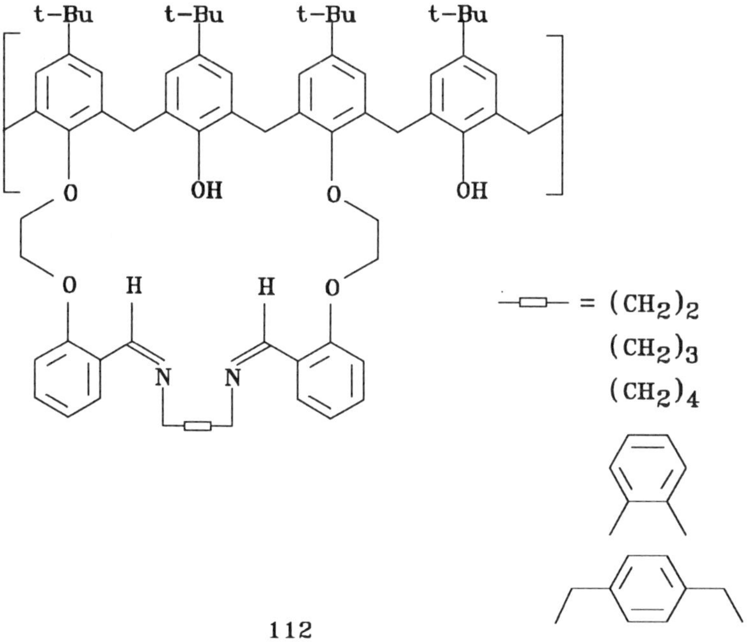

112

However, most of the other alkali, alkaline earth and first row transition metals are also extracted in significant amounts by these derivatives. Another set of calixarene-based ligands **113-116** have either amide, thiocarbamoyl, ester or thioether groups appended.[113] These complexants

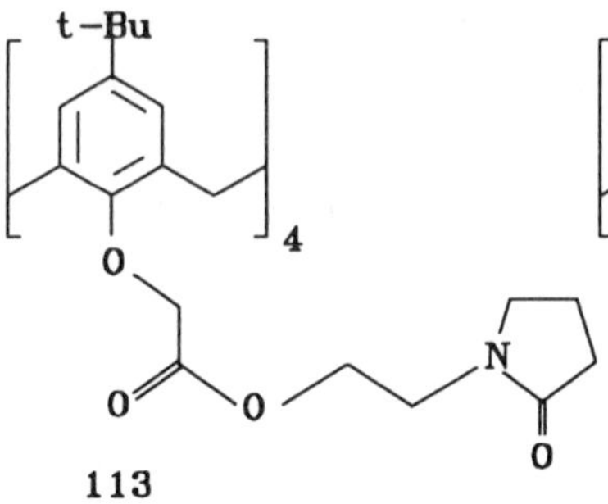

113 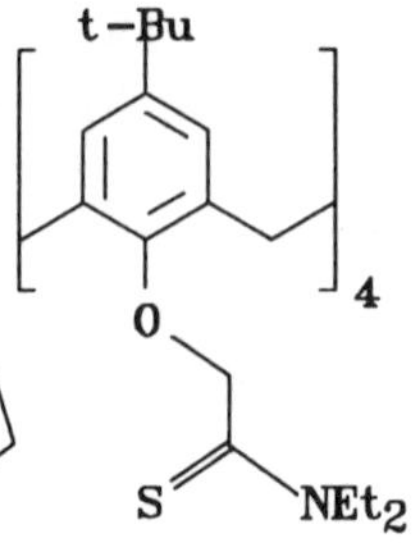 114

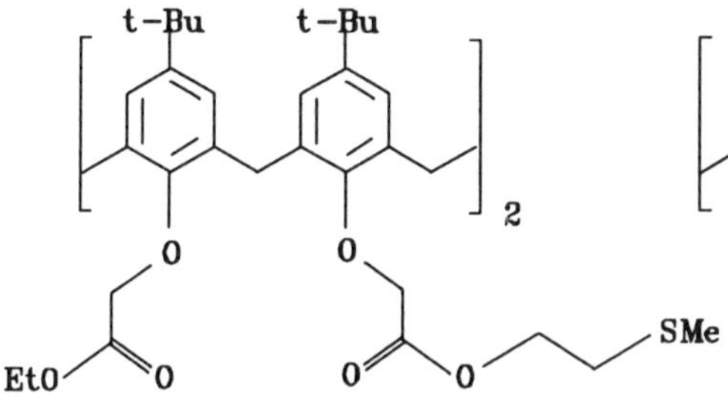

115 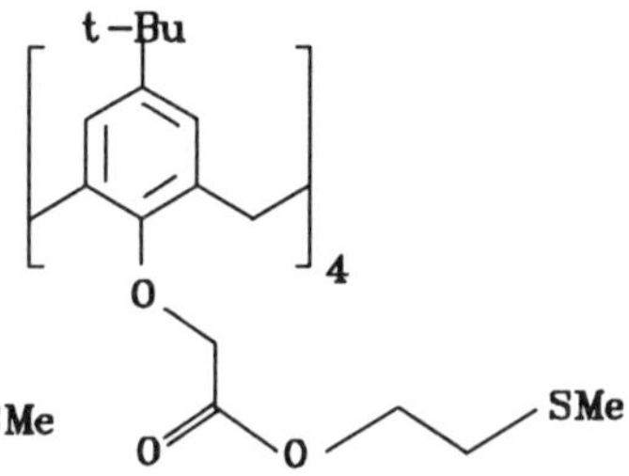 116

have been found to extract silver(I) and lead(II), with sodium(I) and potassium(I) being extracted in smaller amounts. Another new class of derivatized calix[4]arenes bearing thiolate **117** or *N,N*-dimethyldithiocarbamoyl **118** functionalities on the narrow rim have been

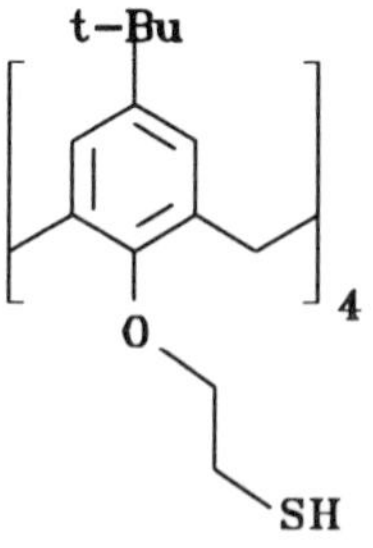

117 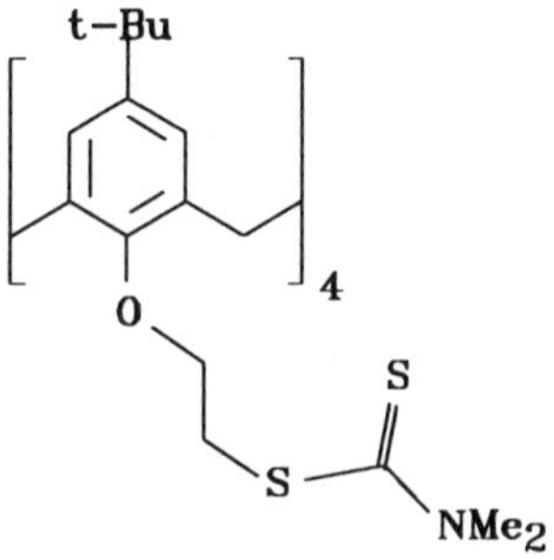 118

synthesized. The calix[4]arene with the thiolate functionalities is unselective and extracts all heavy and precious metals. By contrast the calix[4]arene with *N,N*-dimethyldithiocarbamoyl groups extracts mercury(II), methyl mercury(II), silver(I), palladium(II), gold(III), but not lead(II), cadmium(II), nickel(II) or platinum(II).[114] Under photolytic conditions with a mixture of platinum(II) and platinum(IV), platinum extraction is also observed.[115] Since the extraction selectivities of these substituted calix[4]arenes depend on whether there are hydrogen or *tert*-butyl groups on the upper rim it has been suggested that the size and shape of the calix[4]arene cavity may affect selectivity.[116] These preorganized ligands can be compared to crowns and thiacrowns.[117] A calix[4]arene with a diphenylphosphino group **119** appended to the upper rim has been used as an extractant for heavy metals.

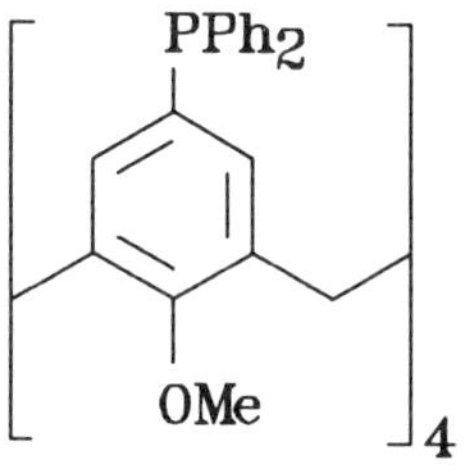

$PPh_2$

OMe

4

119

The extractability for metal ions with this compound follows the sequence mercury(II) > copper(II) > cadmium(II) > zinc(II) > nickel(II) > aluminum(III) > sodium(I) > potassium(I). Since this sequence differs from that observed for triphenylphosphine it has been suggested that the cavity size of the calix[4]arene may be influencing the selectivity.[118]

A family of calix[4]arenes having hydroxamate groups as metal binding sites and either phenol or amide groups as hydrogen donors are extractants for nickel and copper. The calix[4]arenes that have phenolic groups extract nickel(II) in preference to copper(II), while those lacking such H-donors extract copper(II) in preference to both nickel(II) and cobalt(II).[119] A calix[4]arene having alkylammonium substituents on the narrow rim **120** has been used as a "proton-switchable" extractant for chromate and dichromate anion.[120] This system makes use of two amino groups to switch the complexant between it's dicationic and uncharged forms, and to have primary amine groups for hydrogen bonding to the chromium(VI) oxygens.

120

# 5. LIMITATIONS OF CHELATE, MACROCYCLE AND CALIXARENE EXTRACTANTS

It is clear that although there are several relatively inexpensive chelating agents available, and that some have found industrial applications, in many cases they are too unselective for general use. In addition to coordinating with heavy and precious metals they also bind appreciable amounts of alkali, alkaline earth and the lighter transition metals. This lack of selectivity decreases their efficiency and requires the subsequent separation of these metals after the extraction process. Crown ethers and cryptands, by contrast, show remarkable selectivity in their metal ion binding. Unfortunately, aside from some small scale analytical applications, it is problematic for them to be used as large scale extraction agents because of the following limitations:[7] (i) the preparation of crown ethers and cryptands almost always requires multiple-step synthetic procedures that give relatively low yields because of competing polymerization and other side reactions; (ii) typical synthetic procedures are performed under high-dilution conditions ($10^{-4}$ M or less) in order to favor cyclization rather than intermolecular condensation. In industrial terms this translates into very large volumes of organic solvent being required for the preparation of relatively small quantities of the crown ether; (iii) some of the precursor materials are nitrogen and sulfur mustards, and are therefore extremely hazardous for use in large scale synthetic reactions.

# REFERENCES


1. Y. A. Zolotov,"Extraction of Chelate Compounds" Ann Arbor-Humphrey Science Publishers, 1970.
2. P. Cazeneuve, *C. R. Acad. Sci.*, **1900**, *131*, 346.
3. C. F. Bell, "Principles and Applications of Metal Chelation" Clarendon Press, Oxford, U.K., 1977
4. C. J. Pedersen, *J. Am. Chem. Soc.*, **1967**, *89*, 2495.
5. B. P. Hay, *Coord. Chem. Revs.*, **1993**, *126*, 177.
6. J. Gross, G. Harder, A. Siepen, J. Harren, F. Vögtle, H. Stephan, K. Gloe, B. Ahlers, K. Cammann, K. Rissanen, *Chem. Eur. J.*, **1996**, *2*, 1585.
7. J. Rebek, Jr., *Angew. Chem., Int. Ed. Engl.*, **1990**, *29*, 245.
8. E. A. Wintner, M. M. Conn, J. Rebek, Jr., *Accts. Chem. Res.,* **1994**, *27*, 198.
9. F. C. J. M. van Veggel, W. Verboom, D. N. Reinhoudt, *Chem. Rev.,* **1994**, *94*, 279.
10. H.-J. Schneider, *Chem. Soc. Rev.,* **1994**, 227.
11. A. Deratani, B. Sebille, *Anal. Chem.*, **1981**, *53*, 1742.
12. A. T. Yordanov, D. M. Roundhill, *Coord. Chem. Rev.,* **1998**, *170*, 93.
13. D. H. Busch, *Chem. Rev.*, **1993**, *93*, 847.
14. G. Drasch, L. V. Meyer, G. Kauert, Fresenius Z., *Anal. Chem.*, **1982**, *311*, 571.
15. J. M. Lo, Y. P. Lin, K. S. Lin, *Anal. Sci.*, **1991**, *7*, 455.
16. T. Yamane, T. Mukoyama, T. Sasamoto, *Anal. Chim. Acta.*, **1974**, *69*, 347.
17. A. Dornemann, H. Kleist, *Analyst,* **1979**, *104* , 1030.
18. S. Ichinoki, T. Morita, M. Yamazaki, *J. Liq. Chrom.*, **1983**, *6*, 2079.
19. S. Ichinoki, T. Morita, M. Yamazaki, *J. Liq.Chrom.,* **1984**, *7*, 2467.
20. F. L. L. Muller, J. D. Burton, P. J. Statham, *Anal. Chim. Acta.,* **1991**, *245*, 21.
21. G. Bozsai, M. Csanady, *Fresenius Z. Anal. Chem.*, **1979**, *297*, 370.
22. T. Hsieh, L. K. Liu, *Anal. Chim. Acta.*, **1993**, *282*, 221.
23. T. Honjo, *Bull. Chem. Soc. Jpn.*, **1984**, *57*, 591.
24. M. L. Lee, G. Tolg, E. Beinrohr, P. Tschopel, *Anal. Chim. Acta.*, **1993**, *272*, 193.
25. N. Wolf, D. M. Roundhill, *Polyhedron*, **1994**, *13*, 2801.
26. W. H. Chan, S. Y. Lam-Leung, K. W. Cheng, Y. C. Yip, *Anal. Letts.*, **1992**, *25*, 305.
27. *Emerging Technologies in Hazardous Waste Management,* A. P. Hong, T. Chen, R. Okey, ACS Symposium. Ser., No. 607, American Chemical Society. Washington, D.C., 1995.
28. T.-C. Chen, A. Hong, *J. Hazard. Mater.*, **1995**, *41*, 147.
29. A. P. Hong, T. Chen, R. W. Okey, *Proc. Water Environ. Fed. Annu. Conf. Expo.*, 1993, 66 Vol.5, p. 171.
30. R. L. Jenkins, B. J. Scheybeler, M. L. Smith, R. Baird, M. P. Lo, R. T. Haug, *Journal WPCF,* **1981**, *53*, 25.
31. J. L. Howard, J. Shu, *Environ. Poll.,* **1996**, *91*, 89.
32. J. Roca, F. Pomares, Commun. *Soil Sci. Plant Anal.*, **1991**, *22*, 2119.
33. R. Kocjan, M. Garbacka, *Sep. Sci. Technol.*, **1994**, *29*, 799.
34. K. Garnett, P. W. W. Kirk, J. N. Lester, R. Perry, *J. Environ. Qual.*, **1985**, *14*, 549.

35. E. Schonberger, J. Kassovicz, A. Shenhar, *Heavy Met. Environ., Int. Conf., 4th* 1983 vol. 1, 218.
36 W. P Miller, W. W. McFee, J. M. Kelly, *J. Environ. Qual.*, **1983**, *12*, 579.
37. A. K. Guha, P. V. Shanbhag, K. K. Sirkar, C. H. Yun, D. Trivedi, D. Vaccari, *Waste Management*, **1993**, *13*, 395.
38. R. Kocjan, S. Przeszlakowski, *Talanta*, **1992**, 39, 63.
39. R. Kocjan, *Analyst,* **1992**, *117*, 741.
40. K. Hiratani, K. Taguchi, *Bull. Chem. Soc. Jpn.*, **1990**, *63*, 3331.
41. M. Schuster, *Nachr. Chem., Tech. Lab.*, **1992**, *40*, 682.
42. S. Inokuma, K. Hasegawa, S. Sakai, J. Nishimura, *Chem. Lett.*, **1994**, 1729.
43. P. M. van Berkel, W. L. Driessen, J. Reedijk, D.C. Sherrington, A. Zitsmanis, *Reactive & Functional Polymers*, **1995**, *27*, 15.
44. P. Burla, Z. Fresenius, *Anal. Chem.*, **1989**, *306*, 233.
45. I. P. Alimarin, V. M. Ivanov, T. A. Bolshova, E. M. Basova, Z. Fresenius, *Anal. Chem.*, **1989**, *335*, 63.
46. U. B. Talwar, B. C. Haldar, *Indian J. Chem.*, **1969**, *7,* 803.
47. A. Miyazaki, A. Kimura, K. Bansho, Y. Umezaki, *Anal. Chim. Acta*, **1982**, *14*, 213.
48. M. Garcia-Vargas, M. Belizon, M. P. Hernandez-Artiga, C. Martinez, J. A. Perez-Bustamante, *Appl. Spectr.*, **1986**, *40*, 1058.
49. S. Ide, Y. Katayama, H. Nakamura, M. Takagi, *Solv. Extr. Ion Exch.*, **1987**, *5*, 853.
50. T. Kitamori, K. Suzuki, T. Sawada, Y. Gohshi, K. Motojima, *Anal. Chem.*, **1986**, *58*, 2275.
51. V. Leepipatpiboon, *J. Chromat.,* **1995**, *A697*, 137.
52. K. Gloe, H. Stephan, T. Krüger, A. Möckel, N. Woller, G. Subklew, M. J. Schwuger, R. Neumann, E. Weber, *Progr. Colloid Polym. Sci.*, **1996**, *101*, 145.
53. H. Stephan, K. Gloe, T. Krüger, C. Chartroux, R. Neumann, E. Weber, A. Möckel, N. Woller, G. Subklew, M. J. Schwuger, *Solv. Extr. Res. & Dev., Japan,* **1996**, 3, 43.
54. S. Prasad, D. C. *Rupainwar, Poll. Res.,* **1990**, *9*, 15.
55. T. Hayashita, K. Yamasaki, K. Kunogi, K. Hiratani, X. Huang, Y. Jang, R. A. Bartsch, *Abstr. XIX Int. Symp. Macrocycl. Chem.* Lawrence, Kansas 1994, ST5.
56. T. Nabeshima, T. Inaba, N. Furakawa, S. Ohshima, T. Hosoya, Y. Yano, *Tet. Lett.*, **1990**, *31*, 6543.
57. S. S. Lee, D. Y. Kim, J. H. Jung, I. Yoon, S-J. Kim, XXI Symp. on Macrocyclic Chem., Montecatini Terme, Italy, June 1996, Abstr. PB52.
58. G. G. Talanova, A. N. Kozachkova, K. B. Yatsimirskii, M. I. Kabachnik, T. A. Mastryukov, I. M. Aladzheva, O. V. Bikkovskaya, *J. Coord. Chem.*, **1996**, *39*, 1.
59. J. H. Van Zanten, D. S.-W. Chang, I. Stanish, H. G. Monbouquette, *J. Membr. Sci.*, **1995**, *99*, 49.
60. T. Krüger, K. Gloe, B. Habermann, M. Mühlstädt, K. Hollmann, Zeit. für Anorg. und *Allg. Chem.*, **1997**, *623*, 340.
61. K. Gloe, H. Stephan, O. Heitzsch, H. Braun, T. Kind in "*Hydrometallurgy-Fundamentals, Technology and Innovations*", J. B. Hiskey, G. W. Warren, Eds. Littleton, Colorado, 1993, Chapter 52, p. 845.
62. K. Gloe, H. Stephan, R. Jacobi, J. Beger, *Solv. Extr. Res & Dev., Japan*, **1995**,

2, 18.
63. Von E. Hoyer, W. Dietzsch, R. Heber, Wiss. Z., Karl-Marx-Univ., Leipzig, Math.-*Naturwiss. Reihe*, **1975**, *24*, 429.
64. R. W. Frei, G. H. Jamro, O. Navrath, *Anal. Chim. Acta*, **1971**, *55*, 125.
65. A. V. Yazdi, T. A. Hoefling, E. J. Beckman, *Proc. 87th Annu. Meeting of Air, Waste Management Assoc.*, **1994** (Vol. 13, hazard. Waste Management Control), 11, pp. 94-TA42.021.
66. D. E. Leyden, G. H. Luttrell, T. A. Patterson, *Anal. Letts.*, **1975**, *8*, 51.
67. H. Kawamoto, S. Yamazaki, E. Katoh, K-i. Tsunoda, H. Akaiwa, *Anal. Sciences*, **1992**, *8*, 265.
68. A. V. Yazdi, E. J. Beckman, *Met. Mater. Waste Product., Recovery Rem., Proc. Symp.*,1994, 19.
69. J. S. Bradshaw, R. M. Izatt, *Accts. Chem. Res.*, **1997**, *30*, 338.
70. D. M. Roundhill, J. Shen, *CALIX 2001*, J. Vicens, V. Böhmer, J. Harrowfield, eds., Kluwer, Dordrecht, 2001.
71. T. Izumi, S. Oohashi, Y. Tate, *J. Heterocycl. Chem.*, **1993**, *30*, 967.
72. S. R. Cooper, "Crown Compounds: Toward Future Applications", VCH Publishers, New York, 1992.
73. B. G. Cox, H. Schneider, "Coordination and Transport Properties of Macrocyclic Compounds in Solution", Elsevier, 1992.
74. R. M. Izatt, J. S. Bradshaw, S. A. Nielsen, J. D. Lamb, J. J. Christensen, *Chem. Rev.*, **1985**, *85*, 271.
75. M. Kodama, E. Kimura, S. Yamaguchi, *JCS, Dalton Trans.*, **1980**, 2536.
76. A. Bencini, A. Bianchi, M. Micheloni, P. Paoletti, E. Garcia-Espana, M. A. Nino, *JCS Dalton Trans.*, **1991**, 1171.
77. L. Y. Martin, C. R. Sperati, D. H. Busch, *J. Am. Chem. Soc.*, **1977**, *99*, 2968.
78. R. Nagai, M. Kodama, *Inorg.Chem.*, **1984**, *23*, 4184.
79. M. Kato, T. Ito, *Inorg.Chem.*, **1985**, *24*, 504.
80. M. Kato, T. Ito, *Inorg.Chem.*, **1985**, *24*, 509.
81. H. Tsukube, *JCS, Chem.Comm.*, **1983**, 970.
82. H. Tsukube, K. Takagi, T. Higashiyama, T. Iwachido, N. Hayama, *JCS, Perkin Trans. II*, **1985**, 1541.
83. H. Tsukube, *J. Coord. Chem.*, **1987**, *16*, 101.
84. R. M. Izatt, R. L. Bruening, B. J. Tarbet, M. L. Griffin, M. L. Bruening, K. E. Krakowiak, J. S. Bradshaw, *Pure Appl. Chem.*, **1990**, *62*, 1115.
85. D. Cordier, M. W. Hosseini, *New J. Chem.*, **1990**, *14*, 611.
86. H. Tsukube, *JCS, Perkin Trans. I*, **1985**, 615.
87. H. Tsukube, Y. Kubo, T. Toda, T. Araki, *J. Polym. Sci.: Polym. Lett. Ed.*, **1985**, *23*, 517.
88. Y. Inoue, M. Ouchi, T. Hakushi, *Bull. Chem. Soc. Jpn.*, **1985**, *58*, 525.
89. Y. Inoue, F. Amano, N. Okada, H. Inada, M. Ouchi, A. Tai, T. Hakushi, Y. Liu, L. Tong, *JCS, Perkin Trans. II*, **1990**, 1239.
90. Y. Liu, L. Tong, Y. Inoue, T. Hakushi, *JCS, Perkin Trans., II*, **1990**, 1247.
91. Y. Liu, Y. Inoue, T. Hakushi, *Bull. Chem. Soc. Jpn.*,**1990**, *63*, 3044.
92. S. Shinkai, T. Kouno, Y. Kusano, O. Manabe, *JCS, Perkin Trans. I*, **1982**, 2741.
93. S. Shinkai, T. Minami, T. Kouno, Y. Kusano, O. Manabe, *Chem. Lett.*, **1982**, 499.

94. S. Shinkai, Y. Honda, K. Ueda, O. Manabe, *Bull. Chem. Soc. Jpn.*, **1984**, *57*, 2144.
95. S. Shinkai, K. Shigematsu, Y. Honda, O. Manabe, *Bull. Chem. Soc. Jpn.*, **1984**, *57*, 2879.
96. H. Handel, F. R. Muller, R. Guglielmetti, Helv. *Chim. Acta,* **1983**, *44*, 514.
97. M. Zhao, W. T. Ford, *J. Incl. Phenom. Mol. Recogn. in Chem.*, **1994**, *17*, 53.
98. S. Inokuma, S. Sakai, R. Katoh, J. Nishimura, *Bull. Chem. Soc. Jpn.*, **1994**, *67*, 1462.
99. B. Vaidya, J. Zak, G. Bastiaans, M. Porter, J. L. Hallman, N. A. R. Nabulsi, M. D. Utterback, B. Strzelbicka, R. A. Bartsch, *Anal. Chem.*, **1995**, *67*, 4101.
100. T. Kumagai, S. Akabori, *Chem. Lett.*, **1989**, 1667.
101. M-H Cho, H-S. Chun, J-H. Kim, C-H. Rhee, S-J. Kim, *Bull. Korean Chem. Soc.*, **1991**, *12*, 474.
102. R. L. Bruening, B. J. Tarbet, K. E. Krakowiak, M. L. Bruening, R. M. Izatt, J. S. Bradshaw, *Anal. Chem.*, **1991**, *63*, 1014.
103. T. Nabeshima, K. Nishijima, N. Tsukada, H. Furusawa, T. Hosoya, Y. Yano, *JCS, Chem.Comm.*, **1992**, 1092.
104. T. Krueger, K. Gloe, H. Stephan, B. Habermann, K. Hollmann, *XXI Symp. on Macrocyclic Chem.*, Montecatini Terme, Italy, June 1996, Abst. PA 49
105. T. Nabeshima, N. Tsukada, K. Nishijima, H. Ohshiro, Y. Yano, *J. Org. Chem.*, **1996**, *61*, 4342.
106. H. Graubaum, F. Tittelbach, G. Lutze, K. Gloe, M. Mackrodt, *J. Prakt. Chem.* **1997**, *339*, 55.
107. T. Krueger, K. Gloe, H. Stephan, B. Habermann, K. Hollman, E. Weber, *J. Mol. Model.*, **1996**, *2,* 386.
108. O. Heitzsch, K. Gloe, H. Stephan, E. Weber, *Solv. Extr. Ion Exchange*, **1994**, *12*, 475.
109. K. Gloe, O. Heitzsch, H. Stephan, H.-J. Buschmann, R. Trülzsch, R. Jacobi, J. Beger, *Solv. Extr. Res. Devel. (Japan)*, **1994**, *1*, 30.
110. T. Izumi, S. Ooohashi, Y. Tate, *J. Heterocyclic Chem.* **1993**, *30*, 967.
111. M. D. Eley, R. A. Bartsch, M. D. Utterback, *Abstr. XIX Int. Symp. Macrocycl. Chem.* Lawrence, Kansas, 1994, A15.
112. R. Seangprasertkij, Z. Asfari, F. Arnaud, J. Vicens, *J. Org. Chem.*, **1994**, *59*, 1741.
113. K. M. O'Connor, G. Svehla, S. J. Harris, M. A. McKervey, *Anal. Proc.*, **1993**, *30*, 137.
114. A. T. Yordanov, J. T. Mague, D. M. Roundhill, *Inorg. Chem.* ,**1995**, 34, 5084.
115. A. T. Yordanov, J. T. Mague, D. M. Roundhill, *Inorg. Chim. Acta*, **1995**, *240*, 441.
116. A. T. Yordanov, J. T. Mague, D. M. Roundhill, *Inorg. Chim. Acta*, **1995**, *250*, 295.
117. A. T. Yordanov, D. M. Roundhill, *New J. Chem.***1996**, *20*, 447.
118. F. Hamada, T. Fukugaki, K. Murai, G. W. Orr, J. L. Atwood, *J. Incl. Phenom. Mol. Recogn. Chem.*, **1991**, *10*, 57.
119. C. Canevet, J. Libman, A. Shanzer, *Abstr. XX Int. Symp. Macrocycl. Chem.* Jerusalem, Israel, 1995, 35.
120. E. M. Georgiev, N. Wolf, D. M. Roundhill, *Polyhedron*, **1997**, *16*, 1581.

# 6

# PHASE TRANSFER EXTRACTION OF MERCURY

Although many of the heavy metals of the late and post-transition series behave similarly in terms of their extraction properties, mercury is of sufficient importance that we consider it alone. This importance results from the early recognition of the toxicological problems that are caused by mercury, and by the set of environmental problems that are posed by the unique chemistry of the element. Among the chemical properties of mercury that must be considered when addressing its environmental control are the volatility of the element itself with its potential to be carried airborne for considerable distances, the presence of a univalent oxidation state which forms complexes with the mercury in a bimetallic structure $Hg_2$(I), and its ready methylation to give water soluble methylmercury(II) complexes, thereby returning the deposited metal from river and lake bed residues back into the food chain. Nevertheless, the chemistry of mercury does have similarities with those of other heavy metal ions, and throughout this chapter we will compare and contrast the extraction properties of mercury with those exhibited by other metals.

## 1. BACKGROUND

The metal mercury is environmentally important because of its widespread industrial applications. A problem with its use, however, is its toxicity. The combination of these two features means that over the past 100 years considerable amounts of mercury has been dispersed throughout soils and waters, and there is an urgent need for it to be removed and recovered either for further use under more controlled conditions, or for safe storage. The neurological and toxic effects of mercury ingestion have been known for a considerable period of time. The English phrase "mad as a hatter" results from the toxicological properties of the mercury compounds that were used in the hat making industry. In the mid-1800's, hat manufacturers began to use an aqueous mercury nitrate solution to soften, compress, and shape animal furs in a process known as felting. The outcome was skin absorption of mercury solution and inhalation of mercury vapor. The personality changes and tremors that were induced by mercury poisoning led to this English phrase, as well as the expression "Danbury shakes" that refers to the Connecticut town where hatmaking was a major industry. On December 1, 1941, the United States Public Health Service banned the use of mercury by hat manufacturers in 26 states through mutual agreements. The madness in hatters has also received publicity in the novel "Alice in Wonderland". In that book it is noted that both the Hatter and the March Hare are mad. Nevertheless, this diagnosis is rather suspect since it was a rather casual remark by the Cheshire Cat. In the early 20th century it was also common for chemical laboratories to use mercury baths for heating purposes. These baths gave high levels of mercury vapor in their vicinity, and it is likely that the use of such baths had serious effects on the health of, among others, Professor Stock, the German boron chemist. Mercury has also been used in other industrial and agricultural applications. Mercury can be present in waste sites either as complexed ions or as complexed methylated derivatives. The mercury(II) ions in soils will likely be occluded or bound into the clay structures, but methylation by methylcobalamin can still result in their solubilization.

## 1.1. Toxicity and Occurrence

Mercury, like lead, has a long history of industrial use. Also, like lead, its toxicity has long been recognized. Although recognition of the toxicological effects of mercury have resulted in its use being curtailed, there are still various means by which it can become spread throughout the environment. It is estimated that 50 to 70% of mercury in the atmosphere comes from anthropogenic sources. In the United States of America, waste incinerators and coal-fired utilities represent some two thirds of the amount released. A particularly significant problem with mercury is that it can undergo methylation. Methylation results in the mercury being converted into a more lipophilic form, and these methylmercury(II) compounds become accumulated into seafood. This methylation reaction is particularly problematic because mercury that was otherwise considered to be occluded in sediments in the bottom of lakes and rivers can become solubilized and return to the aqueous environment. Warnings for fish begin at levels of 0.16 ppm mercury. This bioaccumulation of mercury by seafood is a persistent problem. As a result these mercury levels must be continually monitored and public warnings issued against consuming seafood from certain locations whenever they reach or exceed these recommended levels. The combination of the airborne movement of elemental mercury, and its solubilization through methylation, means that the problem of mercury in seafood will be a long term problem.

Very recently the food and drug administration have warned women of childbearing age against eating swordfish, shark, king mackerel and tilefish because they may have high levels of methylmercury(II). This warning results from harm that can be caused to the nervous system of an unborn baby, and because of the long term retention of methylmercury(II) in human tissue. Nevertheless, the situation is improving, since seafood canned several decades ago shows higher mercury levels than those found in seafood which is presently entering the food cycle.[1]

Elevated mercury levels in remote aquatic environments have been attributed to long-range atmospheric transport and disposition of anthropogenic mercury. Although geological sources of mercury may be important in some regions, the majority of the global increase in mercury levels is due to human activities.[2] This conclusion is based on analytical data from lake sediments and soils that are good indicators of airborne mercury pollution. In their review, however, these authors caution that care must be taken in interpreting the published data on mercury pollution, and that some of the earlier values may be erroneous.

## 2. COMPLEXANTS

This section is focused on simple complexants that have a single binding site to mercury(II) ions. In some cases coordination through either a single site or multiple sites is possible, but this section includes extractants that usually bind through only a single site. In general these are complexants that are not specifically designed to have multiple coordination positions for binding to mercury(II). This section also includes examples of complexants that are bound to polymeric supports. These polymeric materials are included here because, although they have multiple binding sites for mercury(II), the attachment to the metal will be primarily *via* a single donor atom-mercury interaction. Although conceptually it is to be expected that soft donor atoms such as sulfur will be primarily incorporated into extractants for mercury, there are examples of compounds with oxygen and other donor atoms being used in such applications. Another aspect of mercury extraction that cannot be ignored is the recognition that elemental mercury is a significant contributor to mercury pollution. This results from mercury having a sufficiently high vapor pressure that its airborne migration results in its occurrence being widespread, even into non-industrialized locations and the polar ice caps.

In the presence of chloride ion mercury(II) is present as a combination of the species $HgCl_2$, $HgCl_3^-$, $HgCl_4^{2-}$ and $CH_3HgCl$. Since many environments contain high concentrations of chloride ion, mercury can only be extracted from such a medium in the presence of a better complexant. In the absence of halide ion, the mercury(II) cation is the predominant species in water. Under these conditions, therefore, extractants that are poorer ligands than the chloride ion can be used for its removal from water.

## 2.1. Oxygen Donors

Although soft donor atom extractants are usually preferable, hard oxygen donor ligands can also be used for mercury(II). One such compound is oleic acid.[3] Mercury(II), along with divalent cadmium and zinc, have also been extracted by the use of a series of *n*-dodecyloligo-(oxyethylene) carboxylic acids. The binding is believed to be *via* the ether oxygens, and the extractability of mercury(II) drops drastically upon the addition of sodium chloride because of the formation of hydrophilic chloro complexes.[4] The influence of the length of an alkyl chain incorporated into the alkyltri-(oxyethylene) carboxylic acid has also been investigated, with the shorter chain derivatives exhibiting higher selectivities for mercury(II).[5] The selectivity for divalent ions improves when sodium chloride is added to the aqueous phase. This effect is due to the chemical equilibria involving complexation of chloride ion, hydroxide ion, and the *n*-alkyltri-(oxythylene) carboxylic acid.

## 2.2. Nitrogen Donors

Compounds with amine functionalities have been used for the extraction of mercury(II). This is not unexpected since mercury(II) ammine complexs are well known and are quite stable. However, the complex this is usually extracted, is the alkylammonium salt $[R_3NH]_2HgCl_4$. High molecular weight amines are efficient solvents for its extraction from brine or other chloride ion containing solutions. The amine can be regenerated by stripping the metal with aqueous solutions of nitric acid, ethylenediamine, or propylenediamine.[6] This affinity of long chain amines for mercury(II) has been used to remove both it and other trace metals from nuclear grade uranium. The particular amine used in this application is trioctylamine.[7]

## 2.3. Sulfur Donors

Compounds with sulfur groups are frequently used as extractants for mercury(II) from aqueous solution. In many cases these compounds with sulfur donor functional groups are incorporated into a resin in order that the resulting material can be used as a solid phase extractant for mercury. One such resin contains a hexylthioglycolate functionality that is selective for mercury(II), along with other similar heavy metal ions. The mercury(II) can then be subsequently eluted from the resin with an aqueous hydrochloric acid solution.[8] Thiol-functionalized mono-structured silicas also show a selective affinity for mercury(II), while having no binding affinity for cadmium(II), lead(II) and copper(II).[9] Alternatively, sulfur groups can be appended onto a fluid polymer that can be used as an extractant in a variety of different ways. An example of such a material involves the preparation of a sulfurized jojoba oil that can be used either with the extractant dissolved in a solvent for use in liquid-liquid extraction, or with it adsorbed onto a resin matrix for use in a solid-liquid extraction process.[10]

# 3. CHELATES AND MULTIDENTATES

Chelate and multidentate ligands have been widely used in the extraction of mercury(II) from contaminated waters. For the monovalent mercury(I) ion, which in its coordination chemistry frequently adopts a two-coordinate linear coordination geometry, chelation is not a viable option. Divalent mercury(II), however, commonly adopts a four-coordinate tetrahedral geometry, and now the use of chelate ligands is a useful strategy to follow in order to effect its liquid-liquid extraction from aqueous solution. A somewhat novel approach to separating the monovalent mercury(I) ion from other monovalent cations involves the use of chromatography on paper impregnated with ammonium molybdophosphate, which can itself act as an *O*, *O*-chelate. The technique is extremely sensitive and can be used to detect as little as 10 picogram of mercury that is contained in a 0.1 milligram sample.[11]

## 3.1. Sulfur and Nitrogen Donor Chelates

Although this section is focused on sulfur and nitrogen donor chelates, the majority of chelate extractants that are used for removing mercury(II) from aqueous solution are sulfur derivatives. This choice is a consequence of both the metal ion and the donor ligand being classified as soft. Nevertheless there are some exceptions to this concept. There are several examples of *N*-donor chelates being used as liquid-liquid extractants for mercury(II). One such chelate is the amine base Bindschedler's green, which is 4,4-*bis*-(dimethylamino) diphenylamine. The extraction of mercury(II) with this reagent involves converting the mercury(II) ion into a bromo complex before transferring it into a 1,2-dichloroethane phase using the oxidized form of the base.[12] Another such *N*–donor chelate is a dipyridyl amide-functionalized polymer that is prepared by a ring opening metathesis polymerization method. This polymer shows excellent selectivity toward mercury(II), and it allows for the metal ion to be selectively extracted over a broad range of concentrations. This resin can also be used for the extraction of palladium(II). This extractant can also be used to remove mercury(II) and palladium(II) from complex mixtures of other metal salts. The good stability of the complex allows for high metal loadings on the polymer to be achieved.[13]

Chelates having both nitrogen and sulfur donor atoms have also been used as extractants for mercury(II). An example of such an extractant is a *N*-(hydroxymethyl) thioamide resin that has been used as a stationary phase for the chromatographic separation of a range of heavy metal ions, including mercury(II). The resin is used along with a mineral acid eluent that contains either chloride or sulfate ions.[14]

### 3.1.1. Dithiocarbamates

One of the common extractants for mercury(II) is a monoanionic dithiocarbamate salt. The dithiocarbamate anion **1** has two sulfur donor groups that can bind to a single mercury center. The tertiary amine nitrogen group does not participate in the binding to the metal. Dithiocarbamates are

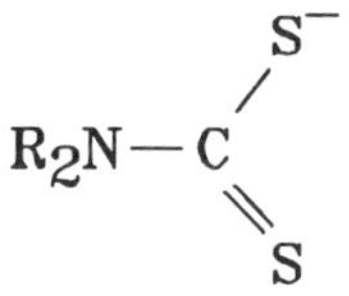

$$R_2N-C(=S)S^-$$

**1**

preferred over xanthates ($ROCS_2^-$) as extractants because xanthates have a tendency to undergo C-O cleavage and eliminate carbon disulfide. Another advantage of dithiocarbamates as extractants is that they can be used in systems that employ liquid carbon dioxide as the non-aqueous solvent. This is a particularly useful property since supercritical carbon dioxide is becoming an increasingly popular medium for carrying out metal extractions. In choosing substituents on the dithiocarbamate for such an application it is noteworthy that the dithiocarbamate complexants with the smaller solubility parameters form metal complexes that have the higher solubilities in supercritical carbon dioxide.[15] The fluorinated dithiocarbamate anion $(CF_3)_2NCS_2^-$ is a particularly effective compound for extracting mercury(II) into supercritical carbon dioxide.[16] This approach of functionalizing chelates to make them compatible with carbon dioxide is particularly attractive because the non-polar carbon dioxide is a very poor solvent for conventional metal chelating agents. The commitment to exploring supercritical carbon dioxide as a fluid for mercury extraction has considerable merit because a simple pressure quench leads to qualitative recovery of the solubilized mercury complex.[17]

The interaction of mercury(II) with diethyldithiocarbamic acid (HDDC) in the presence of sodium chloride in an acidic solution has been studied with the use of a mercury-203 tracer. In the absence of chloride ion, no extraction of mercury is observed if the DCC : mercury(II) molar ratio is <1. By contrast, in the presence of chloride ion, extraction is found for all DCC : mercury(II) molar ratios. The extracted complexes are HgCl(DCC) and $Hg(DCC)_2$.[18] Extraction data for mercury(II) diethyldithiocarbamates have also been obtained by the sub-stoichiometric extraction of the metal in conjunction with radiometry. The method gives data that are in agreement with other methods.[19] A solution that contains a set of heavy metal ions that includes mercury(II) have been separated by liquid chromatography using a cetrimide-dithiocarbamate ion pair loaded onto a precolumn packed with C18-bonded silica. The metal ions complex with the dithiocarbamate, and these apolar complexes are preconcentrated on the precolumn. The metal ions can then be eluted onto a C-18 separation column, and separated with an acetonitrile/water gradient containing cetrimide and a buffer solution at a pH of 6.8.[20] Poorer performance is observed with diethyldithiocarbamates on a liquid chromatography column packed with a PLRP-S styrene-divinylbenzene copolymer.[21] Trace mercury in both aqueous solution and biological samples has been measured by diethyldithiocarbamate extraction in conjunction with gold(III) back extraction. Differential pulse anodic stripping voltammetry is then used for measuring the mercury concentration in solution.[22]

### 3.1.2. Thiocarbazone

Mercury(II) and methylmercury(II) can be extracted from solution by the use of diphenylthiocarbazone-treated polyurethane foams. These materials are effective for aqueous solutions covering a wide pH range, and for mercury concentrations that are less than 10 ppm.[23] Subsequently, however, it has been noted that the mercury dithizone complex is relatively unstable, and if it is identified by visible spectrophotometry the position of the absorption band that is used to analyze the complex is affected by the presence of added anions.[24]

### 3.1.3. Sulfur Containing Metal Complexes

A novel approach to separating nanogram quantities of mercury(I), mercury(II), and methylmercury(II) that may involve their multidentate binding to nitrogen and sulfur donors uses the metal complex $Cr(H_2O)_5NCS^{2+}$ as the complexant. The mercury complexes are subsequently separated by ion exchange.[25] The affinity of mercury(II) for such sulfur donor ligands on other metal centers has led to lithium-intercalated transition metal disulfides $Li_xMS_2$ being used for the extraction of heavy metals, including mercury(II). The materials $Li_xMoS_2$ and $Li_xWS_2$ have been found to bring down the mercury levels from 200 ppm in aqueous solution down to approximately 0.5 ppb. The lithium ion in the material is replaced by ion exchange to give $Hg_yMS_2$. The selectivity for the heavy metal ions follows the sequence mercury(II) > lead(II) > cadmium(II) > zinc(II). Heating the solid $Hg_xMoS_2$ under vacuum at 425°C gives $MoS_2$ and mercury vapor.[26]

# 4. EXTRACTION AND DETECTION TECHNIQUES

Because of the long term recognition of mercury as a toxic pollutant, and because of its unique and varied chemistry, much effort has been devoted to developing techniques for the trace detection of mercury both in the solution phase and the gas phase. Although some of these techniques can be used for other metals, others of them are unique for mercury.

Microemulsions have been used to advantage in the extraction of mercury from an aqueous phase. As an example, a microemulsion containing oleic acid can be used reduce the mercury content of an aqueous phase from 500 ppm down to 0.3 ppm, which represents a 40-fold improvement over equilibrium extraction methods. The kinetics of the process are consistent with film theory predictions for an instantaneous reaction that is mass transfer controlled, and involves mercury(II) migrating to the droplet surface.[27] Methods have been developed for the automated determination of mercury in a continuous flow liquid-liquid extraction process. By this method, preconcentration, reaction with complexant, and detection of mercury, can all be integrated into a single system.[28]

Because of its intense absorption and emission properties, spectrophotometric methods are useful for the quantitative estimation and trace detection of mercury. This was recognized at an early date with atomic absorption spectrometry (AAS) being used to determine the distribution coefficient, D, and the partition coefficient, P, of mercury(II) from an aqueous phase into an organic phase.[29] For trace amounts of the metal, a capillary electrophoretic method has been used for the separation and determination of both mercury(II) and organomercury(II) species. The mercury is present as the water soluble dithizone complex, and coated chromatographic columns are used. Detection limits in the microgram per liter are obtained.[30] An automated online HPLC system has been used for the separation and detection by cold vapor atomic absorption spectroscopy (CVAAS) of methyl, ethyl and phenylmercury compounds. Buffered mixtures containing ammonium tetramethylenedithiocarbamate have been used, and *sub* ng/ml concentrations of mercury in domestic water have been detected.[31] Capillary tube isotachophoresis has been used for the separation of heavy metal ions. A pH in the 7-9 range is used, and the mobility of the leading electrolyte is sufficiently low that cyanide can be used as the terminating ion. The order of the effective mobilities of the metal cyanide complexes that lead to separation follow the sequence: mercury < cadmium < zinc < silver < cobalt < copper.[32]

The trace detection of mercury often requires modifications to be made to standard analytical methods. One such modification involves a reduction in the background noise. For mercury, therefore, a separative column atomizer usable at temperatures up to 1350° C has been employed in conjunction with a gas chromatograph, and direct analysis by AAS. This modification reduces the background in some biological samples, but in the direct atomization of samples in an EDTA matrix complete elimination of the background absorption was not possible.[33]

Radioactive mercury can be separated by using the liquid anion exchanger triisooctylamine. The technique requires less than 5 minutes to complete following combustion, and requires few chemical manipulations, thereby minimizing risks.[34]

## 5. COLUMN MATERIALS FOR MERCURY SEPARATION

A wide range of solid phase absorbents has been used for the extraction and separation of mercury from aqueous solution. The following is a representative list of such materials, although it is in no way comprehensive. Metal ions such as mercury(II) that form anionic chloro complexes are more weakly retained on resins of very low exchange capacity. This property can be used to develop separation methods. For example, mercury(II) is retained on a column comprised of 0.21 mequiv/g XAD-4 anion-exchange resin from a 0.2M HCl solution, while base metals pass through. The mercury can be subsequently eluted from the column using a more concentrated HCl solution.[35] A pulsating column technique based on the resilience of an open-cell polyurethane foam has been developed for the separation of metals, including mercury. A tributyl phosphate-plasticized zinc dithizonate foam is used when the method is used for the separation of mercury.[36]

## 5.1. Organic Adsorbents

The compound α-hydroxyisobutyronitrile reacts with metals in ammoniacal solution to give negatively change cyano complexes. This reaction has been used to advantage in the extraction of heavy metal ions. For mercury(II) and other heavy metals that are present in their chloride form, separation from alkali and alkaline earth metal ions can be achieved on a strongly basic anion exchanger with the metals present as their cyano complexes.[37] Mercury(II), lead(II), silver(I) and thallium(I) can be separated by paper isotachophoresis using 0.1N $HNO_3$ and 0.1 N $LiNO_3$ as leading and terminating electrolytes respectively. The metals are then visualized by spraying with potassium chromate.[38] Thin layer chromatography on Avicel, which is a microcrystalline cellulose, has been used to separate mercury(II) and other heavy metal ions as their anionic EDTA complexes. The complex can be visualized with dithiazone.[39, 40]

## 5.2. Inorganic Adsorbents

Inorganic adsorbents are particularly attractive because they are inexpensive and they do not leave organic residues. In addition these compounds are often minerals which are non-toxic components of soils. Soda lime glass microbeads have been used as an adsorbent for mercury(II). The mercury is 87-9% absorbed from a $10^{-5}$ M solution in pH 7 phosphate buffer containing ethylenediamine (en) when the ratio of en : mercury(II) is in the 4-40 range. The method is satisfactory for the extraction of mercury(II) from organomercurials. Since the extraction is poor at low pH, mercury(II) can be desorbed quantitatively with hydrochloric acid solutions.[41] Heavy metal ions, including mercury(II), have been separated on stannic phosphate and stannic tungstate papers. A range of solvent systems allow for the separation of mercury(II) from mixtures containing over thirty cations.[42] Stannic tungstoarsenate after being dried at 500° C has also been used as a solid phase for the quantitative separation of mercury and lead.[43]

# 6. MACROCYCLES AND OPEN CHAIN ANALOGS

The most common macrocycle types that are used in metal ion extraction are cyclic polyethers, polythioethers, polyamines, and occasionally porphyrins. Since mercury(II) is classified as a soft metal ion, the polyether macrocycles with oxygen donor groups are not generally suitable for either its complexation or extraction. The donor groups that more commonly form strong complexes with mercury(II) are nitrogen and sulfur, therefore in order to develop good extractants for mercury(II), azamacrocycles or thiamacrocycles are usually targeted. Mercury can be extracted by the use of such macrocyclic complexants. These macrocycles can be comprised of sulfur or nitrogen donor atoms, or they can have a combination of these heteroatoms within the macrocycle.

## 6.1. Open Chain Analogs

The cyclic polyethers **2-4** (X = PhN, O, $CH_2$) have been used for the extraction of mercury(II) into an organic solvent. This group of compounds have four or five functional groups that can coordinate to mercury. For each of the sequences NONON, NOOON, NSNSN, NSOSN, NOON and NSSN the end nitrogens are of the pyridine type, with the inner functionalities being aliphatic. The highest extraction is obtained with the system having a NSNSN sequence of donor groups.[44]

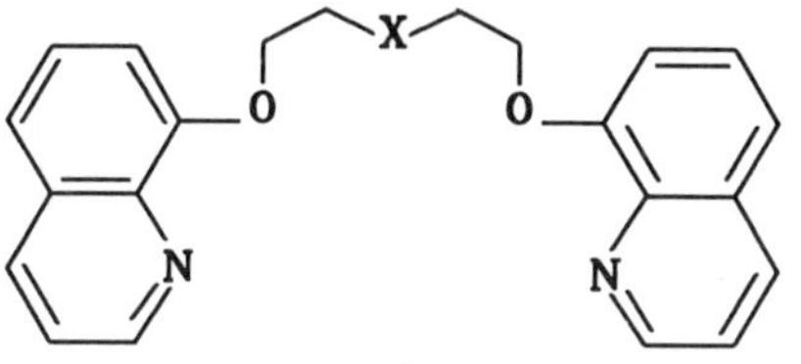

2

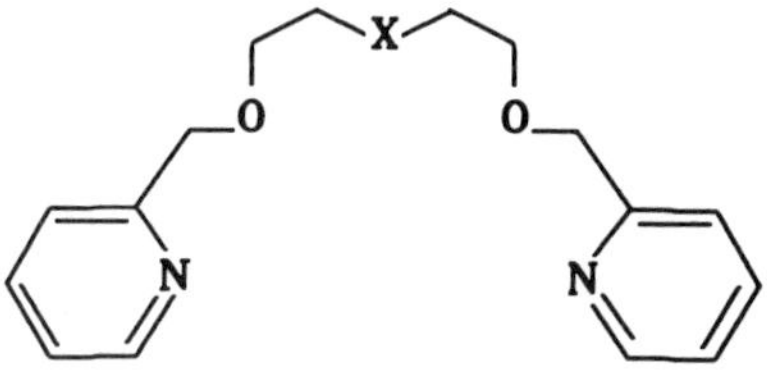

3

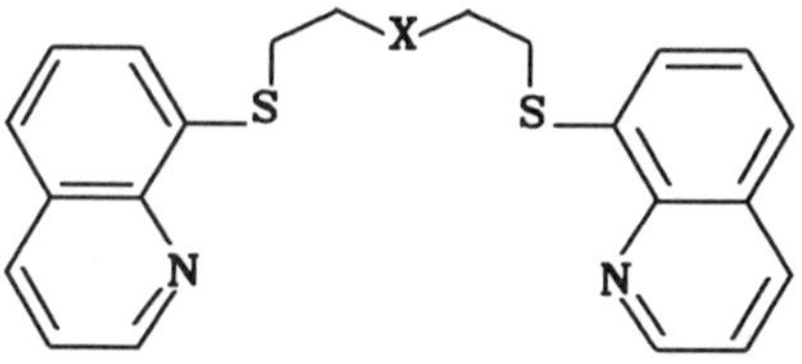

4

## 6.2. N-Donor Macrocycles

Although both linear and cyclic polyamines and polyethers are complexants and extractants for mercury, the polyamines are the more effective. An example of a mercury(II) selective system with nitrogen donor groups in a macrocycle has been identified. The system involves the use of a dibenzo-*bis*-trioazole crown in supercritical carbon dioxide. The system can be used to quantitatively extract mercury(II) over a pH range of 8 to ~ 2. The divalent metal ions cadmium(II), cobalt(II), manganese(II), nickel(II), lead(II), and zinc(II) are virtually non-extractable under these conditions, so good separation can be achieved.[45]

## 6.3. S-Donor Macrocycles

Thiacrowns are also used as complexants and extractants for mercury(II). The strategy behind this concept is that such a soft metal ion will be strongly complexed by the sulfur donor atoms in the macrocycle. In addition, the cavity size can be modified to match the ionic radius of mercury(II). Examples of such thiacrowns are compounds **5**-**9**.[46, 47] Those

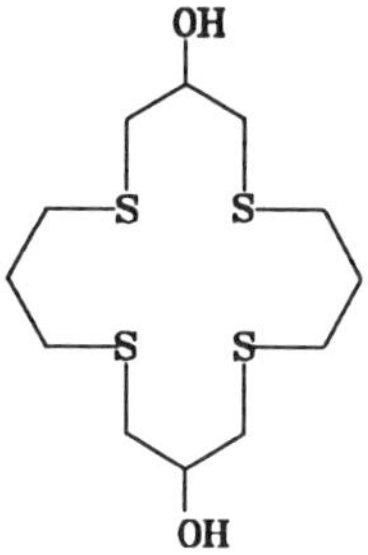

OH S S S S OH

5

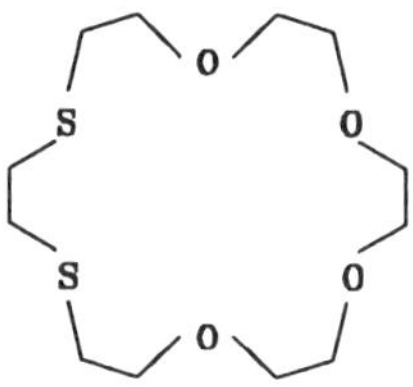

O S O S O O

6

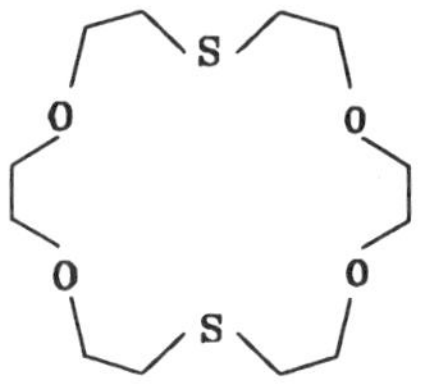

S O O O O S

7

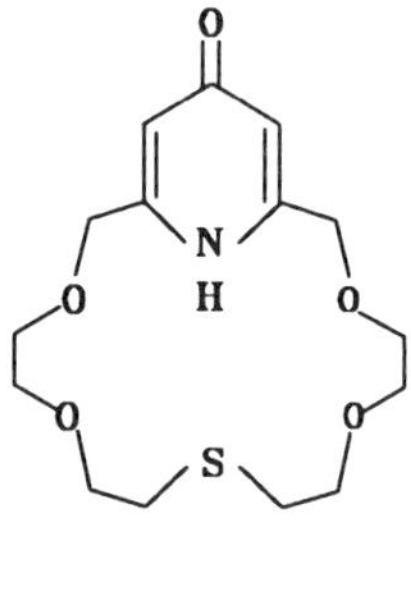

O N H O O O S

8

O⁻ N O O O O S

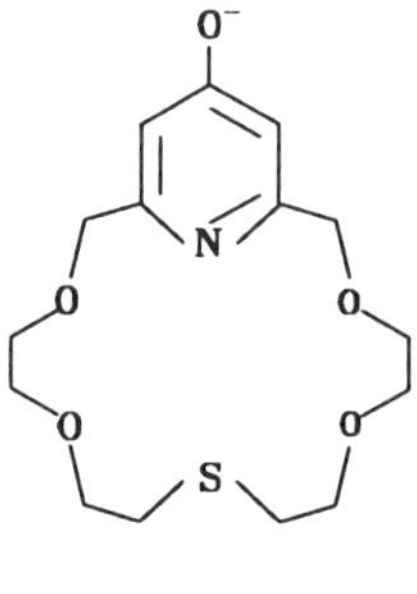

9

macrocycles that have pendant hydroxymethyl groups have the additional feature that they can be chemically bound to a solid support *via* these functionalities. The metal ion binding by these macrocycles is driven by enthalpy, with the entropy changes being consistently negative. Thiacrowns show a greater interaction with mercury(II) than lead(II), which allows for them to be used to separate the two metal ions. A useful feature of thiacrowns is their stability to aqueous acidic conditions. This feature has been used to develop a linked polystyrene material containing pendant thiacrowns that is effective for the extraction of Hg(II) from acidic aqueous solutions.[48]

Methylmercury(II) can also be extracted with thiacrowns and selenacrowns (**10**, X = S; **11**, X = Se). Although both thiacrowns and selenacrowns extract mercury, there are different selectivities for the mercury(II) and the methylmercury(II) salts. For mercury(II), both the dithia and diselenabenzocrowns **10** show high extraction of the metal picrates into

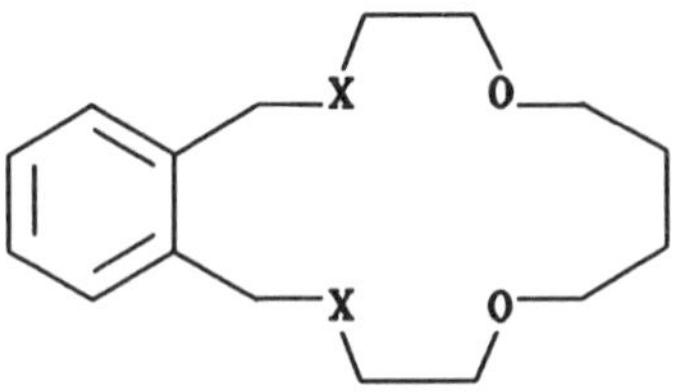

**10** 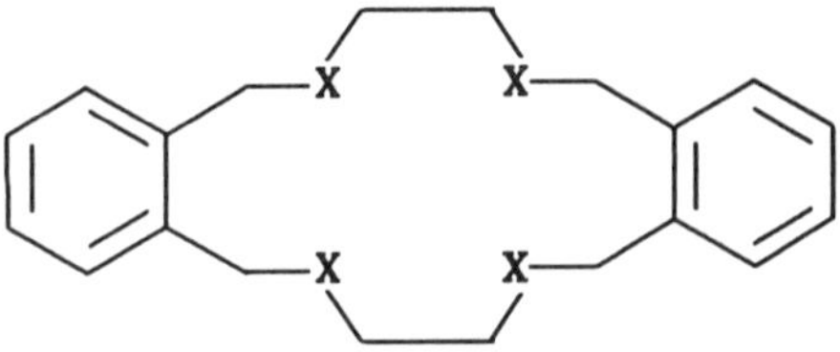 **11**

dichloromethane, but for methylmercury(II) only the diselenabenzocrown is effective. For the tetraselenadibenzocrown **11**, both mercury(II) and methylmercury(II) show good extraction properties.[49]

## 6.4. O-Donor Macrocycles

Although in general macrocycles with oxygen donor groups are poor extractants for mercury(II), an early example of the use of such a polyether macrocycle is the application of the dicyclohexano-18-crown-6 **12** for the preferential transport of cation-anion moieties in emulsion liquid membranes.

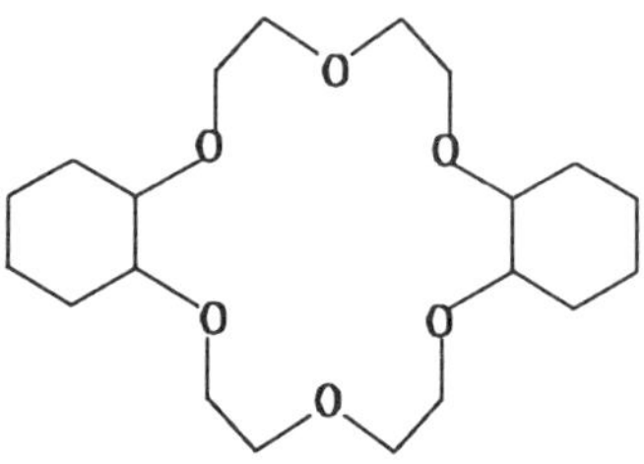

12

The competitive transport of mercury(II), cadmium(II) and zinc(II) with $4\text{x}10^{-3}$ M thiocyanate in the source phase and with nitrate or thiosulfate in the receiving phase are shown in Table 1. This system is selective for mercury(II) over cadmium(II) and zinc(II). The reversed selectivity is found with $4\text{x}10^{-1}$M thiocyanate.[50]

**Table 1. Competitive Transport(%) of mercury(II), cadmium(II) and zinc(II)**

| $NO_3^-$ | | | | $S_2O_3^{2-}$ | | |
|---|---|---|---|---|---|---|
| time, min | Hg(II) | Cd(II) | Zn(II) | Hg(II) | Cd(II) | Zn(II) |
| 3 | 26 | 0 | 0 | 65 | 3 | 0 |
| 5 | 48 | 0 | 0 | 72 | 3 | 0 |
| 10 | 80 | 3 | 0 | 77 | 3 | 0 |

## 6.5. *N,O*-Donor Macrocycles

Mixed *N,O*-macrocycles have been used as extractants for mercury(II). One such family of macrocycles incorporate pendant proton-ionizable functionalites onto the macrocycle. These macrocycles contain either a carboxylic acid or a triazole functional group that can be deprotonated bind with metal ions to form neutral complexes. The extraction of these complexes into organic solvents requires no specific counteranion, and is reversible with respect to pH.[51] Such a pH switch is particularly important if subsequent recovery of the mercury is required. A series of such triazole macrocycles **13-19** ($R_1$ and $R_2$ are aliphatic substituents) have been

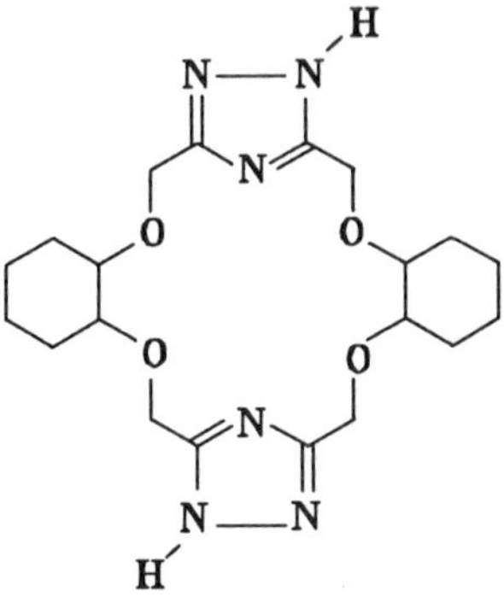

13

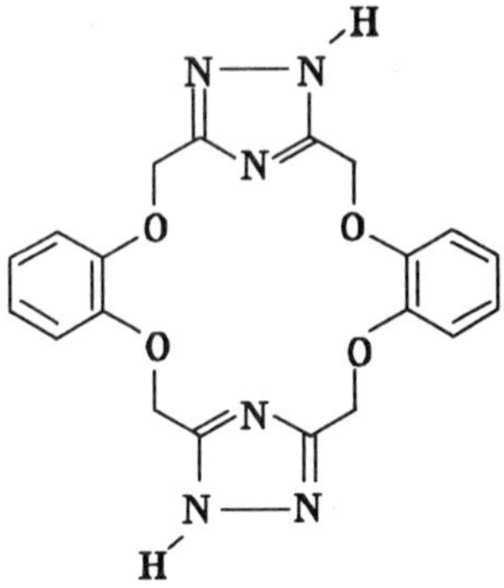

14

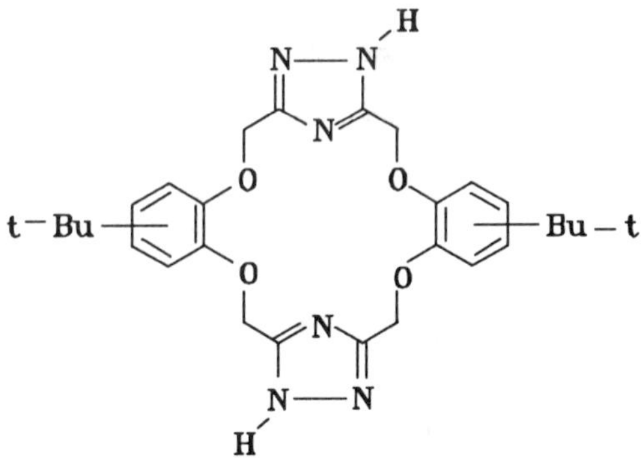

15

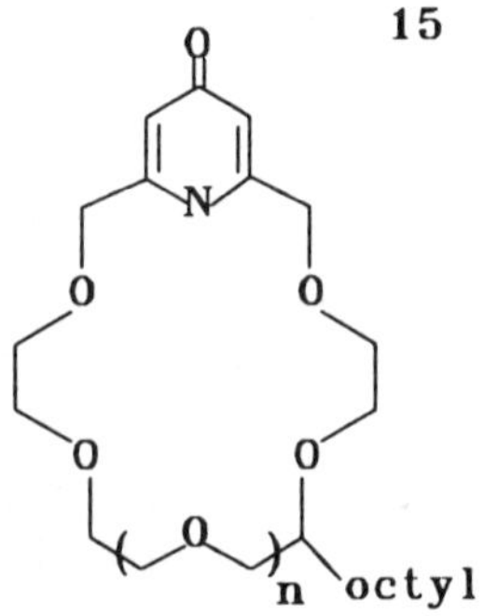

16

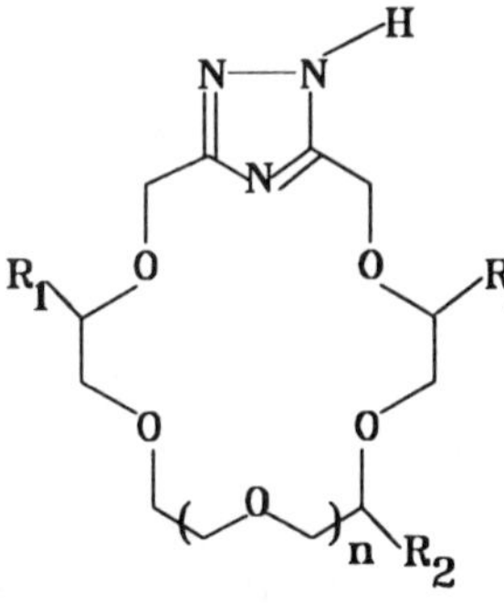

17

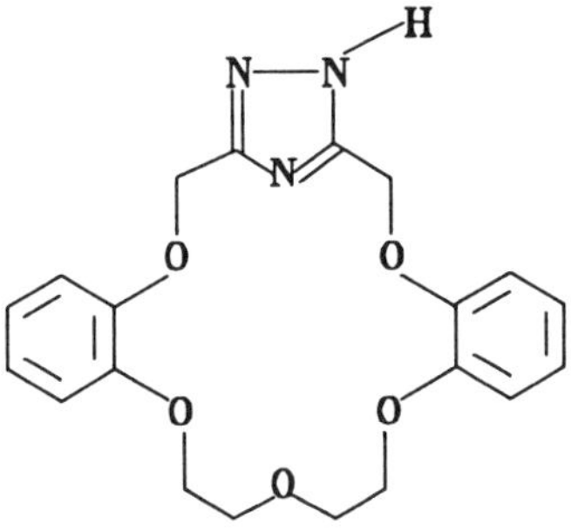

18 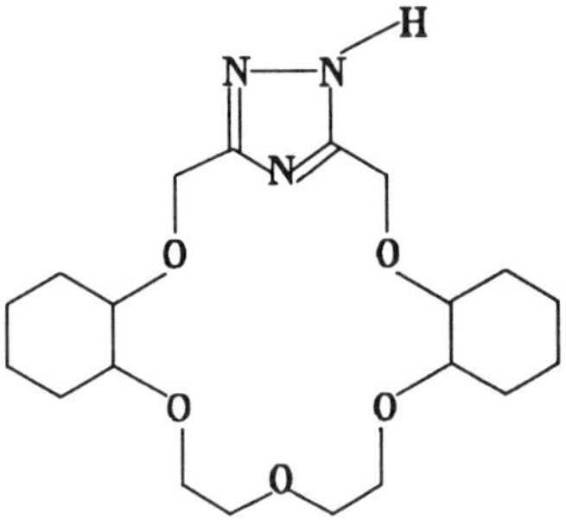 19

used for the extraction of mercury(II) into a solution comprised of methanol (5%) in carbon dioxide. No significant extraction of cadmium(II), cobalt(II), manganese(II), nickel(II), lead(II) or zinc(II) is observed with these complexants, and gold(III) is more poorly extracted than is mercury(II).[52] Immobilization of *N*– or *N,O*-macrocycles allows for their use as column materials for the adsorption and separation of heavy metal ions. As an example, the silica-gel bound diaza-18-crown-6 can be used for the removal and separation of mercury(II) and silver(I).[53]

## 6.6. Chromogenic and Fluorogenic Macrocycles

Chromogenic and fluorogenic crown ethers have been used for the selective extraction and determination of mercury(II). These macrocycles have either 2-hydroxy-5-nitrobenzyl or 7-hydroxy-4-methyl coumarin-8-methylene substituents appended to a 4,13-diazadibenzo-18-crown-6 macrocyclic host.[54] The selectivity for mercury(II) over the next best metal ion is in the $10^6$ to $10^7$ range, making these compounds very highly selective for mercury(II).

# 7. PORPHYRINS

Tetraphenylporphyrin complexes have been investigated as potential compounds for the separation of metal ions. In a study of their chromatographic behavior with four different adsorbents and eleven different solvent systems it has been found that a mixture of the lead(II), cadmium(II) and mercury(II) complexes show isographic behavior.[55]

# 8. CALIXARENES

When suitable functional groups are appended onto either the narrow or wide rim, the calixarene platform can be useful for the synthesis of selective extractants for mercury. Such functionalities usually have soft phosphorus or sulfur donor atoms, but other possibilities exist.

## 8.1. P and S-Donor Calixarenes

Phosphorus and sulfur are potentially useful functionalities for synthesizing calixarenes that are selective extractants for lead, cadmium, and mercury. A calixarene with diphenylphosphino groups on the wide rim extracts both cadmium(II) and mercury(II), but its selectivity against other metals is relatively low.[56] Since this calixarene does, however, show different selectivites than triphenylphosphine itself, it is likely that the three-dimensional cavity of the calix[4]arene host is influencing the complexation properties. Calix[4]arenes with sulfur donor funtionalities have also been investigated as extractants for mercury, lead, and, cadmium. A thiol functionality on the narrow rim **20** leads to good extractability, but low selectivity of heavy metal ions. Better selectivity is obtained with an (*N,N*-dimethyldithiocarbamoyl) ethoxy substituted calix[4]arene **21**, which extracts mercury(II), but not lead(II) and cadmium(II).[57] Changing the wide rim between a *tert*-butyl **21** and a hydrogen **22** leads to differences in the extractability properties of the complexant, although no reason has been offered to explain these observation.[58] Since calix[4]arenes are preorganized complexants, they have been compared to crowns in their coordination properties with mercury(II).[59] Such comparisons are difficult to make, however, because calixarenes are less rigid than crowns. Thioether **23** and **24**, and thiophene **25** substituents on the narrow rim also result in calix[4]arenes that are good extractants for mercury(II).[60, 61] By a combination of $^1$H NMR and electronic absorption spectroscopy it has been concluded that mercury(II) binds to the sulfur functionalities of these calixarenes.[62] The distribution coefficients $[Hg(II)]_{org}/[Hg(II)]_{aq}$ for chloroform as the organic phase for the calix[4]arenes **20-25** are shown in Table 2.[63] These distribution coefficient data show that the (*N*, *N*-dimethyl dithiocarbamoyl) ethoxy substituted calixarenes **20** and **21** have the highest

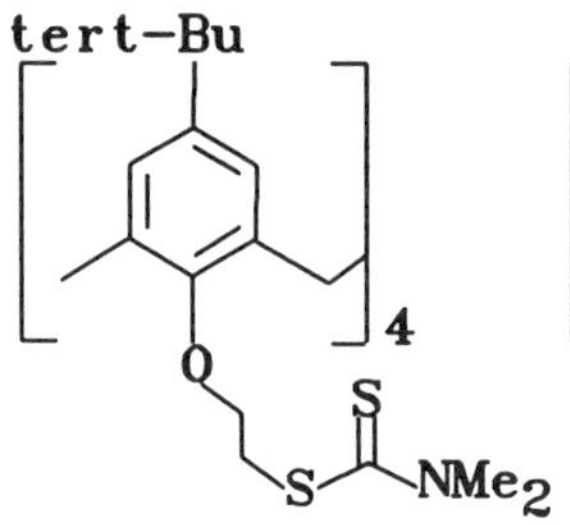

20 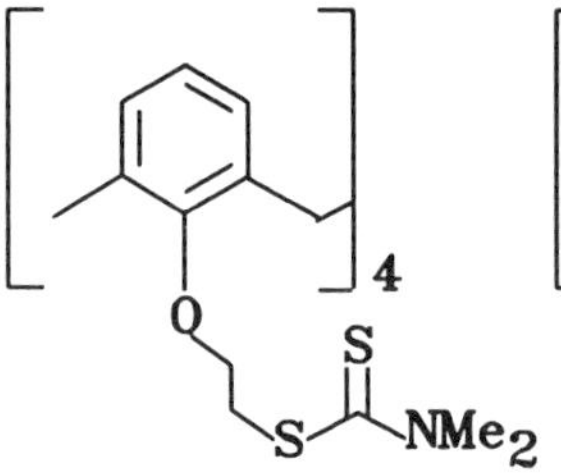 21 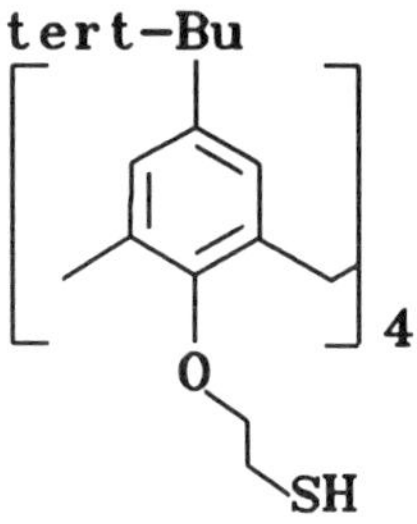 22

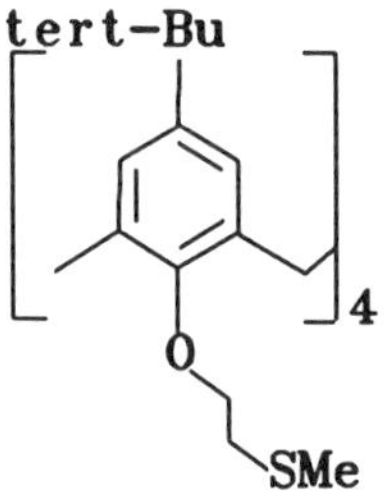

23 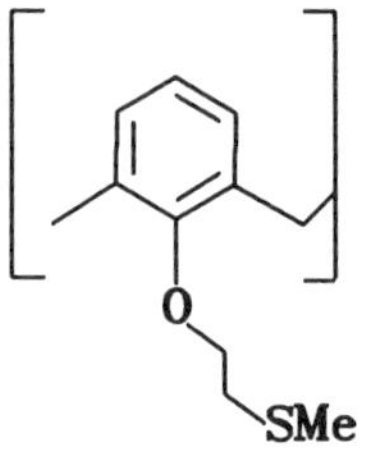 24 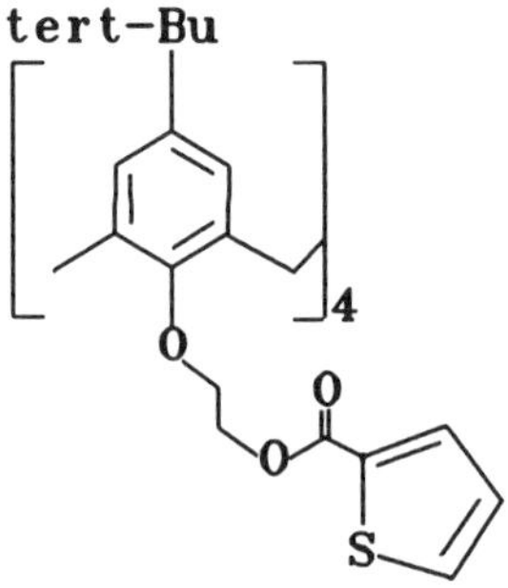 25

distribution coefficients, which when coupled with their selectivity against lead(II) and cadmium(II) makes them useful extractants for mercury(II). By

**Table 2. Distribution Coefficients D, given as $[Hg(II)_{org}]/[Hg(II)_{aq}]$**

| **Compound** | **20** | **21** | **22** | **23** | **24** | **25** |
|---|---|---|---|---|---|---|
| D | 4.3 | 5.7 | 6.1 | 0.06 | 1.6 | 0.85 |

using $^1$H NMR and electronic spectroscopy it has been concluded that the mercury(II) binds to the sulfur groups of these calixarenes.[64] Calix[4]arenes with analogous sulfur functionalities bound to the wide rim have also been found to be effective extractants for mercury(II) into chloroform.[65] These calix[4]arenes are **26-29**, and their distribution coefficients for mercury(II)

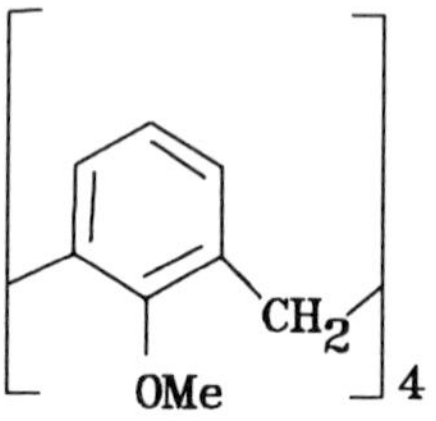

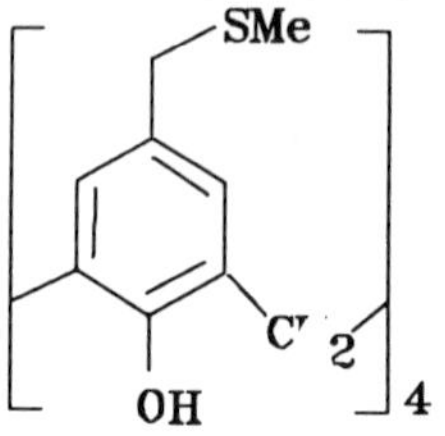

extraction are shown in Table 3. These data show extraction is observed with the methylthioethoxy and (*N,N*-dimethyldithiocarbamoyl) ethoxy substituted derivatives **28** and **29**.

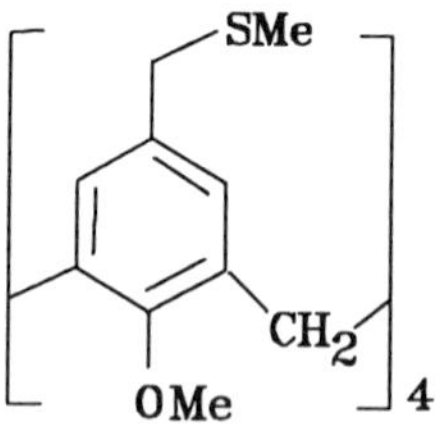

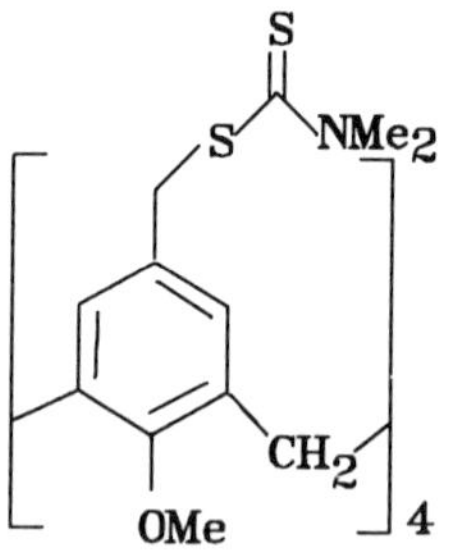

**Table 3. Distribution Coefficients D for Mercury(II)**

| Compound | 26 | 27 | 28 | 29 |
|---|---|---|---|---|
| D | 0 | 0 | 11.5 | 2.7 |

## 8.2. *N*-Donor Calixarenes

Other functional groups can be appended to calixarenes in order to make them be selective extractants for mercury(II). Phenylazo groups appended to the narrow rim of a calix[6]arene have been found to be extractants for both mercury(II) (18%) and mercury(I) (8%) into chloroform.[66] Another such group is the *N*-sulfonylcarboxamide moiety. Such calix[4]arenes efficiently extract mercury(II) from acidic aqueous nitrate solutions with excellent selectivity over alkali, alkaline earth, and many transition metal ions, including lead(II).[67] Calix[4]arenes with fluorogenic *N*-(5-dimethylaminonaphthalene-1-sulfonyl) carbamoylmethoxy groups appended to the narrow rim again show excellent selectivity for mercury(II). The fluorescence quenching due to mercury(II) binding is unaffected by the presence of a 100-fold excess of alkali or alkaline earth metals, or by a group of transition metal and post-transition metal ions.[68]

## 9. ELEMENTAL MERCURY

As was mentioned previously, the volatility of elemental mercury causes it to become widely dispersed into soils and waters. Often it is found at very large distances from where it was initially released. Under certain environmental conditions this mercury can become oxidized to mercury(II), or oxidatively methylated to methylmercury(II). If such oxidations occur, then these charged species can be removed from soils and waters by conventional extractants. Nevertheless, under other circumstances the metal will remain as elemental mercury, and it then needs to be effectively treated in that form.

Although there are early measurements of the solubility of mercury in water at 25° C, the values tend to be lower than those obtained more recently by radioactive mercury detection methods, or by neutron activation analysis. Subsequently the solubility of mercury in water between 0° C and 120° C has been measured using atomic absorption analytical methods.[69] When dissolved in air-free water mercury is monatomic and un-ionized, and is present as the spherically symmetric atom. The mole fraction of mercury in water increases at higher temperature, ranging from 3.73(14) x $10^{-9}$ at 0° C to 31.0(8) x $10^{-9}$ at 90° C. Mercury dissolved in water shows large positive deviations from Raoult's Law, and the Henry Law constants are very much greater than the known mercury saturation vapor pressure. The large positive values for the free energy for mercury transfer into water are predominantly due to the large negative entropy term at 0° C, but to the large positive enthalpy term at 120° C. The un-ionized monatomic mercury is readily lost by volatilization from aqueous solution at ambient temperatures.

Since elemental sulfur has historically been the material of choice for the cleanup of small mercury spills, the retention of mercury vapor by sulfur supported on sepiolite has been studied, and the results compared with materials containing sulfur supported on activated carbon. Samples with 10% sulfur on sepiolite of varying size and shape have been prepared from powders sulfurized by reaction/deposit. The material retains mercury by the irreversible conversion to mercuric sulfide. This chemical reaction is quite rapid and governed by mass transfer. This material also offers a much better capacity than the commercially available sulfurized activated carbon.[70]

The challenge remains, however, to find a molecule or material that will act as a host for elemental mercury such that the metal can be subsequently recovered, and the adsorbent recovered for further use. Possible approaches involve using a host molecule that is selective for mercury, or using an oxidant/complexant combination that will result in it being extracted in a bound mercury(II) form. Since mercury is one of the few metals that forms amalgams with other metals at ambient temperature, it is also possible that this property of the metal could be used to advantage in designing a selective extraction process.

## REFERENCES

1. P. J. Craig in *Organometallic Compounds in the Environment*, P. J. Craig, ed., Wiley, New York, N. Y., 1986, chapter 2.
2. W. F. Fitzgerald, D. R. Engstrom, R. P. Mason, E. A. Nater, *Eviron. Sci. Technol.*, **1998**, *32*, 1.
3. K. A. Larson, J. M. Wiencek, *Ind. Eng. Chem. Res.*, **1993**, *32*, 2854.
4. J. Strzelbicki, W. Charewicz, J. Beger, L. Hinz, *Can. J. Chem.*, **1988**, *66*, 1965.
5. J. Strzelbicki, W. Charewicz, J. Beger, L. Hinz, *Can. J. Chem.*, **1988**, *66*, 2640.
6. F. L. Moore, *Environ. Sci. Technol.*, **1972**, *6*, 525.
7. S. De Moraes, A. Abrao, *Anal. Chem.*, **1974**, *46*, 1812.
8. E. M. Moyers, J. S. Fritz, *Anal. Chem.*, **1976**, *48*, 1117.
9. R. B. Little, C. Burda, S. Link, S. Logunov, M. A. El-Sayed, *J. Phys. Chem.*, **1998**, *102*, 6581.
10. J. Wisniak, G. Schorr, D. Zacovskry, S. Belfer, *Ind. Eng. Chem. Res.*, **1990**, *29*, 1907.
11. G. Alberti, *J. Chromatogr.*, **1967**, *31*, 177.
12. M. Tsubouchi, *Anal. Chem.*, **1970**, *42*, 1087.
13. F. Sinner, M. R. Buchmeiser, R. Tessadri, M. Mupa, K. Wurst, G. K. Bonn, *J. Am. Chem. Soc.*, **1998**, *120*, 2790.
14. C. Y. Liu, M. J. Chen, T. J. Chai, *J. Chromatogr.*, **1991**, *555*, 291.
15. C. M. Wai, S. Wang, J.-J. Yu, *Anal. Chem.*, **1996**, *68*, 3516.
16. S. Wang, C. M. Wai, *Environ. Sci. Technol.*, **1996**, *30*, 3111.

17. A. Yazdi, E. J. Beckman, *Mater. Res. Soc. Symp. Proc.*, **1994**, *344*, 211.
18. A. Wyttenbach, S. Bajo, *Helv. Chim. Acta.*, **1973**, *56*, 1198.
19. S. J. Yeh, J. M. Lo, L. H. Shen, *Anal. Chem.*, **1980**, *52*, 528.
20. H. Irth, G. J. De Jong, V. A. Th. Brinkman, R. W. Frei, *Anal. Chem.*, **1987**, *59*, 98.
21. W. Langseth, *J. Chromatogr.*, **1988**, *438*, 414.
22. J.-M. Lo, J.-D. Lee, *Anal. Chem.*, **1994**, *66*, 1242.
23. A. Chow, D. Buksak, *Can. J. Chem.*, **1975**, *53*, 1373.
24. R. Litman, E. T. Williams, H. L. Finston, *Anal. Chem.*, **1977**, *49*, 983.
25. R. J. Baltisberger, C. L. Knudson, *Anal. Chem.*, **1975**, *47*, 1402.
26. P. K. Dorhout, S. H. Strauss, *ACS Sympos. Ser.*, **1999**, *727*, 53.
27. K. A. Larson, J. M. Wiencek, *ACS Sympos. Ser.*, **1994**, *554*, 124.
28. M. Aguda, A. Rios, M. Valcarcel, *Anal. Chem.*, **1993**, *65*, 2941.
29. M. D. Morris, L. R. Whitlock, *Anal. Chem.*, **1967**, *39*, 1180.
30. P. Jones, S. Hardy, *J. Chromatogr.*, **1997**, *765*, 345.
31. C. Sarzanini, G. Sacchero, M. Aceto, O. Abollino, E. Mentasti, *J. Chromatogr.*, **1992**, *626*, 151.
32. S. Tanaka, T. Kaneta, H. Yoshida, *J. Chromatogr.*, **1988**, *447*, 383.
33. M. Yanagisawa, H. Suzuki, K. Kitagawa, S. Tsuge, *Spectrochim. Acta, Part B*, **1983**, *38B*, 1143.
34. F. W. Wilshire, J. P. Lambert, F. E. Butler, *Anal. Chem.*, **1975**, *47*, 2399.
35. D. T. Gjerde, J. S. Fritz, *J. Chromatogr.*, **1980**, *188*, 391.
36. T. Braun, S. Palagyi, *Anal. Chem.*, **1979**, *51*, 1697.
37. L. Legradi, *J. Chromatogr.*, **1974**, *102*, 319.
38. V. Taglia, *J. Chromatogr.*, **1973**, *79*, 380.
39. Z. Masoomi, D. T. Haworth, *J. Chromatogr.*, **1970**, 48, 581.
40. M. J. Zetlmeisl, D. T. Haworth, *J. Chromatogr.*, **1967**, *30*, 637.
41. W. Haerdi, E. Gorgia, N. Lakhova, *Helv. Chim. Acta.*, **1971**, *54*, 1497.
42. M. Qureshi, I. Akhtar, K. N. Mathur, *Anal. Chem.*, **1967**, *39*, 1766.
43. M. Qureshi, V. Sharma, R. C. Kaushik, T. Khan, *J. Chromatogr.*, **1976**, *128*, 149.
44. H. Sakamoto, S. Ito, M. Otomo, *Chem. Lett.*, **1995**, 37.
45. S. Wang, S. Elshani, C. M. Wai, *Anal. Chem.*, **1995**, *67*, 919.
46. R. M. Izatt, J. S. Bradshaw, S. A. Nielson, J. D. Lamb, J. J. Christensen, D. Sen, *Chem. Rev.*, **1985**, *85*, 271.
47. G. Wu, W. Jiang, J. D. Lamb, J. S. Bradshaw, R. M. Izatt, *J. Am. Chem. Soc.*, **1991**, *113*, 6538.
48. T. F. Baumann, J. G. Reynolds, G. A. Fox, *Abstr. 214th Natl. ACS Meeting*, Las Vegas, Nev., Sep 1997, ENVR 117.
49. T. Kumagai, S. Akabori, *Chem. Lett.*, **1989**, 1667.
50. R. M. Izatt, R. L. Bruening, W. Geng, M. H. Cho, J. J. Christensen, *Anal. Chem.*, **1987**, *59*, 2405 .
51. R. M. Izatt, G. C. Lind, R. L. Bruening, P. Huszthy, C. W. McDaniel, J. S. Bradshaw, J. J. Christensen, *Anal. Chem.*, **1988**, *60*, 1694.
52. S. Wang, S. Elshani, C. M. Wai, *Anal. Chem.*, **1995**, *67*, 919.
53. R. M. Izatt, R. L. Bruening, M. L. Bruening, B. J. Tarbet, K. E. Krakowiak, J.

S. Bradshaw, J. J. Christensen, *Anal. Chem.*, **1988**, *60*, 1825.

54. B. Vaidya, J. Zak, G. J. Bastiaans, M. D. Porter, J. L. Hallman, N. A. R. Nabulski, M. D. Utterback, B. Strzelbicka, R. A. Bartsch, *Anal. Chem.*, **1995**, *67*, 4104.
55. K. S. Hui, B. A. Davis, A. A. Boulton, *J. Chromatogr.*, **1975**, *115*, 581.
56. F. Hamada, T. Fukugaki, K. Murai, G. W. Orr, J. L. Atwood, *J. Incl. Phenom. Mol. Recogn. in Chem.*, **1991**, *10*, 57.
57. A. T. Yordanov, J. T. Mague, D. M. Roundhill, *Inorg. Chem.*, **1995**, *35*, 5084.
58. A. T. Yordanov, D. M. Roundhill, J. T. Mague, *Inorg. Chim Acta.*, **1996**, *250*, 295.
59. A. T. Yordanov, D. M. Roundhill, *New J. Chem.*, **1996**, *20*, 447.
60. A. T. Yordanov, B. R. Whittlesey, D. M. Roundhill, *Supramol. Chem.*, **1998**, *9*, 13.
61. A. T. Yordanov, B. R. Whittlesey, D. M. Roundhill, *Inorg. Chem.*, **1998**, *37*, 3526.
62. A. T. Yordanov, D. M. Roundhill, *Inorg. Chim. Acta.*, **1998**, *270*, 216.
63. A. T. Yordanov, B. R. Whittlesey, D. M. Roundhill, *Inorg. Chem.*, **1998**, *37*, 3526.
64. A. T. Yordanov, D. M. Roundhill, *Inorg. Chim. Acta.*, **1998**, *270*, 216.
65. A. T. Yordanov, O. M. Falana, H. F. Koch, D. M. Roundhill, *Inorg. Chem.*, **1997**, *36*, 6468.
66. E. Nomura, H. Taniguchi, S. Tamura, *Chem. Lett.*, **1989**, 1125.
67. G. G. Talanova, V. S. Talanov, R. A. Bartsch, *JCS, Chem. Comm.*, **1998**, 1329.
68. G. G. Talanova, N. S. A. Nazar, V. S. Talanov, R. A. Bartsch, *Anal. Chem.*, **1999**, *71*, 3106.
69. D. N. Glew, D. A. Hames, *Can. J. Chem.*, **1971**, *49*, 3114.
70. M. I. Guijarro, S. Mendioroz, V. Munoz, *Ind. Eng. Chem. Res.*, **1998**, *37*, 1088.

# 7

# PHASE TRANSFER EXTRACTION OF LEAD AND CADMIUM

The metals lead and cadmium have played important roles in the economic development of nations because of their widespread industrial applications. A problem with their use, however, is their toxicity. The combination of these two features means that over the past 100 years considerable amounts of these metals have been dispersed throughout soils and waters. Now there is an urgent need for them to be removed and recovered, both for safety reasons and for economic reasons so that they can be further used under more controlled conditions. Lead and cadmium have been introduced into the environment from a variety of sources. Cadmium is present in metal coatings and batteries, and lead has found widespread use in both automobile and housing applications. These metals can be present in waste sites as complexed divalent ions, or as complexed methylated derivatives. The simple metal ions in soils are occluded or bound into the clay structures. The methylated derivatives, although soluble in water, are more hydrophobic, and therefore their solubility in organic solvents is higher.

## 1. LEAD

### 1.1. Toxicity and Occurrence

The toxic affects of lead have been widely discussed. This toxicity is particularly problematic for children. Because 74% of all private housing built before 1980 in the United States of America contains some lead paint, one out of nine of all children under the age of six has enough lead in their blood to place them at risk, and children with high lead levels are 6 times more likely to have reading disabilities.[1] The threshold for lead levels is 10 micrograms per deciliter of blood, which puts some four million children at risk for lead poisoning in the United States. Since lead poisoning is mainly silent, most poisoned children have no symptoms. As a result the majority of cases go undiagnosed and untreated.

An early use of lead is a Turkish statue that dates from about 6,500 B.C., and lead water pipes from Roman times that are some 2,000 years old. More recently lead has been used in paints, fuels, batteries, solders, pipes and ammunition, resulting in large amounts being dispersed worldwide.[2,3] Chelation therapy is used in the treatment of lead poisoning. Among the compounds that are used are calcium disodium ethylenediamine tetraacetate, 2,3-dimercapto-1-propanol, 3-mercapto-D-valine (D-penicillamine) and *meso*-2,3-dimercaptosuccinic acid. Although chelation therapy is beneficial, these metals are cumulative toxins, which makes it difficult to keep them at low levels.[4]

## 1.2. Complexants

The complexation chemistry of lead is both similar to and different from that of mercury. Lead primarily exists as its divalent state, but tetraethyllead and its degradation products also represent a significant environmental problem because of its widespread use as an additive to increase the performance of automobile fuels. A significant difference between the complexation chemistry of lead and mercury is that lead(II) has a lesser preference for soft ligands than does mercury(II). Thus lead(II) can be extracted with oxygen-donor atom complexants that are ineffective for mercury(II). This difference frequently allows for this pair of metal ions to be selectively extracted from solutions containing mixtures of the two metal ions.

### 1.2.1. Halides and Acids

Simple halide ions can be used for the extraction of lead(II). Both nitric and hydrochloric acids have been tested as extractants for lead and zinc from both sandy and clay soils. Hydrochloric acid is the most effective because of the complexation properties of the chloride ion, and lead originating from pollution is more easily and completely removed than lead which originates from natural sources.[5] Halides are also particularly useful for carrying out assays of alkyllead compounds. As an example, the presence of sodium chloride increases the quantity of trimethyllead that can be extracted from an aqueous solution into chloroform.[6]

Both oxygen and nitrogen donor ligands have been used for the extraction of lead. One technique for the determination of lead in geochemical samples involves extracting it and other metals with a MAGIC organic extracting solution. In addition to methyl isobutyl ketone, this solution also contains iodide ion and an amine.[7] The mechanisms by which stripping with this MAGIC extraction system is accomplished include poisoning of the amine ion exchange agents with anions that are incompatible with the system, and oxidation of the halide complexing ion.[8] A similar extraction procedure with hydrochloric acid, potassium iodide, and a long chain amine has been used for the removal of trace amounts of lead and other metals from nuclear grade uranium.[9]

Although hydrochloric acid is a good extractant for lead(II) because of the complexation properties of the chloride ion, both nitric acid and acetic acid are also used. However, in an evaluation of the extraction of lead and other trace metals from sediments it is found that extraction with boiling nitric acid solution leads in some cases to low values.[10] Nitric acid has also been used to stabilize the metal complex formed in the extraction of lead and other metals with methyl isobutyl ketone.[11] A mixture of acetic acid and sodium acetate has been used for the extraction of lead and copper from model soils prepared from natural minerals and humic acid. The extractable amounts of the metals and the extent of recovery depends on the soil composition. The presence of dissolved humic acid during extraction influences metal redistribution as evidenced by an increase in the adsorption of copper and a decrease in the adsorption of lead.[12] A broad range of extractants have been used for the extraction of lead, cadmium, copper and zinc bound to aluminum hydroxide gel, gibbsite, or alumina. Alkali and alkaline earth chlorides are the poorer extractants, and acetic, oxalic, nitric and hydrochloric acids are the most effective.[13]

The problem of obtaining reliable quantitative data results in these selective extraction procedures being suspect when applied to real sediments. As a result, a study has been carried out with synthetic models to test the procedures used for the extraction of metals in aquatic sediments. The major sedimentary geochemical phases were prepared and characterized by both X-ray diffraction and scanning electron microscopy. The trace metal is doped into each phase by adsorption or co-precipitation. Lead is present as the calcite phase, and the copper is in either the humic acid or the iron hydroxides.[14] Metal recoveries are calculated by comparing the sum of the contribution from each extraction step to the total amounts available in each sediment. These values fall in the 80-107% range. These sequential extractions are, however, unreliable because of elemental redistribution. This feature becomes apparent when experiments with multiphase model sediments are performed. Thus where extractions perform poorly on model systems, performance on real sediments are questionable.

### 1.2.2. Poly(ethylene glycol)

Poly(ethylene glycol) **1** is a fluid that has been used for the extraction of several different metals, including lead. This fluid has drawn particular attention because it forms a biphasic system in aqueous solution.[15, 16] Aqueous biphasic systems of **1** consist of two immiscible phases that are

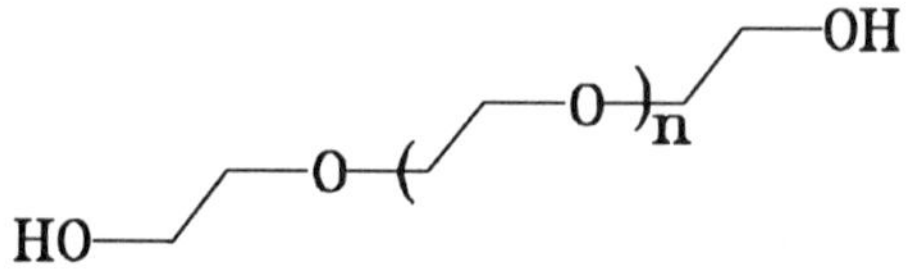

HO–CH₂CH₂–O–(CH₂CH₂–O)ₙ–CH₂CH₂–OH

1

formed when water soluble polymers are combined with inorganic salts. The major component in each phase is water, thereby eliminating the need for the use of volatile organic compounds. Lead(II) has been separated by a poly(ethylene glycol) biphasic system.[17] The metals lead(II), cadmium(II) and mercury(II) partition to the poly(ethylene glycol) rich phase in the presence of a sufficient quantity of halide ion. The D values decrease in the order $I^- > Br^- > Cl^- > F^-$, which correlates with the relative free energies of hydration and the distribution rations for halides in an aqueous poly(ethylene glycol) phase containing ammonium sulfate. This halide ion trend also correlates with the expected stability of the halide complexes $MX_4^{2-}$. Thiocyanate ion has also been used instead of halide ions, although its effectiveness for lead(II) is less than that for mercury(II) and cadmium(II).[18]

## 1.3. Chelates and Macrocycles

Chelate and macrocyclic ligands can also be used for the extraction of lead from soils, waters, and mammalian species. Although the complexation chemistry of lead(II) and cadmium(II) is not as extensive as that of the transition metals, there are still numerous studies that have been carried out on both the complexation chemistry and the liquid-liquid extraction properties of these metals.

### 1.3.1. Chelating Agents for *In Vivo* Extraction Therapy

Chelating agents that have been used as therapy for ingested lead include EDTA,[19] *N*-(2-hydroxyethyl) ethylenediaminetriacetic acid (HEDTA, **2**), [20, 21] sodium 2,3-dimercapto-1-propane sulfonate (DMPS, **3**),[22-26]

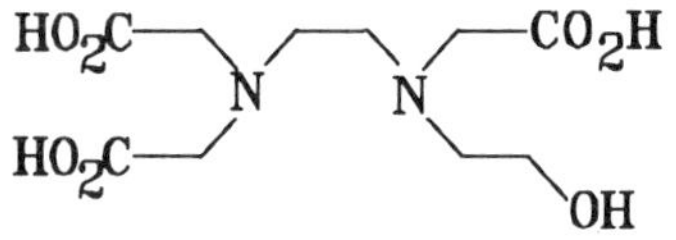

$HO_2C$ $CO_2H$ N N $HO_2C$ OH

2

penicillamine, **4**,[22, 27] 2,3-dimercaptosuccinic acid (DMSA, **5**),[28, 29] (2-mercaptopropionyl) glycine,[30] and 2,3-dimercaptoglycine.[30-32] In 1991 the

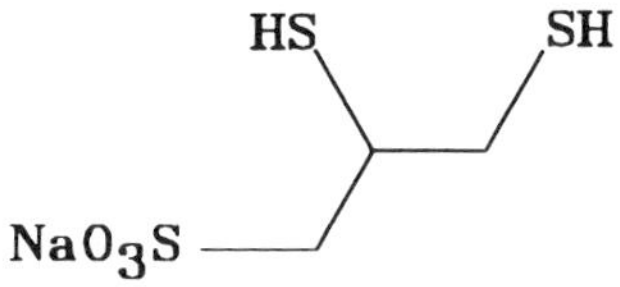

HS SH $NaO_3S$

3

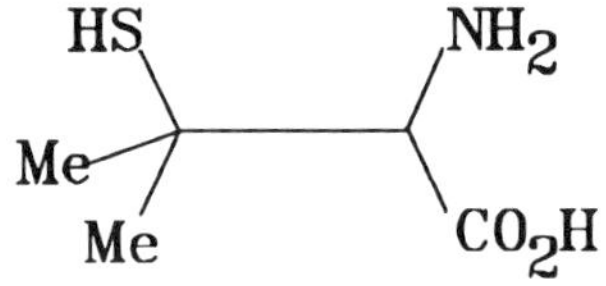

HS $NH_2$ Me Me $CO_2H$

4

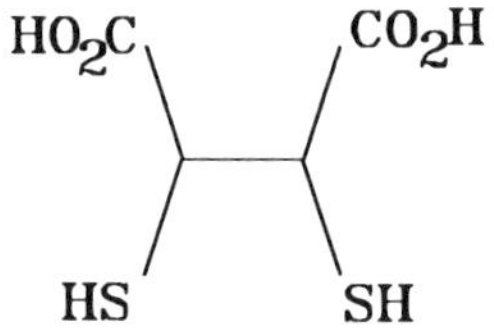

$HO_2C$ $CO_2H$ HS SH

5

FDA approved the use of *meso*-DMSA for lead poisoning. Lead is

completely chelated by DMSA at pH 7.4 for DMSA : lead(II) concentration ratios that are greater than or equal to unity, and the complexes that form have the stoichiometries Pb(DMSA) and HPb(DMSA).[28] The binding sites on the DMSA are the two thiolates and one carboxylate group. The formation constants of the lead chelates of *rac*-DMSA are greater than those of *meso*-DMSA. However, although *rac*-DMSA is better than the *meso* form for mobilizing lead and cadmium, its disadvantage is that its use also causes a greater loss of zinc.[29] A solution to this problem is to use Zn(*rac*-DMSA)$_2$ instead of the free chelate.[33]

### 1.3.2. Oxygen and Nitrogen Donor Extractants

Although chelation therapy uses soft sulfur donor atoms, extractants with oxygen and nitrogen donors can also be effective for the removal of the hard lead(II) and cadmium(II) cations from soils and waters. Such an *O*-donor chelate having a poly(ethylene ether) type backbone is the dibenzopolyether dicarboxylic acid ionophore **6**. This complexant that has two ionizable hydrogens can extract divalent metals into an organic phase as

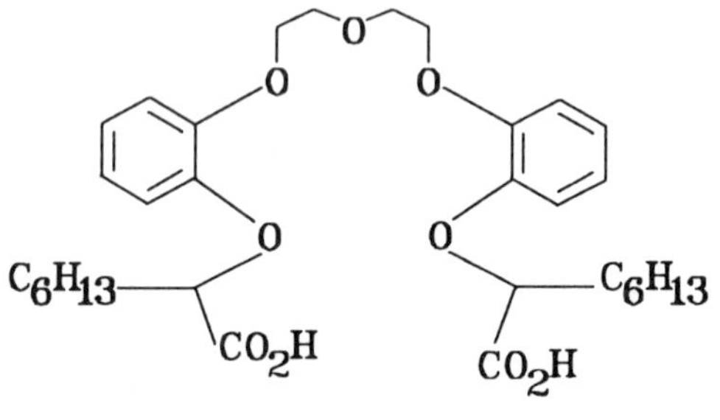

**6**

uncharged complexes. This ionophore is selective for the liquid-liquid extraction of lead(II) from copper(II).[34] The stoichiometry of the extracted complex has a 1 : 1 metal : ligand ratio. When the compound is incorporated into a polymeric plastic membrane the resulting material can be used for the separation of lead(II) from copper(II). For *N*-donors, the simple *N,N*-disubstituted amides have been used for the extraction of lead(II) and other metals into a toluene phase.[35]

A woven polyacryloamidoxime cloth that has a large capacity for lead and copper has been prepared from the polyacrylonitrile precursor by reaction in methanolic hydroxylamine.[36] Also an iminodiacetate cellulose chelating sorbent derived from cotton fabric has also been synthesized and used for the extraction of lead, cadmium and copper. The material adsorbs these metals with a capacity up to 0.38 mmol in metal ion. This material is therefore useful both for preconcentration of metal ions, and for selective metal ion extraction.[37]

Schiff bases also form complexes with lead(II) that can be subsequently extracted into an organic phase. Complexation with two such Schiff bases $H_2$salen and $H_3$saltren leads to the complexes $Pb_3(salen)_2^{2+}$ and Pb(saltren).[38] The coordination geometries about lead(II) are *pseudo* pentacoordinate, with one of the coordination positions being occupied by the lone electron pair on lead. An important aspect of the economics of metal extraction with chelates such as these and with EDTA is the subsequent recovery of both the metal and the chelate. For EDTA with both lead and mercury, an effective recovery system has been developed whereby after separation of the metal complexes that are formed, both the metal and the free EDTA are recovered by electrolysis. The overall system when applied to soils involves soil washing, followed by electrochemical treatment and EDTA recycling. The method is being tested on soils that are contaminated with both lead and mercury.[39]

### 1.3.3. Sulfur Donor Dithiocarbamate Extractants

Although lead(II) and cadmium(II) can be extracted into an organic phase with complexants having oxygen and nitrogen donor atoms, there is also an extensive literature on the use of sulfur donor ligands as extractants for these metals. Dithiocarbamates have received attention, both as simple dithiocarbamate salts or as derivatives having this functionality incorporated into macrocycles or appended onto calix[4]arenes. Radiotracers have been used to determine the extraction constants for diethyldithiocarbamates and pyrrolidinecarbodithioates of lead(II).[40] Ammonium pyrrolidine-dithiocarbamate has also been used as a chelator for lead(II) in its extraction from biological materials. Metal analysis is important in this system, especially when the metals are only present in trace amounts. However, when potassium cyanide is used to mask the effects of iron, zinc, and copper, lead in the 0.05 to 20 μg range can be detected by atomic absorption methods.[41] With ammonium pyrroline-carbodithioate and sodium diethyldithiocarbamate as extractants, the solution acidity must be controlled.

At high acidity both chelating agents decompose. This problem can be circumvented, however, if the lead complex is extracted from the aqueous acid quickly after addition of the complexant.[42, 43] The diethyldithiocarbamate extractant can also be used to extract lead, cadmium and copper that are present at the ppb level from river water into isoamyl alcohol at the ppb level. However, only copper can be quantitatively recovered by this procedure.[44] The extraction of lead and other metals from rocks as their chloro complexes has been accomplished using diethyldithiocarbamate and 8-hydroxyquinoline as chelates, along with a mixture of methyl isobutyl ketone and butyl acetate as the solvent system.[45] The diethyldithiocarbamate complexes of lead and cadmium, after extraction from natural waters, can then be adsorbed onto a Dowex column with a sorption solution comprising of THF/methylglycol/6M HCl. Lead and copper can be eluted with 6M HCl, and cadmium with 2M $HNO_3$.[46] A variation on the dithiocarbamate method has been used to determine trace metals in seawater. The method involves extraction of the metal dithiocarbamate complexes into chloroform at a pH of 4.56, followed by back extraction with a dilute solution of mercury(II). The method preconcentrates trace metals such as lead, cadmium and copper, and is successful because the extraction constant of mercury dithiocarbamate is much greater than those of the other metals.[47] For comparison between the metals lead and cadmium, the extraction constants (log $K_{ex}$) of the *bis*-(diethyldithiocarbamate) lead(II) and cadmium(II) complexes are 7.94(9) and 5.77(5) respectively.[48]

The ammonium pyrrolidine-*N*-carbodithioate and methyl isolutyl ketone liquid-liquid extraction system has been used over a pH range of 0 to 12. The use of either ammonium citrate or tiron effectively masks the effect of iron. For solutions above pH 8.0 citrate is used, and for those above pH 6.5 tiron is used.[49] The addition of sodium chloride as a salting-out agent makes it possible to concentrate the metals. Lead(II), cadmium(II) and copper(II) can then be determined at the ppb level by atomic absorption spectroscopy.

### 1.3.4. Analytical Methods for Lead

A graphite-furnace atomic absorption spectrophotometer has been found to be useful for the determination of lead and cadmium because there are few interferences and sample-size limitations, and the data show good accuracy and consistent precision.[50] Although atomic absorption spectroscopy is the method of choice for the analysis of lead, high-performance liquid chromatography has also been used for the metal. The UV detector is set to the absorbance peak of the dithizone or diethyldithiocarbamate complexants, which allows for the use of small volumes of the extractants.[51] Reversed phase HPLC following extraction with hexamethyleneammonium-hexamethylenedithiocarbamate has been used to determine lead and copper in both citrus leaves and rice flour. The metals are extracted into chloroform as their chelates, separated on a C18 (Cosmosil 5C18) column, and the eluent monitored at 260 nm.[52]

### 1.3.5. Other Sulfur Donor Extractants

Diethyldithiophosphate (DTP, **7**) is one of a series of homologous compounds that can be used as liquid-liquid extractants for heavy metals. The compound can coordinate to metal ions *via* the sulfurs to give 4-membered ring chelate complexes. The compound does, however, have rather limited use because the EtO-P functionalities are susceptible to hydrolysis. Extended contract with water will also result in hydrolysis of the phosphorus-sulfur bonds.

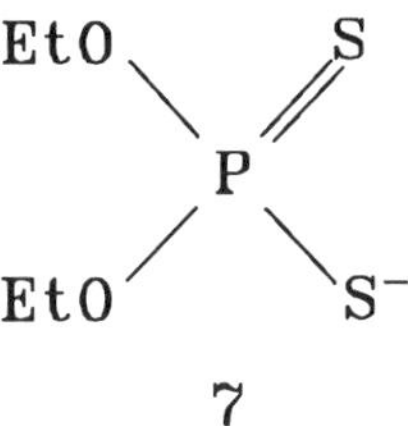


7

Extraction of lead(II) occurs into a chloroform phase, with the metal being extracted as a complex of stoichiometry $Pb(DTP)_2$. This complex can be detected by its UV absorbance band at 293nm.[53] Using octadecyl functional groups bonded to silica as sorbent, and methanol as eluent, dialkyldithiophosphates $(RO)_2PS_2^-$ have been used as extractants for lead, cadmium, and copper. Quantitative extraction is observed for short alkyl chains (R up to propyl) in acidic medium.[54]

Lead and other heavy metals have been extracted by a mixture of diphenylthiocarbazone, **8**, 8-quinolol, **9**, and acetylacetone, **10**, in ethyl propionate as solvent. The system has been used for the metals lead, cadmium, copper, and silver, and found to be highly selective.[55] This

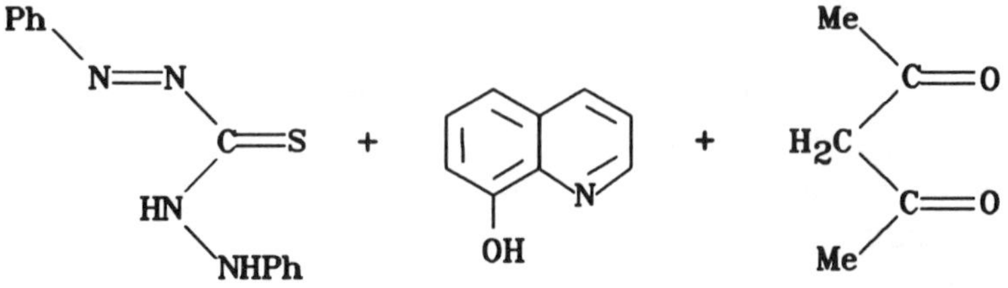

Ph–N=N–C(=S)–NH–NHPh + 8-hydroxyquinoline + Me–C(=O)–$H_2C$–C(=O)–Me

8 9 10

selectivity has been tested with thirty four other ions and compounds. With this method all of the metals can be determined even when they are present at the low parts per billion level in an aqueous solution. The metal analysis is carried out by AAS.

A method for the identification of lead and other toxic metals has been developed that involves extraction of the metals into carbon tetrachloride as their dithizone complexes, followed by separation using thin layer chromatography on silica. The method has application in biological samples, and has a sensitivity of $10^{-7}$ g ion/l.[56] In addition to liquid-liquid extractants, solid phase extractants that incorporate *bis*-(2,4,4-trimethyl) monothio phosphonic acid on a functionalized silica surface have been prepared, and then used for the extraction of lead and other toxic metals. The material shows the selectivity order of cadmium(II) > lead(II) > zinc(II), thereby allowing these three metals to be separated from each other. The metals can be subsequently stripped off the resin by using a dilute solution of nitric acid.[57]

Camphor-3-thioxo-2-oxime has been used as a sensitive analytical reagent for both the extraction and spectrophotometric determination of trace amounts of lead. The reagent rapidly forms a stable yellow-orange chelate complex with lead(II) over the pH range 9.3 to 9.6, and this complex is selectively extracted into carbon tetrachloride. The sensitivity of the method allows for the detection of $5 \times 10^{-3}$ µg lead(II).[58] Another approach to designing selective complexants for the extraction and determination of lead is to target thiohydroxamates. This choice is made because of the very high stability constants of complexes such as lead with compounds such as *N*-phenylthiobenzohydroxamate **11**.[59] The resulting *bis*-(thiohydroxamate) lead complexes have *pseudo* trigonal bipyramidal structures involving *O*, *S*-coordination, with the fifth coordination position being occupied by the lone

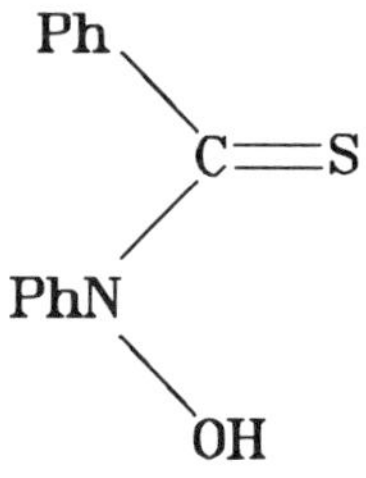

**11**

electron pair on lead(II).[60] The similar hydroxypyridinethione compound **12** forms a *bis*-(ligand) Pb(II) complex that shows an analogous coordination geometry.[61]

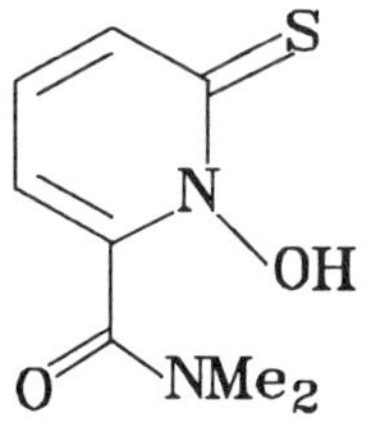

**12**

## 1.4. Macrocycles

Macrocycles are also effective complexants for lead and cadmium. Both oxygen and nitrogen macrocycles have been used for the extraction of lead(II). Similarly both open-chain and macrocyclic ethers are complexants and extractants for cadmium, mercury, and lead. A set of *bis*-coumarin podands with a polyether linkage chain bind lead.[62] For crowns such as 18-crown-6, the inclusion of lead in the cavity results in the formation of an almost hexgonal planar complex.[63] In addition to achieving binding selectively due to hole size, a second mechanism is operable whereby macrocycles select between metal ions in a process that is termed dislocation discrimination.[64] Dislocations occur when a gradual change of properties along a ligand series induces a sudden change in coordination behavior for adjacent complexes in a series. An example of this feature is found for zinc and cadmium with a series of $N_3O_2$-donor macrocycles. For zinc(II) and cadmium(II) with 17-, 18- and 19-membered macrocyles, the first two show the relative stability order of cadmium(II) > zinc(II). For the 9-membered macrocycle the reverse is found, probably because in this case the ether oxygens are not complexed to the metal.[65] This dislocation discrimination effect has been studied by NMR, molecular mechanics, and x-ray diffraction methods.

### 1.4.1. Oxygen Donors

The neutral macrocyclic carriers dibenzo-18-crown-6, dicyclohexano-18-crown-6, 18-crown-6, and polynactin have been used as liquid-liquid extractants for lead(II). The rate constants for the phase transfer process are compatible with what is expected for a diffusion-limited process.[66] Dicyclohexano-18-crown-6 selectively removes lead(II) either from aqueous solutions containing this metal alone, or with it mixed with other metal ions. The lead(II) is extracted as its thiocyanate, nitrate, or chloride salt, and the solvents used for its extraction are nitrobenzene, chloroform, toluene, or an equimolar mixture of nitrobenzene and chloroform.[67] Acyclic polyether dicarboxylic acids having *pseudo*-18-crown-6 frameworks **13-18** have been used for the selective extraction of lead(II). The compound **15** has R = $C_6H_{13}$, $C_8H_{17}$, $C_{12}H_{25}$, $C_{16}H_{33}$. The selectivity

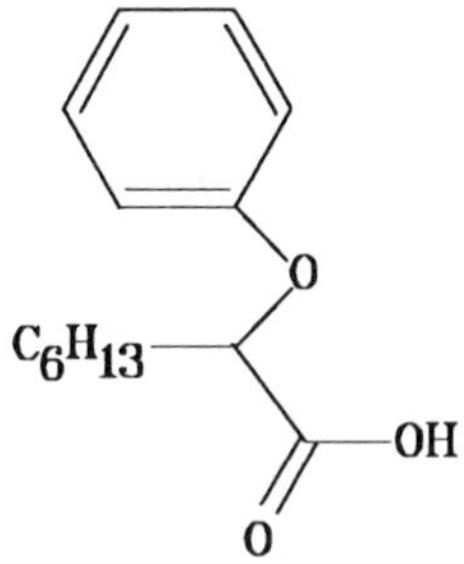

13 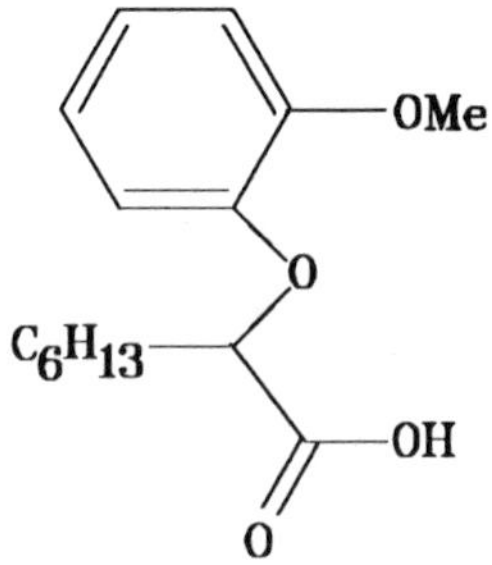 14

order for the extraction of a series of metals with compound **15** (R = $C_8H_{17}$) follows the sequence lead(II) > copper(II) > zinc(II), cadmium(II) > nickel(II). Furthermore, the polyether dicarboxylic acids show higher

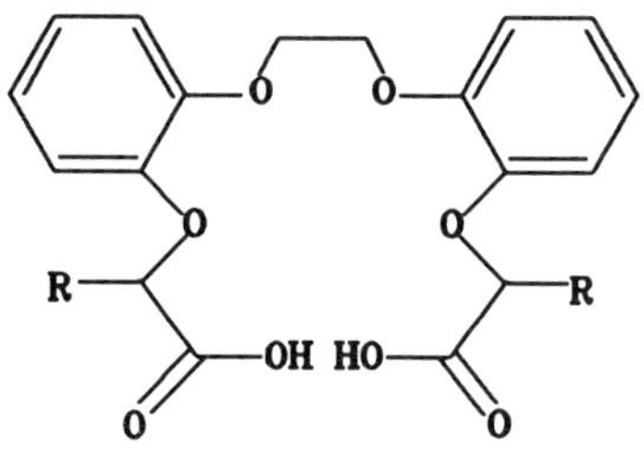

15

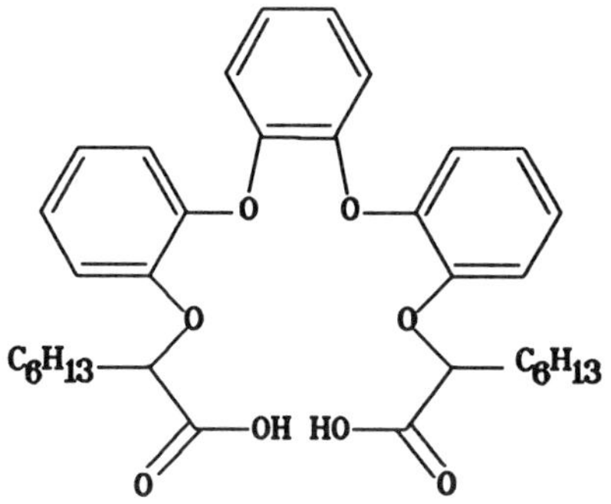

16

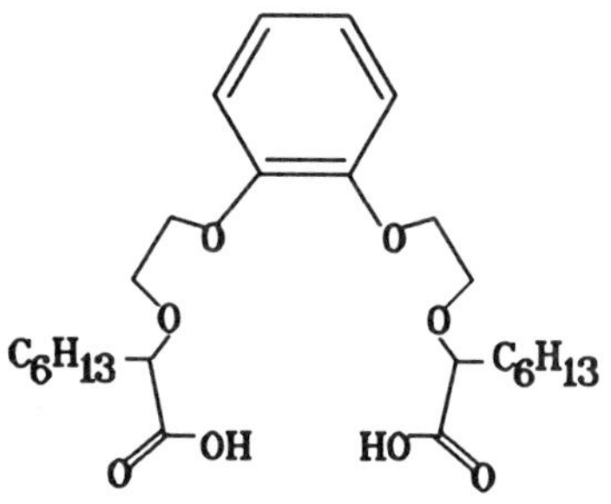

17

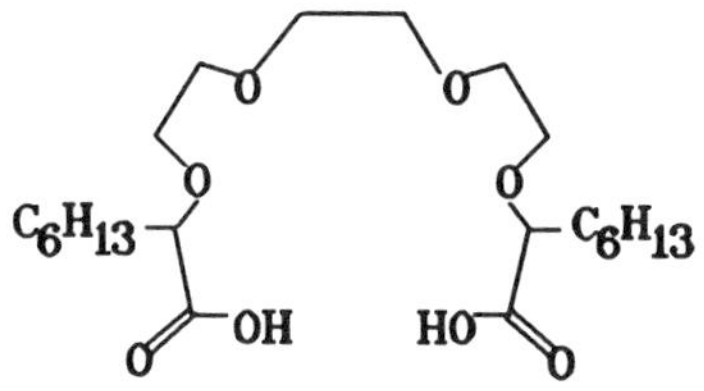

18

relative Pb(II) : Cu(II) selectivities than do the monocarboxylic acids. Alkyl chain length variation along the sequence hexyl, to octyl, to dodecyl, to hexadecyl does not affect the Pb(II) : Cu(II) selectivity.[68] An induced-fit mechanism is invoked to explain these selectivities. In addition to liquid-liquid extractions, an inert polymeric support impregnated with di-*tert*-butylcyclohexano-18-crown-6 has been used for the separation of the lead-210 isotope from both its daughter products and from other impurities.[69]

### 1.4.2. Nitrogen Donors

Tetraazamacrocycles have also been used for the liquid-liquid extraction of lead(II) salts. Three such lipophilic macrocycles that have been investigated are the compounds **19-21**. The extraction efficiency is sensitive to the metal ion that is being extracted, the cavity size of the macrocycle,

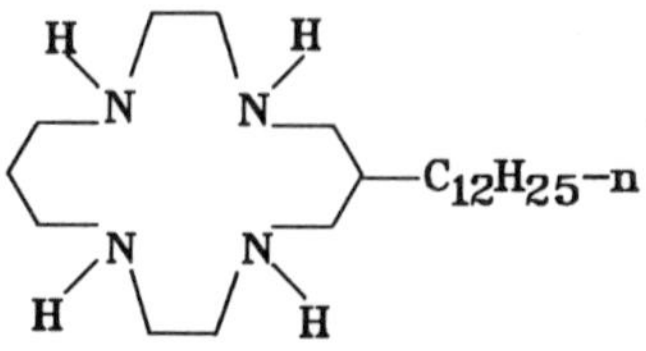

19 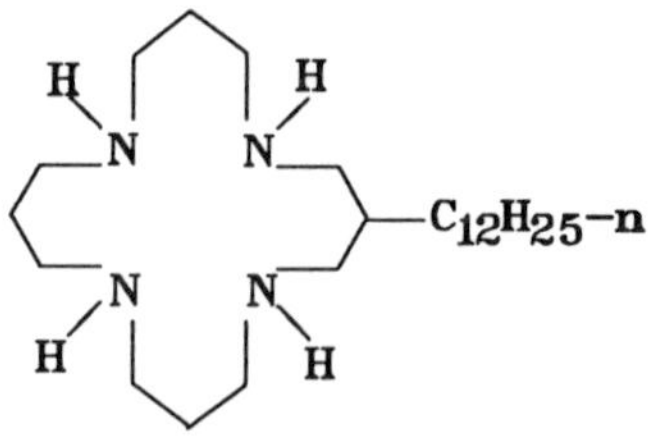 20

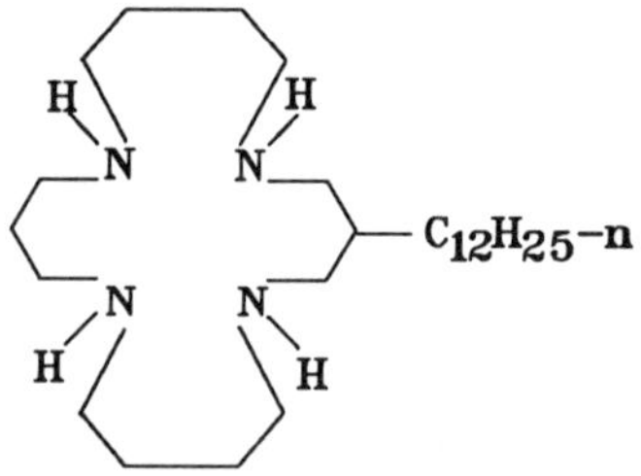

21

and the pH of the aqueous phase. A comparison of the relative extractablilities of lead(II), cadmium(II), copper(II), and silver(I) into a diethyl ether phase as their perchlorate salts are shown in Table 1.[70] These

**Table 1. Extraction (%) of lead(II), cadmium(II), copper(II) and silver(I) Perchlorates into Diethyl Ether**

| Macrocycle | Cu(II) | Cd(II) | Pb(II) | Ag(I) |
|---|---|---|---|---|
| **19** | 67 | 50 | < 5 | 75 |
| **20** | 62 | 66 | 70 | 76 |
| **21** | 44 | 41 | 57 | 95 |

data show a general increase in the extraction percentage as the size of the metal ion increases across a row in the sequence copper(II) (0.69Å), cadmium(II) (0.97Å), lead(II) (1.20Å), silver(I) (1.26Å). The cavity size of the macrocycle follows the sequence **19** < **20** < **21**. For lead(II) and cadmium(II) the extraction is at a maximum for compound **20**. In addition to being used as liquid-liquid extractants, these macrocycles can also be dispersed onto a solid matrix (Amberlite XAD7), which can then be then used for the extraction of lead, cadmium, copper and silver.[71]

Both linear and cyclic polyamines act as complexants for lead(II). The open-chain amines form only mononuclear complexes, but the cyclic amines give both mono-and bi-nuclear lead complexes. The log k stability constants have been plotted against the number of nitrogen atoms for both the open-chain and the macrocyclic ligands.[72] For the open-chain amines, the value of log k remains nearly constant for the range of 4-9 nitrogen atoms in the chain. However, for the cyclic amines, a maximum is observed for 5 nitrogen atoms, with a sharp decrease in log k being observed for either a smaller or a larger number of nitrogens.[72] Cadmium and lead are also complexed by macrocycles having pendant chelates. For example, a tetraazamacrocycle with a phosphinic acid pendant arm shows selectivity for larger cations such as cadmium(II) or lead(II).[73] The size of the macrocycle, the number of 5-membered rings formed in the complex, and the electronic and steric properties of the pendant donor groups are all important in determining the effectiveness of the macrocycle. Another set of macrocyclic ligands have a combination of *O*, *S* and *NH* funtionalities in selected positions in the ring. Although these macrocycles coordinate lead(II), they have higher stability constants for silver(I).[74]

### 1.4.3. Sulfur Donors

Thiacrowns are also used as complexants and extractants for lead(II), cadmium(II) and mercury(II).[75, 76] Those having pendant hydroxymethyl groups can be bound to a support *via* these functionalities. The metal binding by these macrocycles is driven by enthalpy, with the entropy changes being consistently negative. In general, however, thiacrowns show a greater interaction with mercury(II) than they do with lead(II). When choosing a macrocycle as an extractant, it should be remembered that a useful feature of thiacrowns is their stability to aqueous acidic conditions.[77]

A combination of cyclic thioether complexants and electrochemical methods has been used to study lead(II), cadmium(II), and copper(II) at the interface between water and 1,2-dichloroethane. Half-wave potentials and Gibbs energies for direct ion transfer processes follow the order: lead(II) < cadmium(II) < copper(II). The hydrophobic cyclic thioethers that have been used for this purpose are 1,5,9,13-tetrathiacyclohexadecane (**22**), 1,4,7,10-trithiacyclododecane (**23**), 1,4,7-trithiacyclooctane (**24**), 1,4,7-trithiacyclononane (**25**), and trithiane (**26**).[78] Lead(II) ion transfer assisted by **23** and **24** behaves in an electrochemically reversible manner. Cadmium (II)

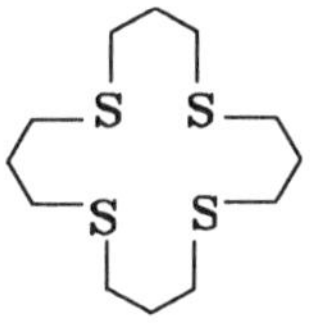

22 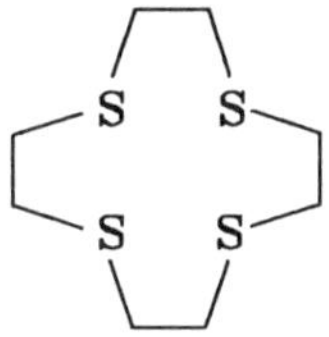 23

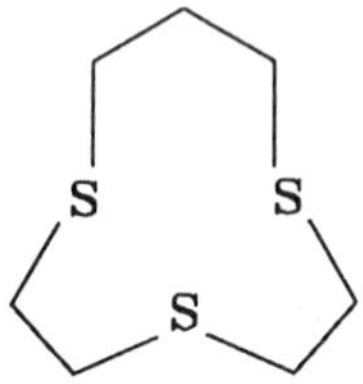

24 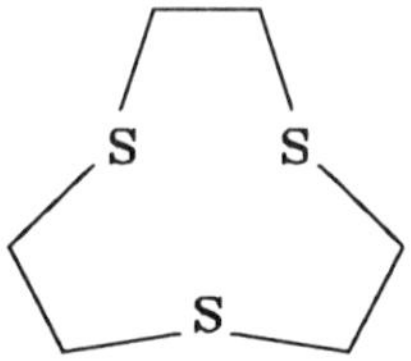 25 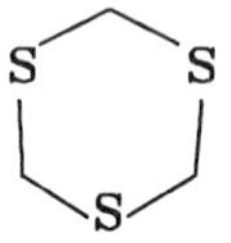 26

shows reversible assisted transfer with **24**. In addition to liquid-liquid extraction, cadmium(II), zinc(II), and lead(II) can each be extracted from aqueous solution using a solid-phase support prepared by attaching *bis*-(2,4,4-trimethyl) monothiophosphonic acid to a functionalized silica surface.[79] The selectivity sequence is cadmium(II) > lead(II) > zinc(II). These metal ions can be simultaneously removed by selective stripping with citric acid and hydrochloric acid solutions.

## 1.5. Calixarenes

Calixarenes, the cyclic oligomers obtained by condensation reactions between phenols and formaldehyde, can be chemically modified to be selective extractants for lead(II). One example is a calixarene carboxylate resin immobilized with polyallylamine that has been used to selectively adsorb trace amounts of lead(II) in the presence of a large excess of zinc(II).[80] Calixarenes have also been used for the complexation and liquid-liquid extraction of lead(II). Calix[n]arenes (n = 4, 5, 6) having thioamide functionalities (R = Et(a), $(CH_2)_4$(b), Pr(c)) appended to their narrow rim **27**

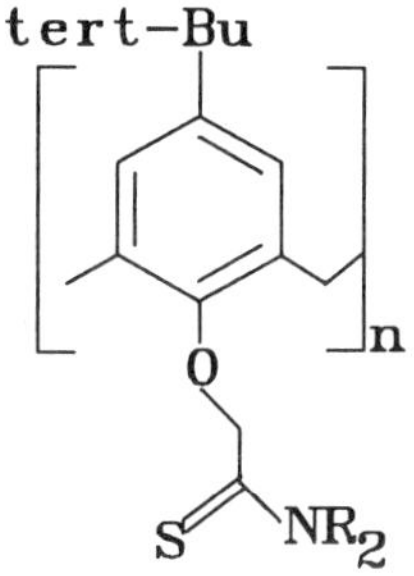

27

effectively extract lead(II), copper(II) and silver(I) picrates from water into dichloromethane. For cadmium(II) only the calix[5]arene derivative is effective. The extraction data collected in Table 2 show higher extractability

**Table 2. Extraction data (%) for Calix[n]arene Thioamides**

| **Cation** | **27a4** | **27b4** | **27a6** | **27b6** | **27c5** |
|---|---|---|---|---|---|
| Na(I) | 7 | 4 | 5 | 5 | - |
| Pb(II) | 56 | 96 | 32 | 46 | 64 |

| | | | | | |
|---|---|---|---|---|---|
| Cd(II) | 8 | 9 | 8 | 58 | 42 |
| Cu(II) | 19 | 47 | 15 | 42 | 38 |
| Ag(I) | 80 | 98 | 95 | 94 | 96 |

for these metal ions than is observed for the sodium ion.[81] Calix[4]arenes with *N*-(X)-sulfonylcarboxamide groups of tunable acidity appended to the narrow rim **28**, (R = $CH_3$, $CF_3$, Ph) are also effective extractants for lead(II), cadmium(II), and copper(II). These extractants show a high selectivity for lead(II) over most alkali, alkaline earth and transition metal ions, although

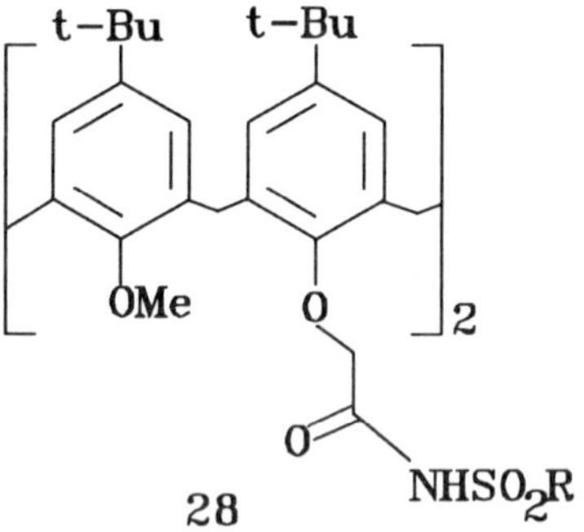

28

they fail to extract lead(II) in the presence of mercury(II).[82] The high selectivity of these calix[4]arenes with hard donor ligands for mercury(II) over lead(II) has been confirmed in further studies.[83] Nevertheless, this selectivity is unexpected since it is soft donor atoms that are expected to show the higher selectivity for mercury(II).

## 1.6. Liquid Carbon Dioxide as the Non-Aqueous Phase

In a series of publications, liquid carbon dioxide has been reported as a potential extractant for lead. Carbon dioxide is a particularly attractive solvent for carrying out extractions. This solvent under supercritical conditions is becoming of increasing importance in metal extraction technology because it is inexpensive and non-toxic. When used in conjunction with a selective complexant, it can be effectively used to extract lead(II). As was discussed in an earlier chapter, an important consideration with this solvent system is always the solubility of the extractant in supercritical carbon dioxide. The solubility of a series of dithiocarbamate salts following 30 min static equilibration at 45°C and 17.18 Mpa with carbon dioxide are shown in Table 3.The complexant $Bu_4N^+[SC(S)NBu_2]^-$,

**Table 3. Solubilities of Dithiocarbamate Ion Pairs in Supercritical Carbon Dioxide**

| Ion Pair | Solubility (μg / ml) |
|---|---|
| $Bu_4N^+[SC(S)NBu_2]^-$ | 23.24 |
| $Bu_4N^+[SC(S)NEt_2]^-$ | 2.91 |
| $Na^+[SC(S)NEt_2]^-$ | 1.09 |
| $NH_4^+[SC(S)NC_4H_8]^-$ | 0.68 |

which has the highest solubility, effects lead(II) and cadmium(II) removal into carbon dioxide at levels of > 73% and > 94% for 5 and 15 min extraction times respectively.[84] This approach of targeting extractants that are carbon dioxide "philic"is becoming a favored method, and various synthetic strategies are being considered to achieve this goal.[85] One approach to making complexants that are more compatible with liquid carbon dioxide is to introduce fluoro substituents onto them. Two such fluorinated chelates that have been synthesized for extracting lead(II) into liquid carbon dioxide are a fluoroether picolyl amine and a fluoroether *bis*-(picolyl amine). The percent extraction is 20 and 33 respectively. Introduction of a hydrocarbon carbon "spacer" in the form of a propyl group reduces the electron-withdrawing effect of the tail on the head group. The resulting fluoroether propyl *bis*-(picolyl amine) has a higher basicity at nitrogen, and gives 67% extraction of lead(II) into carbon dioxide, a value that is significantly higher than is found with the derivatives that do not have such a spacer.[86] This approach of tailoring the synthesis of complexants so that they are compatible with the solvent system is a new development in liquid-liquid extraction technology. Previously the focus has been on creating selective binding centers for metal ions. This new focus will result in these binding centers being incorporated into systems that are specifically designed for a chosen extraction technologies.

## 1.7. Biological

In evaluating methods for the environmental removal of lead the entire ecosystem must be considered. One aspect of metal removal is that the metal transfers through the system until it eventually becomes bound into the biomass. One approach for lead removal therefore is to stimulate the enhanced production of biomass and consequent lead recovery by enrichment of the microbial environment.[87] Although the biomethylation of mercury(II) is well known, both lead and tin compounds can potentially undergo methylation under conditions that model the natural environment. This premise has been tested by using methyl iodide, trimethylsulfonium iodide, and an *N*, *N*, *N*-trimethylglycinium salt (betaine, $Me_3N^+CH_2CO_2^-$) as possible methylating agents. The reactions have been carried out at ambient temperatures in aqueous media.[88] For the reactions of lead(0) and lead(II) salts with methyl iodide, the $Me_2Pb^{2+}$ and $Me_3Pb^+$ products are formed in much higher yield than is $Me_4Pb$. With trimethylsulfonium iodide, the formation of $Me_2Pb^{2+}$ is observed from elemental lead, but with betaine no methylation of lead is observed for either elemental lead or from lead(II) salts.

## 1.8. Ingestion from Paper

Although many environmental sources of lead are immediately apparent, an overlooked source is printed paper. Up to 200 μg of lead can be extracted from printed paper at pH values in the range of human gastric fluid. Although lead is not extracted at pH values in the range of human saliva, if printed matter is swallowed, there is a possibility that excessive amounts of the metal can be absorbed.[89]

## 1.9. Analytical Methods for Lead

An important aspect of determining the quantity of lead and other metals in samples from a wide range of sources is to continually develop analytical methods that have better sensitivities, consistencies, and tolerance for other elements than are presently available. Indeed the public awareness of heavy metals in the environment is in large part due to improved analytical methods that have allowed for their detection. In this section are covered a range of analytical methods that are used for lead in particular, but can in some cases be used for other heavy metals. Although some of these methods are applicable to a wide range of metals, others are specific for lead. Also, the technique that is chosen for use may be dependent on the origin of the sample. Because of the widespread industrial use of lead it is found in a wide range of sample types. Thus different analytical methods may be required for lead in water, soil, air, or tissue samples. Because the toxicological aspects of lead have been known for a considerable period of time, much effort has been devoted to detecting it accurately in trace amounts. This section therefore consists of analytical methods that have been mainly used for the specific analysis of lead.

### 1.9.1. Biological Samples

The toxicological aspects of lead, and the fact that the metal is a widespread contaminant, has led to considerable effort been devoted to estimating its amount in biological samples. The physiologically based extraction test (PBET) is an *in vitro* test system for predicting the bioavailability of metals from a solid matrix, and it incorporates gastrointestinal tract parameters representative of a human. For lead, the results of the PBET are linearly correlated with results from a Sprague-Dawley rat model between *in vitro* and *in vivo* results. The PBET is not designed to supplant bioavailability studies in animal models, but rather to screen for lead bioavailability when animal results are not available.[90] Atomic absorption spectroscopy is widely used to analyze metal samples. However, because of interferences, liquid chelate extraction is employed in preparing samples for graphite furnace atomic absorption spectroscopy. Filtration of these solutions can then be effected by using a porous Teflon film.[91] This technique allows for the rapid analysis of both lead and cadmium in biological samples.

### 1.9.2. Sediment Samples

Sediment analysis is also an important consideration for lead. In order to obtain meaningful analytical data for lead and other metals in sediments it is important to be aware of their partitioning. This partitioning of metals within the sediment can be affected by the techniques used to preserve the sediments before analysis, and also the presence or absence of atmospheric oxygen during the extraction steps. Drying of the sediment should be avoided. For anoxic sediments the maintenance of oxygen-free conditions during the extraction is of critical importance.[92] For example, treatment of a naturally occurring mixture of manganese oxides with acidified hydroxylamine releases into solution almost all of the calcium, and one-third of the lead. The other two-thirds can be released with oxalate. Examination of the solids by electron microscopy, coupled with electron probe microanalysis, shows that the dissolution behaviors of calcium and lead do not reflect their distributions in the original mixture, where calcium had been associated with both the manganese and iron oxides, but lead largely with the manganese oxide alone. The discrepancies occur because the conditions of the hydroxylamine treatment are sufficiently acidic for the iron oxide to release calcium, but sufficiently basic for it to take up a substantial amount of lead that had been released by dissolution of the manganese oxide.[93] An analytical procedure involving sequential chemical extractions with salt and acid solutions has been developed for the partitioning of particulate trace amounts of lead, cadmium and copper into five fractions. These fractions are exchangeable, bound to carbonates, bound to iron-manganese oxides, bound to organic matter, and residual. A possible application is the prediction of trace metal behavior in estuarine waters or in anoxic lacustrine systems.[94]

A characteristic of all sequential extraction schemes for metals from sediments is that the procedures are complex and time-consuming. The Tessier method requires approximately 24 hr of extraction and centrifugation time to complete.[94] An alternative to conventional methods is to use microwave heating. Use of the microwave technique reduces the time required for fractionation of metals from 24 hr to 4 hr, and is effective for lead.[95] Trace element extractions of atmospheric particulates from glass-fiber filters typically use boiling or refluxing acids or a sonication step. The conditions for ultrasonic extraction of lead have been optimized, and the procedure archived as an EPA designated equivalent method.[96]

### 1.9.3. Microextraction

Recent advances in solid phase microextraction (SPME) methods have been reviewed.[97] The method employs a small fused silica fiber coated with a polymeric stationary phase for analyte extraction from various matrices. This SPME method has been used for the analysis of tetraethyllead and lead(II) in water. Before the analysis, lead is derivatized with sodium tetraethylborate, and the tetraethyllead formed is extracted from the sample headspace. Sub-ppb detection limits have been achieved, and the procedure can be easily performed in the field with the use of a portable gas chromatograph.

# 2. CADMIUM

## 2.1. Toxicity and Occurrence

Cadmium is close to lead and mercury with respect to its being a metal of toxicological concern. Cadmium is only poorly absorbed from the gastrointestinal tract, but it is absorbed from the respiratory tract. After absorption, cadmium is transported in the blood, and becomes bound mainly to blood cells and albumin. After distribution, 50% of the cadmium resides in the liver and kidney. The half-life of cadmium in the body is 10 to 30 years. The metal binds to metallothionein, a low molecular weight protein with a high affinity for cadmium. This very long biological half-life results in its being a poison which accumulates. Acute cadmium poisoning usually results from inhalation of cadmium dusts or the intake of cadmium salts. When the concentration of cadmium in the kidney reaches 200 μg/g there is renal injury. With more severe exposure, glomerula injury occurs. Other health problems that have been attributed to cadmium are hypertension and *itai-itai* disease.[98]

In addition to cadmium being emitted from volcanic eruptions, its use in metal smelting and processing operations leads to it being found in the environment. Tobacco smoke is also a source of cadmium ingestion, since each cigarette contains 1 to 2 μg of cadmium. As a result, the blood cadmium levels for smokers can be up to 4 times as high as those for non-smokers. With 10% pulmonary absorption, the smoking of one pack of cigarettes per day results in approximately 1 mg of cadmium being ingested each year from smoking. When this consideration is taken along with the long biological half life of cadmium in the body, smoking can yield to a significant buildup over a lifetime.[4] Shellfish and animal liver and kidney are among the foods that can be expected to contain 0.05 μg/g cadmium.

## 2.2. Chelates and Macrocycles

Chelate and macrocyclic ligands can be used for the extraction of cadmium from soils, waters, and mammalian species. Many of the systems used are similar to those chosen for lead, but there are other cases where high specificity for cadmium is observed.

### 2.2.1. Chelating Agents for Extraction Therapy

Carbodithioates such as dithiocarbamates have been used for the *in vivo* mobilization of cadmium.[99, 100] This successful application has led to the development of other analogs having more than a single dithiocarbamate functionality.[101] Evaluation of a series of such dithiocarbamates with the formulation $(CH_2)_n[N(CS_2Na)CH_2(CHOH)_4CH_2OH]_2$ (n = 7, 8, 10, 12) shows that those analogs with n = 7, 8, 10 are effective for the *in vivo* removal of cadmium. The rapid reduction in liver cadmium levels after injection indicates that these anions rapidly gain access to intracellular hepatic sites.[102] The composition of such cadmium and mercury dithiocarbamate complexes have been probed by electrospray mass spectrometry.[103]

### 2.2.2. Nitrogen and Oxygen Donor Extractants

Cadmium(II) salts can be extracted from aqueous solution into acetonitrile. The cadmium can be subsequently recovered by salting out with an aqueous solution of sodium sulfate.[104] Cadmium(II) in the form of its complex with salicylaldehyde is extracted from aqueous solution into butanol over a pH range of 4.5 to 6.0. Extraction into non-hydroxylic solvents, however, does not occur.

Nitrogen heterocycles are particularly effective complexants for the liquid-liquid extraction of cadmium(II). One use of these compounds is the enhancement of the effectiveness of other complexants for cadmium(II). As a consequence, the addition of nitrogen heterocycles which do not themselves in the absence of other liquids result in the extraction of cadmium(II), lead to extraction of the metal ion into the hydrophobic solvents benzene and chloroform. The order follows the sequence β-picoline > pyridine >

quinoline. The proposed explanation is that the nitrogen heterocycle binds to cadmium with displacement of coordinated water, thereby increasing the hydrophobicity of the initial complex.[105] Similarly, cadmium(II) anthranilates are only extracted into non-polar solvents in the presence of these same nitrogen heterocycles.[106] For cadmium(II), two molecules of the nitrogen base are complexed, whereas for zinc(II) only one such molecule is bound.

The 6,6-disubstituted-2,2-bipyridines **29** (X = Br, Y = $OC_6H_{13}$; X = Y = $OC_6H_{13}$; X = $NH_2$, Y = $NHC_{12}H_{25}$; X = Y = $NHC_{12}H_{25}$) both extract

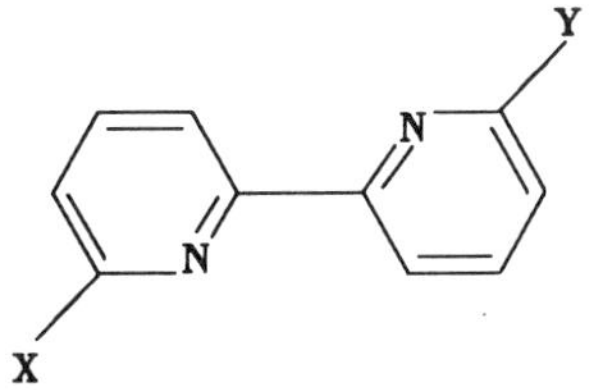

29

and transport heavy metal ions. For example, the 6-amino-6-dodecylamino-2,2-bipyridine derivative is an excellent transport carrier for cadmium(II) and copper(II) through a liquid membrane. Transport of cadmium(II) is observed in this system, although this metal ion is less effectively extracted into an organic phase than are the ions copper(II) and zinc(II). This bipyridine derivative also effects the up-hill transport of both cadmium(II) and copper(II) ions when a proton gradient is available in the system.[107] The mechanism of the transfer of cadmium(II) ion from an aqueous phase into a dichloroethane solution of 1,10-phenanthroline involves the sequential diffusion of 1,10-phenanthroline to the aqueous phase, formation of a 1 : 1 complex, phase transfer of this complex into dichloroethane, and further reaction of this complex with 1,10-phenanthroline in the organic phase.[108] The removal of cadmium(II) and copper(II) from 10M sodium hydroxide solution can be carried out using a three step sequence that involves complexation with phenyl-2-pyridyl ketoxime, filtration of the solution, then extraction into a mixture of ethanol and isopentanol. Zinc(II) is not removed with this reagent.[109]

Amines can also be used for the liquid-liquid extraction of cadmium. Thus cadmium(II) can be extracted from a solution containing iodide ion with a series of high molecular weight amines. All of these amines give over 90% extraction of cadmium(II) from an acidic solution. Cadmium can be separated from mercury by extraction with the amine tricaprylmethylammonium chloride (Aliquat 336-S-I), followed by selective stripping of the cadmium from the amine with ammonium hydroxide or ethylenediamine.[110] Better than 99% separation of cadmium and zinc has also been achieved by first extracting the metals into a 5% solution of Aliquat 336-S-I in xylene, followed by stripping with 1M sodium sulfite. Zinc transfers to the aqueous phase, but cadmium remains in the organic phase. The cadmium can be subsequently stripped from the organic phase by washing with concentrated ammonia.[111]

Cadmium and copper are among a series of metals that can be separated by means of reversed-phase chromatography. The distribution coefficients are then determined by extraction with a tri-*n*-octylamine solution in benzene in a 1 : 1 ratio.[112] Trace amounts of cadmium in uranium can be extracted by treating solutions of the metals in 0.4M hydrochloric acid - 0.01M potassium iodide with a 5 volume % solution of tri-*n*-octylamine in benzene.[113] An integrated process has also been developed based on the application of two membrane technologies for the selective recovery of the cadmium and uranium in wet phosphonic acid. The process involves removal of uranium by ion exchange using Purolite S940 as the ion-exchange resin, followed by the selective removal of cadmium by means of a membrane-assisted solvent extraction with Aliquat 336 as the extractant.[114]

Other miscellaneous complexants have been used for the extraction of cadmium(II). Some of these are oligomers, such as a series of *n*-dodecyloligo (oxyethylene) carboxylic acids that have been used for the separation of cadmium(II), mercury(II), and zinc(II).[115] Also an iminodiacetate cellulose-type chelating absorbent has been synthesized and used for the selective extraction of cadmium(II), lead(II) and copper(II).[116] The macrocyclic oligomer dicyclohexano-18-crown-6 can also be effectively used for the separation of cadmium(II) from mercury(II) and zinc(II) in the presence of the thiocyanate, iodide, bromide, and chloride anions.[117]

## 2.3. Sulfur Donors

Compounds having sulfur donor functionalitiess are also good extractants for cadmium(II). As with mercury(II), dithiocarbamates are excellent complexants and extractants, and a considerable amount of work has been published regarding their use in this context. Again, these extractants can act of *S, S*-donor chelates, or as monodentates bound *via* the anionic sulfur center.

### 2.3.1. Dithiocarbamates

Two dithiocarbamates that have been used for cadmium extraction are diethylammonium diethyldithiocarbamate and ammonium pyrrolidinedithiocarbamate.[118] The extraction constant for cadmium diethyldithiocarbamate into chloroform from water is given by log K = 5.77(5). Solutions of the *bis*-(diethyldithiocarbamate) cadmium(II) complex are stable for at least 40 days.[119] A method is available for the simultaneous determination of extraction constants of simple metal diethyldithiocarbamates. The approach uses a sub-stoichiometric extraction method incorporated with a radiometric technique. When this extraction is applied to metal ion solutions containing excess chloride ion, mixed metal chloride diethyldithiocarbamate complexes are frequently formed, together with the simple metal diethyldithiocarbamates. The extraction constants of these complexes can be determined by this method, along with those of the simple metal diethyldithiocarbamates. The method uses a metal ion that does not form mixed metal chloride diethyldithiocarbamates, and where the extraction constant of its simple metal diethyldithiocarbamate is known.[120]

Metal dithiocarbamates other than the sodium salts have been used for the extraction of cadmium(II). One such salt is the thallium(I) compound,[121] and another is the zinc(II) salt.[122] The toxicity of thallium will, however, preclude the widespread use of this particular compound Dithiocarbamates have also been used in the extraction of cadmium from biological samples. These extracted samples have then been analyzed by isotope dilution (ID) inductively coupled plasma mass spectrometry (ICPMS). For each reference material the measured isotope ratios are corrected for instrumental bias and compared to natural abundance cadmium isotope ratios. Non-extracted samples have large isotope ratios, but interferences have been noted that must be considered. A major problem for non-extracted samples is the presence of sodium, calcium and phosphorus. Quantitative recovery is not critical for IDICPMS analysis, and recoveries appear to be somewhat lowered for matrices having the higher cadmium concentrations, even in the presence of excess complexant. These IDICPMS procedures commonly use cadmium-111 and cadmium-114 for the analysis, with cadmium-111 as the ID spike, and cadmium-114 as the natural abundance reference. Isotope ratios are suitable for use in IDICPMS application if the ratios can be measured to within 1.5% of the accepted ratios.[123] Metal extraction from powdered biological samples into the liquid phase can be effected with ultrasonic agitation. The metals cadmium, copper and lead are quantitatively extracted, and can then be analyzed either by graphite furnace atomic absorption spectroscopy (GFAAS), or by microwave induced plasma-mass spectrometry (MIP-MS).[124] Detection limits for cadmium are 0.012 $\mu g\ g^{-1}$. The mixed ammonium pyrrolidinedithiocarbamate and methyl isobutyl ketone complexant and solvent extraction system has been modified for the determination of cadmium and lead in foods. The samples are digested into a mixture of nitric acid and perchloric acid, extracted into methyl isobutyl ketone, stripped into a solution of nitric acid (3%) and hydrogen peroxide, and the stripping solution then modified with a solution of ammonium hydroxide and ammonium hydrogen phosphate. The detection limit for cadmium in these samples by GFAAS using this method is 1 ng. [125] Supercritical carbon dioxide has also been used as a solvent for the extraction of cadmium, in conjunction with these dithiocarbamate salts as extractants.[126,127]

### 2.3.2. Other Sulfur Donors

Other sulfur donor complexants have been used for cadmium(II) extraction. The extraction processes involving diphenylthiocarbazone (dithizone, Dz) and its metal chelates have been studied using the ascending water electrode. Only for cadmium(II) is a second wave characteristic of the charged mixed ligand complex species $Cd(OAc)Dz_2^-$ observed. For lead(II), only the neutral extractable complex is observed, thereby allowing for a differentiation between the two metals to be made.[128]

Both mono- and di-thiophosphonic acid derivatives have been used for the separation of cadmium(II) and lead(II). A solid-phase extractant for such a purpose has been prepared from such compounds by attaching *bis*-(2,4,4-trimethyl)-monothiophosphonic acid onto a functionalized silica surface. Cadmium(II), lead(II), and zinc(II) have then been separated on this solid resin.[129] Similarly, resins having octadecyl (C-18) functional groups bonded to silica gel as the sorbent, with methanol as the eluent, and with dialkyldithiophosphates with short alkyl chains in acid medium as the complexants, extract cadmium(II), lead(II), and copper(II) quantitatively.[130] Solid-phase extraction with covalent affinity chromatography using thiol-disulfide gels as the absorbent has been used for the preconcentration of cadmium-thionein and copper-thionein proteins from water and plasma. The results correlate with those obtained with metallothionein proteins as determined by AAS.[131]

As noted earlier, in a study of thioamide substituted calixarenes it is found that whereas lead(II), copper(II), and silver(I) are extracted efficiently by all of the thioamides studied, cadmium(II) is only extracted by the calix[5]arene derivative.[132]

## 2.4. Cyclodextrins and Carbohydrates

Cyclodextrins have been used for the simultaneous elution of organic compounds and heavy metals. An example is found with carboxymethyl-β-cyclodextrin which simultaneously increases the apparent aqueous solubilities of anthracene, trichlorobenzene, biphenyl, and DDT, while complexing with cadmium(II). This bifunctional cyclodextrin **30** coordinates organic molecules such as anthracene within the cyclodextrin host, and cadmium(II) chelated by the carboxylate groups. This complexation is not

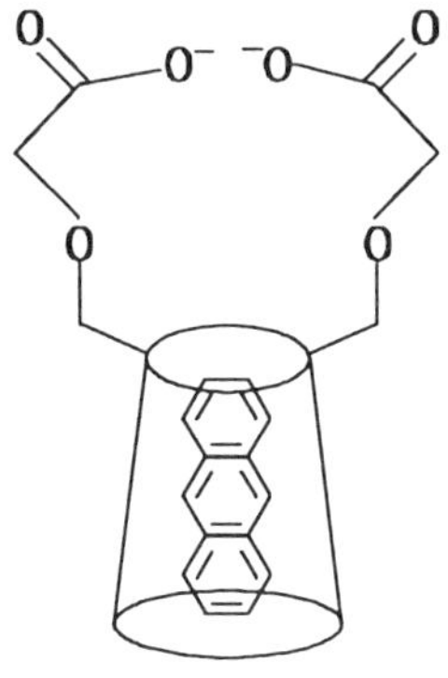

30

significantly altered by changes in pH, or by the presence of excess calcium(II).[133] Such properties have allowed for these compounds to be used in the cleanup of several soil samples.[134]

Carbohydrates can also be used as complexants for lead(II) and mercury(II), with the former having the higher stability in aqueous solution.[135] Similarly, immobilized poly (L-cysteine) has been used for both cadmium(II) chelation and preconcentration.[136]

## 2.5. Cadmium Incorporation into the Food Chain

Toxic heavy metals such as cadmium and mercury can enter the diet of farm animals by a variety of routes, and subsequently become a part of the human food chain. There is therefore a need to be able to predict the levels of contamination in animal tissues if exposed to a contaminated diet, and to estimate how rapidly an animal will decontaminate once the source of contamination is removed from the diet. A study with sheep given a single oral administration of cadmium-109 and mercury-203 involves following the radioisotope concentrations for a year. Feeding cadmium at a constant daily rate over a period of 1000 day period on an uncontaminated diet shows that concentrations in both liver and kidney increase throughout the period on a contaminated diet. However, whereas the cadmium level in the liver declines after the animal is given uncontaminated feed, the cadmium level in the kidney continues to rise during the remaining 1000 days. Thus concentrations of many heavy metal contaminants in animal tissues will not reach equilibrium with the diet within realistic time scales.[137]

# REFERENCES


1. S. Waldman, *Newsweek*, July 15, 1991.
2. M. A. Jandreski, *Clinical Chem. News*, January 1994.
3. A. P. Morton, S. Partridge, A. Blair, *Chem. In Britain*, October 1985, p. 923.
4. R. D. Field, *Clinical Chem. News*, October 1992.
5. C. W. Randall, C. D. Donaldson, P. J. Wigington, Jr., T. J. Grizzard, W. R. Knocke, *Water Sci. Technol.,* **1985**, *17*, 1461.
6. W. R. A. De Jonghe, W. E. Van Mol, F. C. Anson, *Anal. Chem.*, **1983**, *55*, 1050.
7. J. R. Clark, J. G. Viets, *Anal. Chem.*, **1981**, *53*, 61.
8. R. J. Clark, J. G. Viets, *Anal. Chem.*, **1983**, *55*, 166 .
9. S. de Moraes, A. Abrão, *Anal. Chem.*, **1974**, *46* 1812.
10. S. A. Sinex, A. Y. Cantillo, G. R. Helz, *Anal. Chem.*, **1980**, *52*, 2342.
11. T. K. Jan, D. R. Young, *Anal. Chem.*, **1978**, *50*, 1250.
12. T. Qiang, S. Xiao-Quan, Q. Jin, N. Zhe-Ming, *Anal. Chem.*, **1994**, *66*, 3562.
13. J. Slavek, W. F. Pickering, *Can. J. Chem.*, **1987**, *65*, 984.
14. C. Kheboian, C. F. Bauer, *Anal. Chem.*, **1987**, *59*, 1417.
15. G. E. Tolten, N. A. Clinton, P. L. Matlock, *J. M. S. - Rev Macromol. Chem. Phys.*, **1998**, *C38*, 77.
16. R. D. Rogers, J. Zhang in Ion Exchange and Solvent Extraction, J. A. Marinsky and Y. Marcus, eds, Marcel Dekker, New York, 1997, Volume 13, p.141.
17. R. D. Rogers, A. H. Bond, C. B. Bauer, J. Zhang, M. L. Jezl, D. M. Roden, S. D. Rein, R. R. Chomko, *Aqueous Biphasic Separations: Biomolecules to Metal*

*Ions*, R. D. Rogers, M. A. Eiteman, eds., Plenum, New York, 1995, p.1.
18. T. Sotobayashi, T. Suzuki, S. Tonouchi, *Chem. Lett.*, **1976**, 585.
19. E. Friedheim, C. J. Gorvi, *Pharm. Pharmacol.*, **1975,** *27*, 624.
20. S. K. Tandon, J. R. Behari, S. Singh, *Bull. Environ. Contam. Toxicol.*, **1983**, *30*, 552.
21. S. K. Tandon, S. J. S. Flora, S. Singh, *Bull. Environ. Contam. Toxicol.*, **1986**, *37*, 317.
22. R. A. Peter, L. A. Stucken, J. J. S. Thompson, *Nature,* **1945,** *156*, 616.
23. J. J. Chisolm, Jr. *Mod. Treat.*, **1971**, *8,* 593.
24. H. L. Haust, H. Ali, D. S. M. Haines, *J. Biochem.*, **1986**, *2*, 897.
25. T. A. Twarog, M. G. Cherian, *Bull. Environ. Contam. Toxcicol.*, **1983**, *30*, 165.
26. H. L Haust, H. Ali, D. S. M. Haines, *Int. J. Biochem.,* **1980**, *12*, 897.
27. A. Goldberg, J. A. Smith, A. C. Lockhead, *Br. Med. J.,* **1963**, 1270.
28. X. Fang, Q. Fernando, *Chem. Res. Toxcicol.*, **1995**, *8*, 525.
29. X. Fang, Q. Fernando, *Chem. Res. Toxcicol.*, **1994**, *7*, 882.
30. S. K. Tandon, S. J. S. Flora, *Biochem. Int.,* **1986**, *12*, 1963.
31. E. Friedheim, *Lancet,* **1978**, *2 (8102)*, 1234.
32. J. H. Graziano, J. K. Leong, E. Friedheim, *J. Pharmacol. Exp. The.,* **1978**. *206*, 699.
33. X. Fang, Q. Fernando, *Chem. Res. Toxicol.*, **1994**, *7*, 770.
34. T. Hayashita, T. Fujimoto, Y. Morita, R. A. Bartsch, *Chem. Lett.*, **1994**, 2385.
35. J. S. Fritz, G. M. Orf, *Anal. Chem.*, **1975**, *47*, 2043.
36. M. E. McComb, H. D. Gesser, *J. Appl. Polym. Sci.*, **1997**, *65*, 1175.
37. W. H. Chan, S. Y. Lam-Leung, W. S. Fang, F. W. Kwan, *J. Appl. Polym. Sci.*, **1992**, *46*, 921.
38. J. Parr, A. T. Ross, M. Z. Slaurin, *JCS, Dalton Trans.*, **1996**, 1509.
39. S. B. Martin, Jr., H. E. Allen, *CHEMTECH*, **1996**, *26,* 23.
40. L. H. Shen, S. J. Yeh, J. M. Lo, *Anal. Chem.*, **1980**, *52*, 1882.
41. J. Cholak, D. W. Yeager, E. W. Henderson, *Environ. Sci. Technol.*, **1971**, *5*, 1020.
42. R. J. Everson, H. E. Parker, *Anal. Chem.*, **1974**, *46*, 1966.
43. J. D Kinrade, J. C. Van Loon, *Anal. Chem.*, **1974**, *46*, 1894.
44. T. N. Tweeten, J. W. Knoeck, *Anal. Chem.*, **1976**, *48*, 64.
45. P. Hannaker, T. C. Hughes, *Anal. Chem.*, **1977**, *49*, 1485.
46. J. Korkish, *Pure Appl. Chem.*, **1978**, *50*, 371.
47. J. M. Lo, J. C. Yu, F. I. Hutchison, C. M. Wai, *Anal. Chem.,* **1982**, *54*, 2536.
48. S. Bajo, A. Wyttenbach, *Anal. Chem.*, **1979**, *51*, 376.
49. M. C. Williams, E. J. Cokal, T. N. Niemczyk, *Anal. Chem.*, **1986**, *58*, 1541.
50. R. W. Dabeka, *Anal. Chem.*, **1979**, *51*, 902.
51. R. E. B. Edward-Inatami, *J. Chromatogr.*, **1983**, *256*, 253.
52. S. Inchinoki, M. Yamazaki, *Anal. Chem.*, **1985**, *57*, 2219.
53. M. H. Jones, J. T. Woodcock, *Anal. Chem.*, **1986**, *58*, 1845.
54. R. Ma, F. Adams, *Spectrochim. Acta, Part B*, **1996**, *51B*, 1917.
55. S. L. Sachdev, P. W. West, *Environ. Sci. Technol.*, **1970**, *4*, 749.
56. P. Baudot, J. L. Monal, M. H. Livertoux, R. Truhaut, *J. Chromatogr.*, **1976**, *128*, 141.

57. N. V. Deorkar, L. L. Tavlarides, *Ind. Eng. Chem. Res.*, **1997**, *36*, 399.
58. S. Ninan, A. Varadarajan, S. B. Jadhav, A. J. Kulkarni, S. P. Malve, *Spectrochim. Acta Part A.*, **1999**, *55*, 825.
59. R. Dietzel, Ph. Thomas, *Z. anorg. allg. Chem..*, **1971**, *381*, 214.
60. K. Abu-Dari, F. E. Hahn, K. N. Raymond, *J. Am. Chem. Soc.*, **1990**, *112*, 1519.
61. K. Abu-Dari, T. B. Karpishin, K. N. Raymond, *Inorg. Chem.*, **1993**, *32*, 3052.
62. N. Karsli, C. Erk, *Synth. Dyes Pigm..* **1996**, *32*, 85.
63. H. von Arnim, K. Dehnicke, K. Maczek, D. Fenske, *Z. anorg. allg. Chem.*, **1993**, *619*, 1704.
64. K. R. Adam, A. J. Leong, L. F. Lindoy, H. C. Lip, B. W. Skelton, A. H. White, *J. Am. Chem. Soc.*, **1983**, *105*, 4645.
65. K. R. Adam, K. P. Dancey, A. J. Leong, L. F. Lindoy, B. J. McCool, M. McPartlin, P. A. Tasker, *J. Am. Chem. Soc.*, **1988**, *110*, 8471.
66. S. Yoshida, T. Watanabe, *J. Coord. Chem.*, **1988**, *18*, 63.
67. S. Dernini, A. Scrugli, S. Palmas, A. M. Polcaro, *J. Chem. Eng. Data,* **1996**, *41*, 1388.
68. T. Hayashita, H. Sawano, T. Higuchi, M. Indo, K. Hiratani, Z. -Y. Zhang, R. A. Bartsch, *Anal. Chem.*, **1999**, *71*, 791.
69. M. L. Dietz, P. E. Horwitz, *J. Chem. Educ.*, **1996**, *73*, 182.
70. H. Handel, F. R. Muller, R. Guglielmetti, *Helv. Chim. Acta.,* **1983**, *66,* 514.
71. F. R. Muller, H. Handel, R. Guglielmetti, *Helv. Chim. Acta.*, **1983**, *66*, 1525.
72. A. Andrés, A. Bencini, A. Carachalios, A. Bianchi, P. Dapporto, E. Garcia-España, P. Paoletti, P. Paoli, *JCS, Dalton Trans.*, **1993**, 3507.
73. K. Bazakas, I. Lukes. I. *JCS, Dalton Trans.*, **1995**, 1133.
74. K. R. Adam, D. S. Baldwin, P. A. Duckworth, L. F. Lindoy, M. McPartlin, A. Bashall, H. R. Powell, P. A. Tasker, *JCS, Dalton Trans.*, **1995**, 1127.
75. R. M. Izatt, J. S. Bradshaw, S. A. Nielson, J. D. Lamb, J. J. Christensen, D. Sen, *Chem. Rev.*, **1985**, *85*, 271.
76. G. Wu, W. Jiang J. D. Lamb, J. S. Bradshaw, R. M. Izatt, *J. Am. Chem. Soc.*, **1991**, *113*, 6538.
77. T. F. Baumann, J. G. Reynolds, G. A. Fox, *Abstr. 214th Natl. ACS Meeting*, Las Vegas, Nev., Sep 1997, ENVR 117.
78. G. Lagger, L. Tomaszewski, M. D. Osborne, B. J. Seddon, H. Girault, *J. Electroanal. Chem.*, **1988**, *451*, 29.
79. N. V. Deorkar, L. L. Tavlarides, *Ind. Eng. Chem. Res.*, **1997**, *36*, 399.
80. K. Ohto, Y. Tanaka, K. Inoue, *Chem. Lett.*, **1997**, 647.
81. F. Arnaud-Neu, G. Barrett, D. Corry, S. Cremin, G. Ferguson, J. G. Gallagher, S. J. Harris, M. A. McKervey, M.-J. Schwing-Weill, *JCS, Perkin Trans.*, **1997**, *2*, 575.
82. G. G. Talanova, H.-S. Hwang, V. S. Talanov, R. A. Bartsch, *JCS, Chem. Comm.*,**1998**, 419.
83. G. G. Talanova, H.-S. Hwang, V. S. Talanov, R. A. Bartsch, *JCS, Chem. Comm.*,**1998** 1329.
84. J. Wang, W. D. Marshall, *Anal. Chem.*, **1994**, *66*, 1658.
85. A. Yazdi, E. J. Beckman, *Mater. Res. Soc. Symp. Proc.*, **1994**, *344*, 211.
86. A. V. Yazdi, E. J. Beckman, *Ind. Eng. Chem. Res.*, **1996**, *35*, 3644.
87. J. A. Bender, E. R. Archibald, V. Ibeanusi, J. P. Gould, *Water Sci. Technol.*,

**1989**, *21*, 1661.
88. P. J. Craig, S. Rapsomanikis, *Environ. Sci. Technol.*, **1985**, *19*, 726.
89. J. D. Bodgen, M. M Joselow, N. P. Singh, *Arch. Environ. Health*, **1975**, *30*, 442.
90. M. V. Ruby, A. Davis, R. Schoof, S. Eberle, C. Sellstone, *Environ. Sci. Technol.*, **1996**, *30*, 422.
91. K. Yasuda, S. Toda, C. Igarashi, S. Tamura, *Anal. Chem.*, **1979**, *51*, 161.
92. F. Rapin, A. Tessier, P. G. C. Campbell, R. Carignan, *Environ. Sci. Technol.*, **1986**, *20*, 836.
93. E. Tipping, N. B. Hetherington, J. Hilton, D. W. Thompson, E. Bowles, J. Hamilton-Taylor, *Anal. Chem.*, **1985**, *57*, 1944.
94. A. Tessier, P. Campbell, M. Bisson, *Anal. Chem.*, **1979**, *51*, 844.
95. K. I. Mahan, T. A. Foderaro, T. L. Garza, R. M. Martinez, G. A. Maroney, M. R. Trivisonno, E. M. Willging, *Anal. Chem.*, **1987**, *59*, 938.
96. S. L. Harper, J. F. Walling, D. M. Holland, L. J. Pranger, *Anal. Chem.*, **1983**, *55*, 1553.
97. T. Górecki, A. Body-Boland, Z. Zhang, J. Pawliszyn, *Can. J. Chem.*, **1996**, *74*, 1297.
98. C. D. Klaasen in *Goodman and Gilman's The Pharmacological Bassi of Therapeutics*, Section XVII, 69, A. G. Gilman, L. S. Goodman, T. W. Rall, F. Murad, Eds., Seventh Ed., Macmillan, New York, 1985.
99. S. G. Jones, P. K. Singh, M. M. Jones, *Chem. Res. Toxicol.*, **1988**, *3*, 248.
100. M. M. Jones, P. K. Singh, S. G. Jones, M. A. Holscher, *Pharmacol. Toxicol.*, **1991**, *68*, 115.
101. P. K. Singh, C. Xu, M. M. Jones, K. Kostial, M. Blanusa, *Chem. Res. Toxcicol.*, **1994**, *7*, 614.
102. P. K. Singh, M. M. Jones, K. Kostial, M. Blanusa, M. Piasek, *Chem. Res. Toxicol.*, **1996**, *9*, 313.
103. A. M. Bond, R. Colton, J. C. Traeger, J. Harvey, *Inorg. Chim. Acta.*, **1993**, *212*, 233.
104. T. Fujinaga, Y. Nagaosa, *Chem. Lett.*, **1978**, 587.
105. J. M. Singh, S. N. Tandon, *Can. J. Chem.*, **1978**, *56*, 2922.
106. B. Jain, J. M. Singh, R. N. Goyal, S. N. Tandon, *Can. J. Chem.*, **1980**, *58*, 1558.
107. N. Kishii, K. Araki, S. Shiraishi, *JCS, Dalton Trans.*, **1985**, 373.
108. Z. Yoshida, H. Freiser, *Inorg. Chem.*, **1984**, *23*, 3931.
109. D. Reiner, D. P. Poe, *Anal. Chem.*, **1977**, *49*, 889.
110. C. W. McDonald, F. I. Moore, *Anal. Chem.*, **1973**, *45*, 983.
111. C. W. McDonald, T. Rhodes, *Anal. Chem.*, **1974**, *46*, 300.
112. B. Neef, H. Grosse-Ruyken, *J. Chromatogr.*, **1973**, *79*, 275.
113. S. De Moraes, A. Abrão, *Anal. Chem.*, **1974**, *46*, 1812.
114 I. Ortiz, A. I. Alonso, A. M. Urtiaga, M. Demircioglu, N. Kocacik, N. Kabay, *Ind. Eng. Chem. Res.*, **1999**, *38*, 2450.
115. J. Strzelbicki, W. Charewicz, J. Beger, L. Hinz, *Can. J. Chem.*, **1988**, *66*, 1695.
116. W. H. Chan, S. Y. Lam-Leung, W. S. Fong, F. W. Kwan, *J. Appl. Polym. Sci.*, **1992**, *46*, 921.
117. R. M. Izatt, R. L. Bruening, W. Geng, M. H. Cho, J. J. Christensen, *Anal. Chem.*, **1987**, *59*, 2405.

118. J. D. Kinrade, J. C. Van Loon, *Anal. Chem.*, **1974**, *46*, 1894.
119. S. Bajo, A. Wyttenbach, *Anal. Chem.*, **1979**, *51*, 376.
120. S. J. Yeh, J. M. Lo, L. H. Shen, *Anal. Chem.*, **1980**, *52*, 528.
121. G. Soundararajan, M. Subbalyan, *Anal. Chem.*, **1983**, *55*, 910.
122. S. Bajo, A. Wyttenbach, *Anal. Chem.*, **1977**, *41*, 158.
123. R. A. Vanderpool, W. T. Buckley, *Anal. Chem.*, **1999**, *71*, 652.
124. H. Minami, T. Honjyo, I. Atsuya, *Spectrochim Acta Part B*, **1996**, *51B*, 211.
125. R. W. Dabeka, *Anal. Chem.*, **1979**, *51*, 902.
126. J. Li, E. J. Beckman, *Ind. Eng. Chem. Res.*, **1998**, *37*, 4768.
127. J. Wang, W. D. Marshall, *Anal. Chem.*, **1994**, *66*, 1658.
128. W-H. Yu, H. Freiser, *Anal. Chem.*, **1989,** 61, 1621.
129. N. V. Deorkar, L. L. Tavlarides, *Ind. Eng. Chem. Res.*, **1997**, *36*, 399.
130. R. Ma, F. Adams, *Spectrochim. Acta, Part B*, **1996**, *51B*, 1917.
131. A. K. M. Kabzinski, *J. Chromatogr. A*, **1997**, *766*, 121.
132. F. Arnaud-Neu, G. Barrett, D. Curry, S. Cremin, G. Ferguson, J. F. Gallagher, S. J. Harris, M. A. McKervey, M.-J. Schwing-Weill, *JCS, Perkin 2,* **1997**, 575.
133. X. Wang, M. L. Brusseau, *Environ. Sci. Technol.*, **1995**, *29*, 2632.
134. M. L. Brusseau, X. Wang, W. -Z. Wang, *Environ. Sci. Technol.*, **1997**, *31*, 1087.
135. M. Palma, Y. L. Pascal, *Canad. J. Chem..*, **1995**, *73*, 22.
136. H. A. Jurbergs, J. A. Holcombe, *Anal. Chem.,* **1997**, *69*, 1893.
137. N. A. Beresford, R. W. Mayes, N. M. J. Crout, P. J. MacEachern, B. A. Dodd, C. L. Barnett, C. S. Lamb, *Environ. Sci. Technol.*, **1999**, *33*, 2395.

# 8

## PHASE TRANSFER EXTRACTION OF COPPER, SILVER AND GOLD

These three coinage metals are grouped together because of the similarities of their chemistries and because each of them has a high commercial value. These metals have long been recognized for their high electrical conductivities and their ease of manipulation in manufacturing processes. In addition, both silver and gold have been used in jewelry. Copper is also used in the manufacture of bronzes, brass, other copper alloys and ammunition. Previously the metal has been used for water pipes and as a roofing material in buildings. Silver is used for plating other metal surfaces in addition to being used as a photographic material. Gold is also used for plating metals and as a decorative metal. These metals are also used as alloys, including those alloys that are comprised of all three of them together.

As a result of these uses these metals are widely dispersed in nature, and because of their economic value improved methods for their extraction and recovery are always being sought. Gold has been mined for centuries, and many of these operations have used cyanide as the extractant to obtain the metal. Clearly this method cannot be used in present times, therefore more environmentally acceptable methods are required for mining this metal. These new methods are also needed for cleaning up the residues from earlier operations. The newly emerging electronics industry is also generating a need for the safe disposal of these metal residues.

The toxicological effects of copper have been the subject of some debate. The metal is one of the essential elements for life, a fact that has been recognized since around 1920. Since, however, the metal is also beneficial to plant growth, the normal diet of humans contains sufficient copper to meet the daily needs. The two main diseases that are associated with copper are Wilson's disease and Menkes' disease. In addition the soluble salts of copper(II), notably copper sulfate, are strong irritants to skin and mucous membranes. Furthermore copper oxide fumes can cause metal fume fever. Both Wilson's disease and Menkes' disease are disorders of copper metabolism. If untreated the former condition can lead to a fatal accumulation of copper in the liver, brain, and kidney. Silver and gold both exhibit little or no toxic effects, other than silver(I) salts which are found to be irritating to skin and mucous membranes.

# 1. COPPER

## 1.1. Monodentate Ligands

These ligands comprise of pseudohalides, mineral acids, and amines, although micelles and vesicles are also included in this section. In solution copper exists primarily in the form of divalent salts and complexes. These compounds can have tetrahedral, square planar, trigonal bipyramidal, or octahedral geometries.

### 1.1.1. Thiocyanate

Ammonium thiocyanate has been used as an extractant for copper(II). With propylene carbonate being used as the organic phase the distribution ratio is 93. Large quantities of mercury, zinc and cobalt interfere because they also form extractable thiocyanate complexes. This extraction technique has been used for the analysis of copper in both steel and wrought aluminum alloy samples.[1]

### 1.1.2. Cyanide

Cyanides can also be used in copper extraction. One application is in the redox extraction from molten cyanides where oxidation of cyanide to dicyanamide occurs.[2] Another application involves the extraction of copper(II) from aqueous solution into dichloroethane. The method is based on the formation of the complex ion $Cu(CN)_2^-$ and its generation of an ion-pair with methylene blue.[3]

### 1.1.3. Mineral Acids

In the extraction of copper and other metals from sludge with dilute acids it has been found that sulfuric acid is preferable to hydrochloric acid. An advantage of sulfuric acid is the precipitation of the excess calcium in the sludge as its sulfate.[4] Acids can also be used to extract copper from soils. Extraction with a mixture of acetic acid and sodium acetate leads to a redistribution of metals within the soils. Humic acid and pyrolusite show the highest binding affinity for copper and lead. The extractabilities are therefore dependent on the types of soils. The presence of dissolved humic acid during extraction has a significant influence on metal redistribution, as evidenced by an increase in the adsorption of copper, and a decrease in the adsorption of lead.[5]

### 1.1.4. Amines

Amines are effective extractants for copper(II). In many cases the ones that are chosen are long-chain hydrophobic amines. More commonly though oximes are preferred, especially those of the LIX family of compounds.

### 1.1.5. Micelles and Vesicles

Copper(II) can be extracted by micelles and vesicles. A series of 6-[(alkylamino)methyl]-2-(hydroxymethyl) pyridine type micellar extractants in combination with cetyltrimethylammonium bromide (CTAB) and $C_{12}EO_6$ micelles have been used. The removal of copper(II) is achieved by ultrafiltration.[6] The selective extraction of copper(II) has been carried out using hydrophobic ligands solubilized in vesicles. The vesicle systems consist of dimethyl-di-*n*-alkylammonium bromides. The complexants are a single chain ligand 6-[(hexadecylamino)methyl]-2-(hydroxymethyl)pyridine, and a double chain one, 6-[(di-*n*-dodecylamino)methyl]-2-

(hydroxymethyl)pyridine. The amount of copper extracted is lower than that with classical micelles under similar conditions.[7]

## 1.2. Chelate Ligands

The chemical kinetics and mechanisms involved in the solvent extraction of copper chelates has been reviewed in 1977. This review focuses on copper extraction with oximes and hydroxyquinolines. These include the LIX family of extractants, along with 2-hydroxy-5-nonyl phenyl methyl ketone oxime (SME 529), 2-hydroxy-5-nonyl phenyl benzyl ketoxime (P17), 4-thia-1-azabicyclo [3.2.0.]-heptane-2-carboxylic acid (P50) and 7-(4-ethyl-1-methyloctyl)-8-quinolinol (Kelex 100).[8] These compounds were selected because they are widely used in the liquid-liquid extraction of copper(II).

### 1.2.1. Oxygen Donors

Copper(II) can be extracted from aqueous solution into chloroform using salicylate and methyltrioctyl ammonium chloride (Aliquat 336). The complex species in the chloroform phase are $Cu(HL)_3^-$ and $Cu(HL)L^-$ (where HL is the monovalent salicylate anion). In solutions of pH greater than 9 the complex $CuL_2^{2-}$ is also formed in the system.[9] Acetylacetone is effective for extracting copper(II) into dichloromethane. The chelate complex can then be analyzed by liquid chromatography on a Cosmosil 5 SL (silica gel) stationary phase.[10] Phosphates are other oxygen donor chelates that have been used as extractants for copper(II). Methyldiphenyl phosphate, for example, extracts copper(II) nitrate in the form of the complex $[Cu(MePh_2PO_4)_3(H_2O)_3](NO_3)_2$.[11] The compound *bis*-(2-ethylhexyl) phosphoric acid is also effective for the liquid-liquid extraction of copper(II) into kerosine.[12]

Acyclic polyethers have also been used as extractants for copper(II), although many of these extractants are also effective for lead(II).[13] Complexants that can form *N,O*-chelates with copper(II) have also been used as extractants for copper(II). Trace amounts of copper can be extracted from edible fats and oils using a solution that is 0.01% EDTA in 18% aqueous hydrochloric acid solution. The extraction is close to quantitative.[14] Copper(II), as well as lead(II) and cadmium(II), can be extracted by an iminodiacetate cellulosic chelating sorbent derived from cotton fabric. This sorbent can be used for the recycling, preconcentration, and selective extraction of these metal ions.[15] Solvent extraction with *N, N*-ethylene-*bis*-(salicylaldimine) as complexant has been used for the determination of copper in ore samples and rock residues. The complex is eluted with a ternary mixture of methanol-acetonitrile-water, and then analyzed by HPLC.[16] A family of *bis*-(4-acylpyrazol-5-one) derivatives ($H_2BPn$, **1**) having two 1-phenyl-3-methyl-4-acylpyrazol-5-one subunits linked by polymethylene chains - $(CH_2)_n$- of various lengths (n = 0-8, 10, 20) are good extractants for copper(II). In complexation with copper(II) this quadridentate ligand is influenced by the length of the polymethylene chain. The stoichiometries of the extracted complexes are Cu(BPn) (n = 8, 10, 20) and $Cu(HBPn)_2$ or $Cu_2(BPn)_2$ (n = 1-7). The odd-numbered polymethylene

Me Me N N N N $(CH_2)_n$ O O O O H H

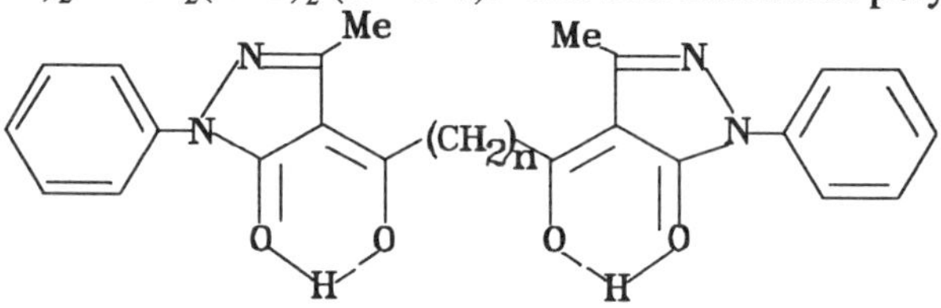

1

chains have higher extraction constants than do the even-numbered chains. This situation occurs because the former exists as a staggered conformational isomer where the hydrogen atoms in the methylene groups can be as far as possible from each other. Thus steric hindrance is less with the odd-numbered polymethylene chains.[17]

Kelex 100 is a material produced by Ashland Chemicals for the extraction of copper from acidic solution. The active component is an alkenyl derivatized 8-hydroxyquinol **2**.[18] Dipentyl pyridine carboxylates (L)

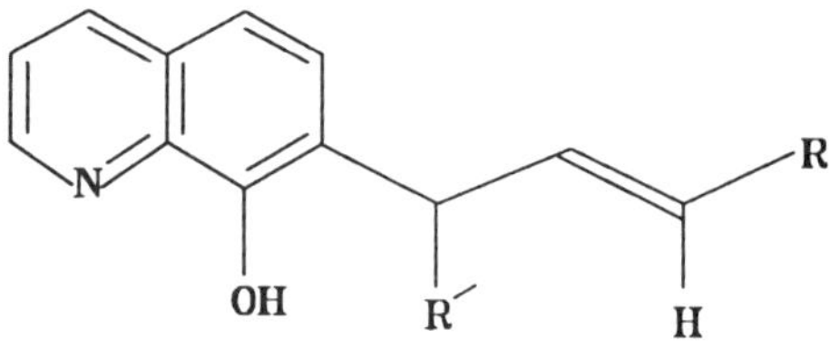

N OH R' H R

2

having ester groups in different positions have been used as extractants for copper(II) in the presence of chloride ion. The most effective compound is dipentyl pyridine-3,5-dicarboxylate. All of the compounds except dipentyl pyridine-2,6-dicarboxylate give complexes of stoichiometry $CuCl_2L_2$.[19]

### 1.2.2. Oximes

Oximes are the most commonly used extractants for copper(II). For this purpose a series of 2-hydroxybenzophenone oxime reagents **3** have been

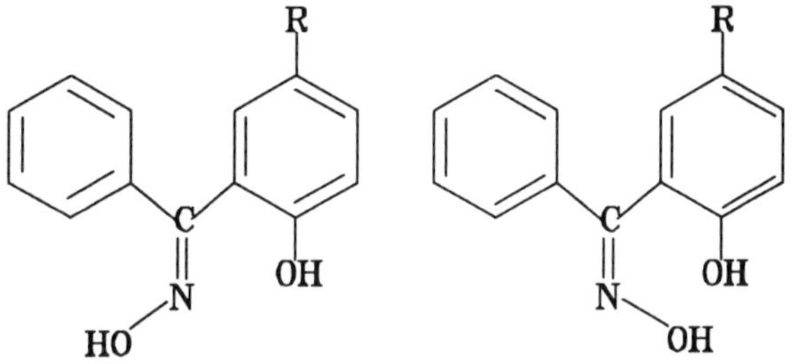

anti syn

3

developed and marketed under the name LIX Reagents. These commercial reagents contain both the *syn* and *anti* isomer, with the latter being the active extractant.[20] Both copper(II) and cadmium(II) can be recovered from 10M sodium hydroxide solution by precipitation and extraction with phenyl-2-pyridyl ketoxime.[21] Several examples of 5,8-diethyl-7-hydroxy-6-dodecanone (LIX 64N) being used for copper(II) extraction can be found in the literature. The reagent as been used to separate copper(II) and nickel(II) ammines with the copper being first extracted in the acid pH range, followed by nickel(II) extraction from an ammoniacal solution.[22, 23] Separation of copper(II), nickel(II) and cobalt(II) from an ammonia-ammonium carbonate solution results in LIX 64N binding copper(II) preferentially to nickel(II).[24] The compound 2-hydroxy-5-nonyl phenyl methyl ketone oxime (SME 529) has also been used for the separation of copper(II), nickel(II) and cobalt(II) ammines. Copper(II) is extracted at low pH before nickel(II) is extracted from the ammoniacal solutions.[25] Copper(II) and nickel(II) can be extracted from an ammonia-ammonium carbonate solution with 20% LIX 64N in kerosene as the organic phase. Copper(II) can be subsequently removed with sulfuric acid. The loading capacity of the reagent is higher for copper than for nickel, and preferential extraction of copper occurs when the organic phase is successively contacted with fresh aqueous solutions.[26] Copper(II) has also been extracted from ammoniacal solutions with a mixture of the strong extractant *anti*-2-hydroxy-5-nonylacetophenone oxime (LIX 84) and the weaker extractant 1-phenyl-1,3-isodecane dione (LIX 54). The correlation of the equilibrium extraction constant is weak for LIX 84 but very strong for LIX 54.[27] The synergistic extraction and separation of two cations such as copper(II) and zinc(II), or a cation and an anion such as copper(II) and chromium(VI), or two cations and an anion such as copper(II), zinc(II), and chromium(VI), have been accomplished using a hydrophobic microporous hollow fiber membrane-based (HFM) extraction technique. The extraction selectivity of copper(II) and zinc(II) by LIX 84 and *bis*-(2-ethylhexyl) phosphoric acid respectively in such a two-fiber-set HFM extractor is enhanced due to competitive extraction.[28] Kelex 100 supported on a macroporous resin has also been used for the removal of copper(II) from aqueous solution.[29] The rate of extraction of copper(II) into an organic phase with oxime chelates can be monitored with a copper(II) ion selective electrode. Reproducible results are obtained if the two-phase mixture is stirred at a constant rate.[30] The kinetics of the extraction of copper(II) into chloroform solutions of 2-hydroxy-5-nonylbenzophenone oxime shows a first order dependence in metal ion, a second order dependence in extractant, and an inverse dependence on hydrogen ion concentration. These data support a mechanism where the rate-determining step occurs in the aqueous phase.[31] In a study with the oxime ligand Acorga P50 (**4**),however, the reaction takes place at the liquid-liquid interface. This

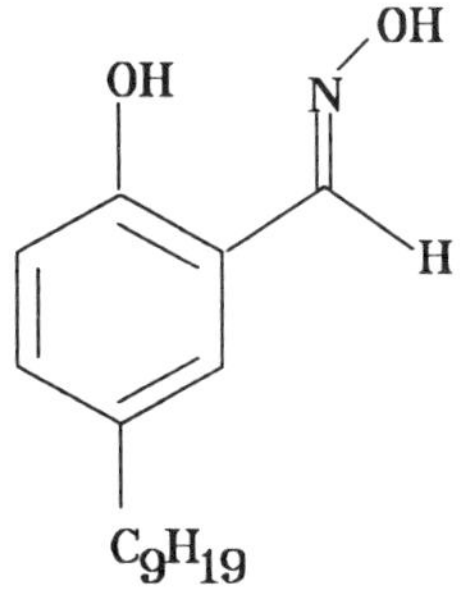

OH N OH H $C_9H_{19}$

**4**

particular study has been carried out using heptane as the organic phase, and a rotating diffusion cell in both the extraction and stripping directions. The data fit a pathway that involves the sequential addition of two oxime ligands in the extraction process, with the stripping step involving the reverse (equation 1).[32] These authors have also considered the rates of processes that occur in a system of two liquid phases. The rate may be limited (i) by the

$$Cu^{2+} (aq) + 2HL (org) = CuL_2 (org) + 2H^+ (aq) \qquad (1)$$

kinetics of a bulk homogeneous reaction, (ii) by the transport of the reactant to the bulk phase, (iii) by the kinetics of a homogeneous reaction in a thin reaction layer, and (iv) by the kinetics of a reaction taking place at the liquid-liquid interface. These considerations need to be applied to each system because it is not always clear whether a reaction is an interfacial one or whether it takes place in the aqueous phase.[8,33,34] Furthermore, it needs to be recognized that results from an experiment where the interfacial area is small and the bulk volume large, will give a valid mechanism under different experimental conditions where the interfacial area compared to the homogeneous volume is much larger. The problem has been treated in terms of four characteristic lengths, $Z_D$, $Z_K$, $Z_I$, and $Z_B$.[35] With a species A in phase $\alpha$ reacting with a species B in phase $\beta$, for diffusion from a droplet of $\alpha$ phase, $Z_D$ will be given by the radius of the droplet $r_\alpha$. The length $Z_K$ is the thickness of the reaction layer. $Z_I$ compares the free energies of the homogeneous and heterogeneous transition states, and $Z_B$ is a length that depends on the geometry of the reactor. Which of the four cases ((i)-(iv)) is observed is dependent on these four characteristic lengths. For $Z_I < Z_D$, reduction of $Z_B$ (macroscopic to microscopic) will not change the case, but for $Z_I > Z_D$, reduction of $Z_B$ may change the mass transport, bulk, interfacial sequence. Increasing [B] can give four different sequences. Also, a small amount of stabilization (~ 20 kJ mol$^{-1}$) of the interfacial transition state by the second solvent will lead to the interfacial reaction being the preferred route. In another physical study, the selectivities between salicylaldoximes and both 2-hydroxyacetophenone and 2-hydroxybenzophenone oximes as copper(II) extractants have been investigated in terms of the electronic and structural differences within the ligand. From both X-ray structural data and a series of different computational models it has been concluded that steric factors are primarily involved in determining extraction behavior.[36]

For the chelate *anti*-2-hydroxy-5-nonylbenzophenone oxime in a rotating diffusion cell the reaction is first order in [$Cu^{2+}$] and [oxime], and inverse first order in $[H^+]$.[37] In a study of the extraction of copper from acidified sulfate solutions, it has been found that dimerization of hydroxyoxime extractants in the organic phase can be neglected except for conditions where its concentration is 20% or higher.[38] In a further report from this same research group using a computer simulation it has been found that the extraction of copper(II) decreases in the following order with respect to the following extraction technologies: systems with cross-current flow, systems with combined flow, and classical systems with sequential stages of extraction and stripping.[39] Copper(II) has also been extracted from a micellar media using two isomeric hydroxyimes that differ in the location of their hydrophobic alkyl chain. These compounds are [(*E*)-1-(2-hydroxy-5-nonylphenyl)-1-ethanone oxime (**5**) and (*E*)-1-(2-hydroxy-5-methylphenyl)-1-decanone oxime (**6**)]. The removal of copper(II) from dilute aqueous solution at pH 6.2 is close to 100% with both cetyltrimethylammonium bromide (CTAB) and *n*-dodecyl-hexa-(ethylene glycol) ether ($C_{12}EO_6$). For the ligand : metal ratio of 5 : 1 used in this system, compound **6** is slightly more efficient than **5**. This result correlates with the faster complexation of **6** with copper(II). The lower complexation rate with **5** may be due to its deeper penetration in the amphiphilic palisade, thereby inhibiting the approach of the metal ion. In the case of CTAB, this means that it must overcome electrostatic repulsions, or in the case of $C_{12}EO_6$ the steric barrier posed by the oxyethylene chains.[40] Further studies on the complexation kinetics of copper(II) with hydroxyoximes in CTAB micelles show that for three 2-hydroxy-5-alkylbenzophenone oximes the complex formation decreases in the order for the alkyl groups methyl > *tert*-butyl > *tert*-octyl. These observations are best fit to an interfacial mechanism.[41] In a

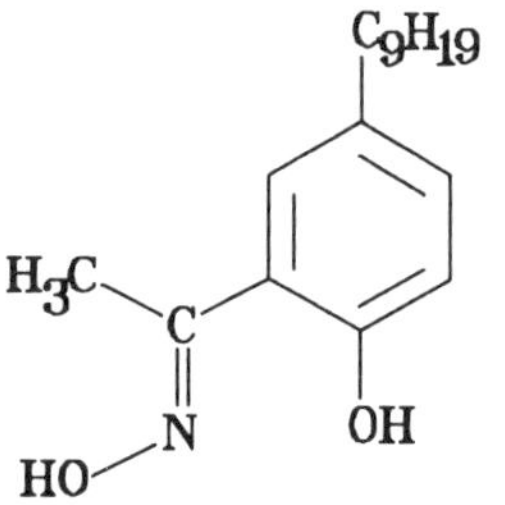

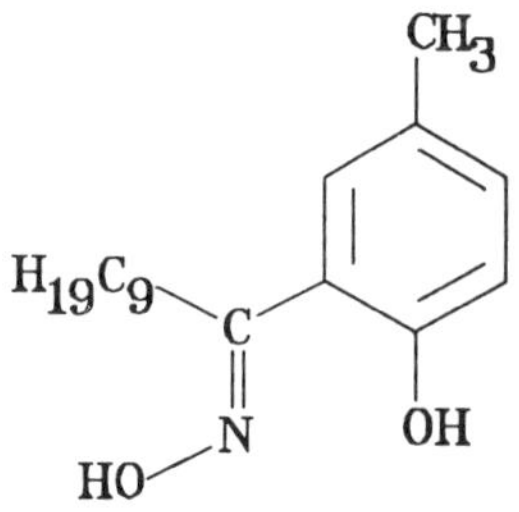

comparative study of copper(II) extraction with hydroxyoximes and dialkylphosphoric acids by a rotating diffusion cell, the thickness of the reaction zone for the oximes is estimated to be $1.3 \times 10^{-3}$ μm, while for dialkylphosphoric acids it is 0.7 μm.[42]

### 1.2.3. Aliphatic and Aromatic Nitrogen Chelates

The aromatic chelate 2,9-dimethyl-1, 10-phenanthroline extracts copper(I) into propylene carbonate from an acetate buffered solution.[43] Triethylenetetramine copper(II) chelates are extracted into a solution of stearic acid in a mixed cyclohexane-isobutanol organic phase.[44] The mixed aliphatic-aromatic nitrogen compound 6-amino-6-(dodecylamino)-2,2-bipyridine is an excellent carrier for the specific transport of copper(II) and cadmium(II) through a liquid chloroform membrane.[45] As a complementary application, cationic copper(II) phenanthroline complexes as ion pairs with anionic detergents have been used for the determination of these detergents. The method shows good agreement with the results obtained by the classical methylene blue method.[46]

Dibutyl-*N,N-bis*-(8-quinolyl) malonamide is a selective extractant for copper(II). This selectivity among copper(II), nickel(II), cobalt(II) and zinc(II) is observed in both competitive and single metal ion extraction. A 1:1 complex **7** is formed where two protons of the amide group have been displaced by copper(II).[47] In a study of the effect of substituents on the

n-Bu, n-Bu, C-NH, C-NH, O, O, N, N $\underset{+2H^+}{\overset{+Cu^{2+}}{\rightleftharpoons}}$ n-Bu, n-Bu, C-N, C-N, O, O, Cu, N, N

**7**

extraction of copper(II) by *N,N-bis*-(8-quinolyl) malonamide, benzyl groups are particularly effective for increasing the extractability.[48]

Micellar particles can solubilize lipophilic extractants in a similar manner to the organic phase in classical biphasic extractions. With a series of 6-(alkylamino)-methyl-2-(hydroxymethyl) pyridines, the rate of complex formation decreases by more than 20 times in CTAB, and 10 times in $C_{12}EO_6$ for alkyl chains up to 16 congeners. The extraction of copper(II) in the micellar *pseudo*-phase is more effective with $C_{12}EO_6$ than it is with CTAB.[49, 50]

### 1.2.4. Other *N,O* Chelates

Oligoethylene glycol *bis*-(hydrazones) are highly selective extractants for copper(II). The extraction constants (expressed as log $K_{ex}$) for ligand : copper(II) (2 : 2 complexes) range from -7.89 to -9.30 for the *bis*

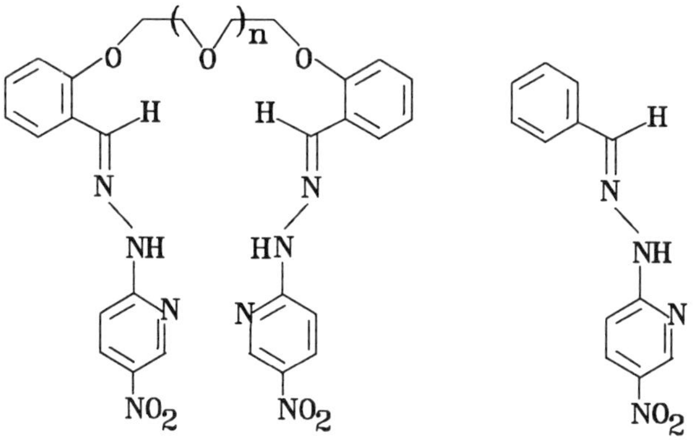

8–11 12

-(hydrazones), with the *mono*-hydrazone showing minimal extraction of copper(II). The extractabilities decrease in the order **8** (n = 1) > **9** ( n = 2) > **10** (n = 3) > **11** (n = 4) >> **12**.[51] In a role reversal technique, the enantioselective solvent extraction of neutral dl-amino acids has been carried out in a two-phase system containing *N*-*n*-alkyl-*L*-proline derivatives and copper(II) ions. Significant enantioselectivity is observed with *n*–butyl, *n*–amyl or *n*-octyl alcohol as solvent. The separation factors (α) for *N*-*n*-dodecyl-L-proline and *N*-n-dodecyl-L-hydroxyproline are in the 1.5 and 1.9 range respectively. This correlates with the stability difference of the mixed ligand complexes in the aqueous phase.[52]

### 1.2.5. Sulfur-Donor Chelates

Dithiocarbamates and other sulfur-donor chelates have been used for the extraction of copper(II). For a solution containing zinc(II), copper(II), iron(II), lead(II), cadmium(II) and manganese(II) diethyldithiocarbamates in river water, only copper can be quantitatively recovered into isoamyl alcohol by solvent extraction.[53] Diethyldithiocarbamates can also be used to extract copper(II) into chloroform from acid solutions in the 1N-5N range.[54] The pyrrolidinedithiocarbamate and hexamethylenedithiocarbamate ions are also effective for removing copper(II) from natural waters.[55, 56] Similarly a dithiocarbamate-functionalized copolymer has been used for the concentration of copper(II). At low loadings the preferred site is pentacoordinate, but at higher loadings the coordination is the tetracoordinate $CuS_2N_2$.[57]

A series of other *S*-donor complexants have been used for the extraction of copper(II). Among these are *mono*-thiothenoyltrifluoroacetone, which as its *bis*-ligand complex can be extracted into cyclohexane.[58] *Mono*-thiodibenzoylmethane derivatives can be used to extract copper(II) into carbon tetrachloride, although the transfer is slow since it requires approximately two hours.[59] The compound 4-dodecyl-6-(2-thiazolylazo) resorcinol is an extractant for copper(II) which can be used in strongly alkaline solutions without the addition of quaternary ammonium salts.[60] Dialkyldithiophosphates are also effective.[61] A triad of complexants with the heteroatom sequences NSNSN (**13**, X = NH), NSOSN (**14**, X = O) and NSSN (**15**, X = $CH_2$) are effective extractants for copper(I), silver(I) and

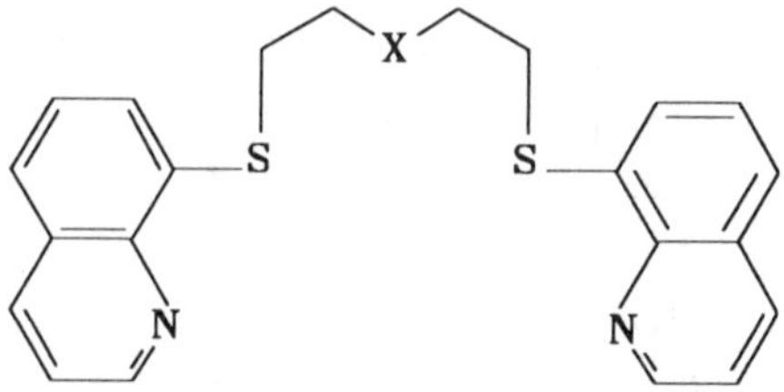

X
S
S
N
N

13–15

mercury(II). The complexants are ineffective for copper(II), and their extraction ability follows the order: copper(I) > silver(I) > mercury(II).[62]

## 1.3. Macrocycles and Calixarenes

### 1.3.1. Azamacrocycles

Copper(II) forms strong complexes with a wide range of nitrogen donor ligands. Azamacrocycles are therefore a particularly attractive set of complexants for use as extractants for copper(II). An early example is the use of the dianion of 1,4,8,11-tetraazacyclotetradecane grafted onto chloromethylated polystyrene. This material binds copper(II) in a mixed solution of copper(II), nickel(II), and cobalt(II)[63] For a series of lipophilic tetraazamacrocycles bound to the polyacrylate XAD7, the extraction order follows the sequence copper(II) > cadmium(II) > lead(II).[64] By comparison, lipophilic open chain tetraamines (**16**, **17**, R = H, $C_{16}H_{33}$) also show a

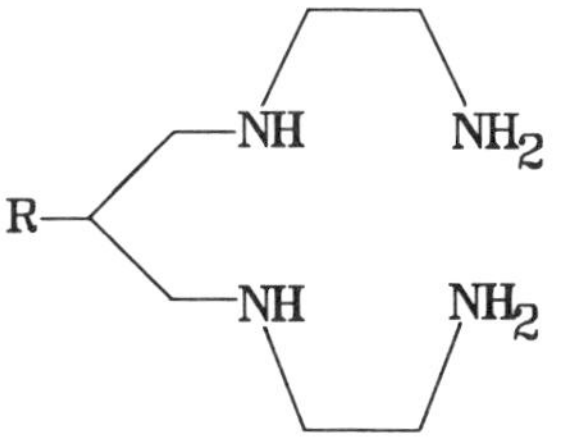

16 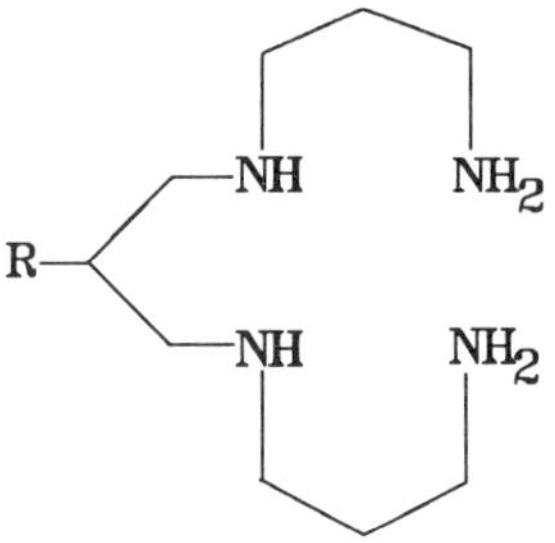 17

preference for the transport of copper(II) across a liquid membrane. The metal is transported as the complex $MX_2L$, where L is the tetraamine and $X^-$ is an anion. The extraction decreases in the relative anion order $Br^-$ > $Cl^-$ > $ClO_3^-$ > $NO_3^-$ > $ClO_4^-$, which correlates with their coordination properties to copper(II).[65] Both a tetraaza- (**18**) and a hexaaza-macrocycle (**19**) that have appended ferrocenyl moieties have been used both as extractants and

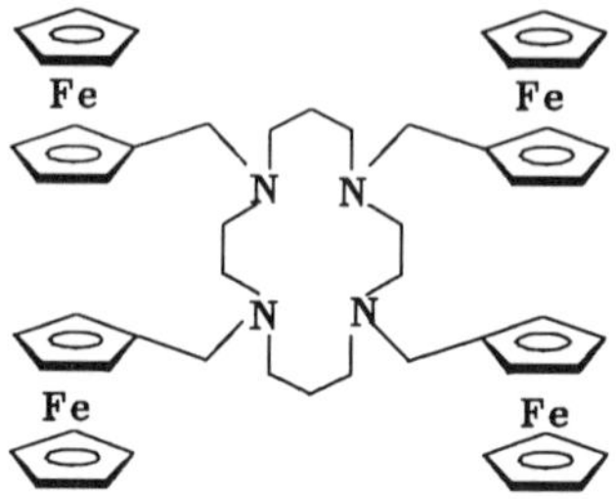

18 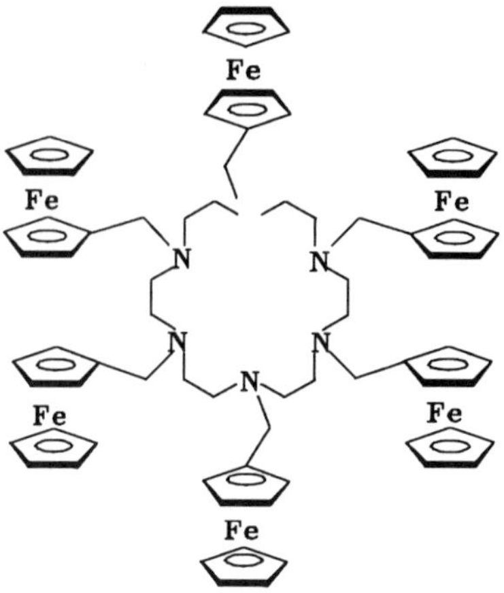 19

electroactive sensors for heavy metal ions such as copper(II).The hexaaza-macrocycle quantitatively extracts copper(II), even though the tetraaza-macrocycle has the better electrochemical receptor response. This difference in behavior is likely due to the higher lipophilicity of the former compound.[66]

### 1.3.2. Thiamacrocycles

Thiamacrocycles have been used for the selective extraction of copper(II). This application is to be anticipated because of the strong complexation of this metal ion with sulfur donor ligands. In one example, the tetrathia-14-crown-4 has been used. The application of this macrocycle involves impregnating it into a sulfonated poly-(styrene-divinylbenzene) resin. This material is effective for the removal of copper(II) from sulfuric acid because of synergism between the macrocyclic binding site and the ion exchange properties of the sulfonate groups. By contrast, the material obtained by impregnating tetrathia-14-crown-4 into the unfunctionalized poly-(styrene-divinylbenzene) resin does not extract copper(II).[67] Interestingly, the affinity of thiacrowns for copper is sufficiently high that they will extract the metal from a mixed copper-ruthenium cluster compound.[68] Thiamacrocycles also assist the electrochemical extraction of both copper(I) and copper(II) from an aqueous phase to an organic one.[69]

## 1.4. Calixarenes

Calixarenes have been used for the liquid-liquid extraction of copper(II). In an early paper the simple unsubstituted *tert*-butyl calix[6]arene has been used as an extractant for copper(II) from alkaline ammonia into organic liquids such as chloroform or benzene. Rapid extraction of 90% of the copper(II) from the solution occurs with the formation of a 1:1 complex.[70] A more planned approach to designing extractants involves attaching thioamide functionalities onto the calixarene in order to make it selective for copper(II), silver(I), cadmium(II), and lead(II).[71]

## 1.5. Porphyrins

Porphyrins have also been used as extractants for copper(II). As an example, *sub*-nanogram quantities of copper(II) have been determined by using 5,10,15,20-*tetrakis*-(1-methylpyridinium-4-yl) porphyrin to extract it into acetonitrile. The method involves the addition of sodium chloride, and it shows no interferences from a wide-range of other metal cations.[72, 73]

## 1.6. Carbon Dioxide

As for other metal ions, carbon dioxide is a fluid that can be used for the liquid-liquid extraction of copper(II) from aqueous solution. Supercritical carbon dioxide containing lithium *bis*-(trifluoroethyl) dithiocarbamate has been used for the extraction of copper(II), both from an aqueous solution and a silica surface.[74] For copper(II) and other metal ions, the solubility of the metal complex shows a strong correlation with the solubility parameters of the complexants as calculated by a group contribution method.[75] An ammonium pyrrolidinedithiocarbamate analog that is soluble in carbon dioxide is a good extractant for both copper(II) and cadmium(II).[76] Copper(II) as its nitrate salt can be removed from water into supercritical carbon dioxide with hexafluoracetylacetone. Extractions range from 14 to 60%, with the highest values being obtained under high pressures of carbon dioxide, and high concentration ratios of hexafluoroacetylacetone : copper(II).[77] 1,1,1-Trifluoroacetylacetone and 2,2-dimethyl-6,6,7,7,8,8,8-

heptafluoro-3,5-octanedione are also effective extractants.[78]

## 1.7. Others

Copper(II) chloride, and other metal chlorides, can be intercalated into graphite from a refluxing carbon tetrachloride solution. After 48 hours refluxing, 22% of the copper is intercalated. This percentage is only exceeded by aluminum(III), which is 40% intercalated afer 3 hours reflux time. Copper(II) can still be selectively extracted into graphite, however, since this high level of intercalation with aluminum(III) is only obtained in the presence of chlorine.[79] Activated carbon can be used to remove trace amounts of copper from water,[80] and copper(II) is also extracted from water by reduction onto a thin mercury film deposited on a wax-impregnated graphite rod.[81]

A biohydrometallurgical extraction process has been developed using a mixed culture of *Thiobacillus thiooxidans* and *Thiobacillus ferrooxidans*. The process is effective for solubilizing both copper and cadmium. Cadmium is solubilized by the biotically formed sulfuric acid, however the copper that is present in fly ash as chalcocite ($Cu_2S$) or cuprite ($Cu_2O$) is dependent on the metabolic activity of the thiobacillus ferroxidans.[82]

Isooctyl thioglycolate is a complexant that quantitatively extracts copper(II), mercury(II), silver(I) and gold(III) from aqueous nitric acid. The organic phase is chloroform, cyclohexane, or ethyl acetate. The copper, mercury, and silver can be subsequently back-extracted into aqueous hydrochloric acid.[83]

# 2. SILVER AND GOLD

Silver and gold are considered together because of their close chemical similarities, and also because as precious metals they have a high commercial value. As a result, recovering them is of commercial importance. In addition to their presence as salts, the metals can be present in their elemental state. This presence of the element is especially true for gold which is very resistant to oxidation, a situation that is exemplified by the fact that aqua regia (a mixture of nitric and hydrochloric acids) takes its name from its ability to dissolve the metal. Silver in solution is usually in the monovalent form, whereas gold can be either monovalent of trivalent. Under the oxidizing conditions of an air environment, gold salts will be present in their trivalent gold(III) form, which commonly adopts the four coordinate square planar coordination geometry that is expected for a $d^8$ metal ion.

## 2.1. Monodentates

The monodentate extractants that are commonly used for silver(I) and gold(III) are similar to those that are used for copper(II), mainly pseudohalide ions and amines.

### 2.1.1. Cyanide

A long time used extractant for gold is the cyanide ion. Although this anion is a good complexant for most metals, it is particularly effective for soft metals such as silver and gold. A problem with this extractant, however, is its high toxicity. Since cyanide is one of the best complexants for these metals, extraction occurs even in the presence of other good ligands. Much of the recent interest in cyanide complexes, therefore, is in the techniques used to separate them and recover the silver and gold. Capillary tube isotachophoresis within the pH range of 7-9 can be used to separate the cyanide complex of silver from other metal ions.[84] Reversed phase ion-interaction HPLC can also be used to separate the cyanide complexes of silver and gold from those of other metals. The effect of the organic modifier, the nature and concentration of the ion-pair reagent, the pH, and the ionic strength of the mobile phase is presented and discussed.[85] Capillary zone electrophoresis has also been used for the separation of cyanide complexes of gold form those of other metals. A phosphate-triethanolamine buffer at pH 8.5 is preferred.[86] Another application of HPLC for the separation and determination of silver and gold cyanide complexes uses a mixture of tetrabutyl ammonium hydrogen sulfate and methanol as the mobile phase.[87] In all of these techniques, however, a problem with the use of cyanide as extractant is its loss by oxidation to cyanate. An effective way of assaying for cyanate in the presence of large concentrations of metal cyanide complexes is ion chromatography. The method uses an anion-exchange column with an anthranilic acid eluent.[88]

Radioactive silver in animal tissue can be assayed by exchange with a solution of silver(I) in basic cyanide. The silver is directly plated onto platinum strips, stripped with nitric acid, precipitated as silver chloride, and counted with a gamma-ray spectrometer.[89]

### 2.1.2. Amines

Long chain amines can be used as extractants for silver and gold. Among those used is tri-*n*-octylamine, which is particularly effective.[90] Amides can also be used for gold extraction. Among those that show complete extraction are *N*-phenylacetamide, *N*-(methylphenyl) acetamide, *N*-(dimethylphenyl) acetamide, *N*-2-pyridylbenzamide, and *N*, *N*-diphenylbenzamidine.[91] The complexes formed between these extractants and gold halides can be spectroscopically observed in the 320-400 nm wavelength range. Amide functionalized XAD-1 (polystyrene-divinylbenzene) strongly retains the noble metals from aqueous or methanolic hydrochloric acid solutions. Gold can be eluted with a solution containing hydrochloric acid and acetone.[92] The ion exchange resin XAD-4 itself can also be used for the separation of gold(III).[93]

### 2.1.3 Oxygen donors

The phosphate moiety can be employed in the removal of silver. As an example, traces of silver(I) can be separated from a large excess of sodium(I) by ion exchange on a column of crystalline cerium(IV) phosphate sulfate.[94] Gold(III) can be extracted into chloroform as its $AuCl_4^-$ ion with 2-nonylpyridine-1-oxide. Selective separation of this complex from other platinum metal chlorides can be effected by adjusting the concentration of the hydrochloric acid in the aqueous phase.[95]

### 2.1.4. Particles

A novel approach to the separation of gold and other precious metals involves their photoreduction onto semiconductor titanium dioxide particles. The rate of photoreduction of gold(III) is nearly independent of the concentration of the semiconductor.[96] This approach, in addition to being novel, is also potentially of commercial value since titanium dioxide is very inexpensive.

### 2.1.5 Detergents and Salts

Detergent extraction of both silver and gold can be carried out in a reversed micellar system comprised of cetyltrimethyl ammonium chloride-water (buffered with sodium carbonate)-chloroform-cyclohexane.[97] The dissolution behavior of gold in washing solutions has been investigated by electrochemical methods. Gold requires a very high oxidation potential and low solution pH in order to be dissolved in chloride or bromide solutions. For iodide solutions lower potentials are required, but for the case of silver, the metal precipitates as silver iodide. Both silver and gold dissolve in ammonia and thiosulfate ion at low oxidation potentials in alkaline solutions.[98]

## 2.2. Chelates

As with other metals, silver and gold can be extracted with chelates.

### 2.2.1. *N, O*-Chelates

A series of podant ionophores have been prepared that show excellent silver(I) selectivity. These complexants are either tridentates or bidentates (X = N, R = H, **20**; X = N, R = Me(*S*-CH)(OC(O)$(CH_2)_4$Me), **21**; X = CH, R = Me(*S*-CH)(OC(O)$(CH_2)_4$Me, **22**); **23**. Podand **21** shows the

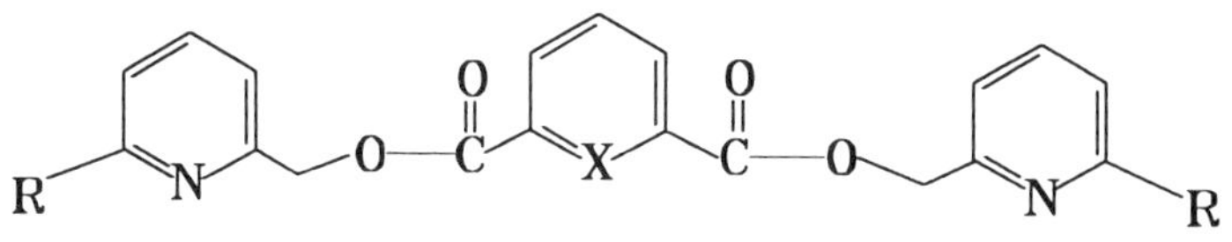

R N O C(=O) X C(=O) O N R

20–22

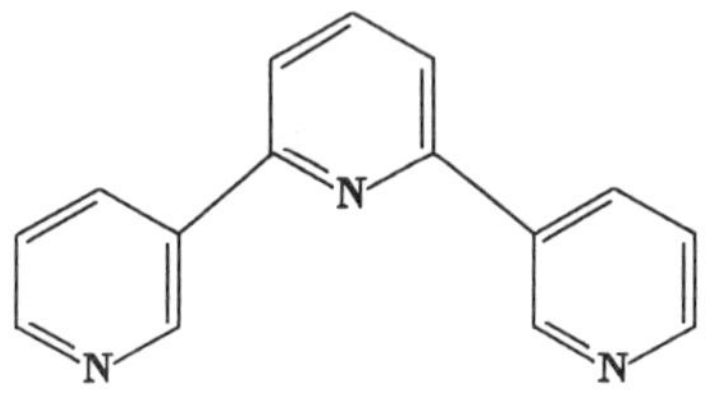

23 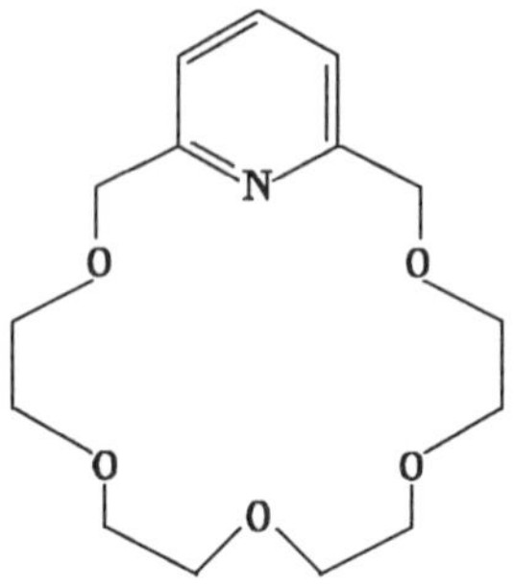 24

highest selectivity for silver(I) transport, even higher than the macrocycle **25**. The selectivities for the different podands with a series of metal perchlorates are shown in Table 1.[99] These data confirm the high selectivity for silver.

**Table 1. Cation Transport Properties of Pyridine Podands**

| Transport Rate x10$^6$ (mol/h) | | | | | | |
|---|---|---|---|---|---|---|
| **Podand** | **$Ag^{+}$** | **$Pb^{2+}$** | **$Cu^{2+}$** | **$Ni^{2+}$** | **$Co^{2+}$** | **$Zn^{2+}$** |
| **20** | 0.6 | <0.1 | <0.1 | <0.1 | <0.1 | <0.1 |
| **21** | 6.4 | <0.1 | <0.1 | <0.1 | <0.1 | <0.1 |
| **22** | 0.2 | <0.1 | <0.1 | <0.1 | <0.1 | <0.1 |
| **23** | <0.1 | <0.1 | <0.2 | <0.1 | <0.1 | <0.1 |
| **24** | 4.6 | <0.1 | <0.1 | <0.1 | <0.1 | <0.1 |

Another *N*, *O*-chelate that shows a high extraction for gold(III) is a reactive fiber containing an amidoxime group. High selectivity is observed from a solution containing a mixture of gold(III), copper(II), zinc(II), and chromium(III). During adsorption the gold(III) is partially reduced into metallic gold, with concurrent oxidation of the amidoxime group into a carboxyl functionality.[100]

### 2.2.2. *S*-Chelates

Extractants of the *O*, *S*- and *N*, *S*-chelate types have been used for silver(I), in addition to copper(II) and mercury(II). For extraction from water into 1,2-dichloroethane, acyclic polyethers incorporating oxygen, nitrogen and/or sulfur atoms in the chain, along with two heterocyclic groups at both ends of this ether chain, have been used. Silver(I) is also extracted into 1,2-dichloroethane with both acyclic and cyclic mono-azatetrathioethers containing a substituted hydrazone group.[101]

Sulfur chelates are also widely used as extractants for silver(I) and gold(III) from aqueous solution. Dithiocarbamates can be used.[102-106] Other sulfur-containing extractants that have been used as liquid-liquid extractants for silver(I) and gold(III) are thiolglycolates,[107,108], xanthates,[109] thiourea,[110] thioamides,[111] and a series of amine-thioether podands **25-28**. These podands

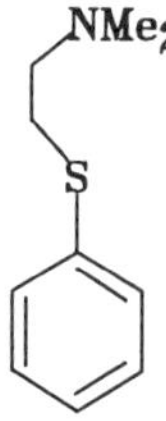

25 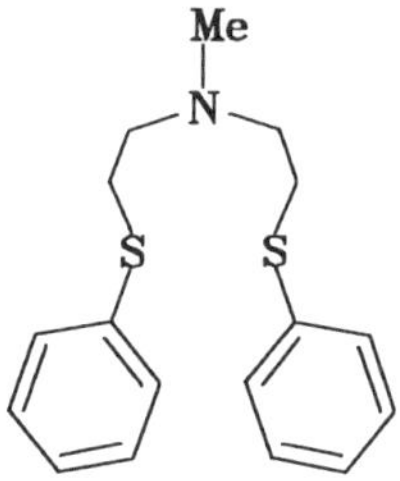 26 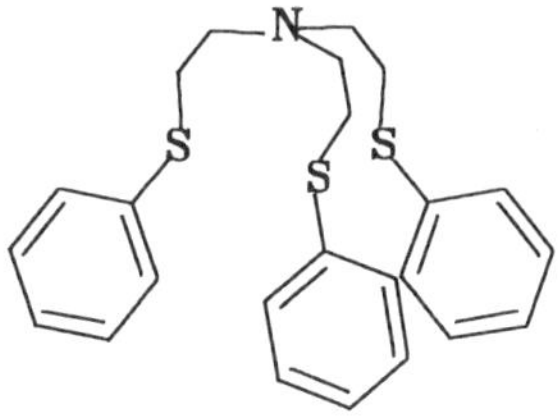 27

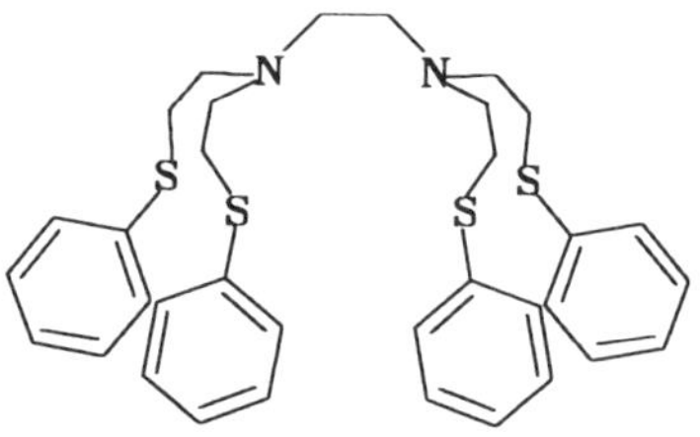

28

show the relative extraction order silver(I) > lead(II) > mercury(II).[112]

## 2.3. Macrocycles

The majority of macrocycles that have been used in the extraction of silver(I) and gold(III) have nitrogen or sulfur atoms within the ring. This choice is made because these particular heteroatoms atoms have stronger binding properties to these soft metal ions.

### 2.3.1. Azamacrocycles and Thiamacrocycles

For the range of available macrocycles, those containing *N, O*; *S*; *S, O;* and *S, Se-* functionalities in the ring have been employed for the extraction of silver(I) and gold(III). Selective transport of silver(I) has been achieved in liquid membrane systems using macrocycles of the proton-ionizable and triazole types (**29**, n = 0, 1. **30**, n = 0, $R_1$ = H, $R_2$ = octyl; n = 1, $R_1$ = H, $R_2$ = octyl; n = 1, $R_1$ = octyl, $R_2$ = H. **31**, ring is aromatic or

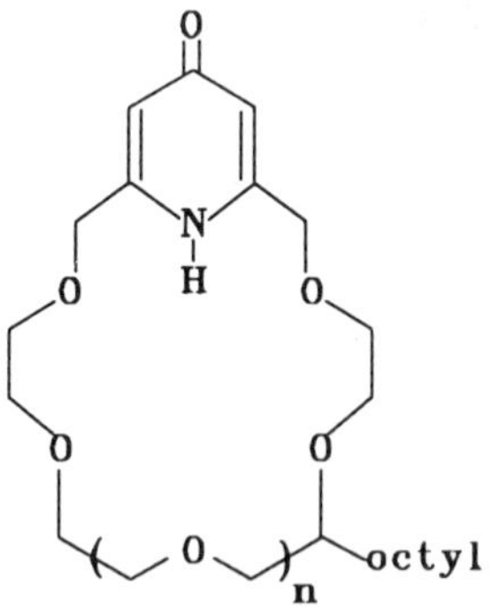

29 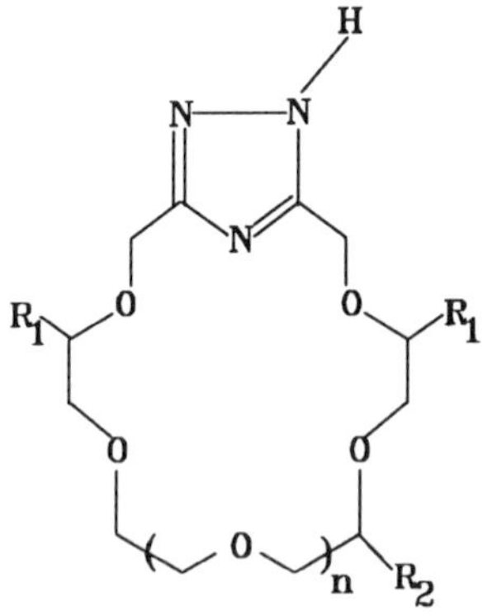 30

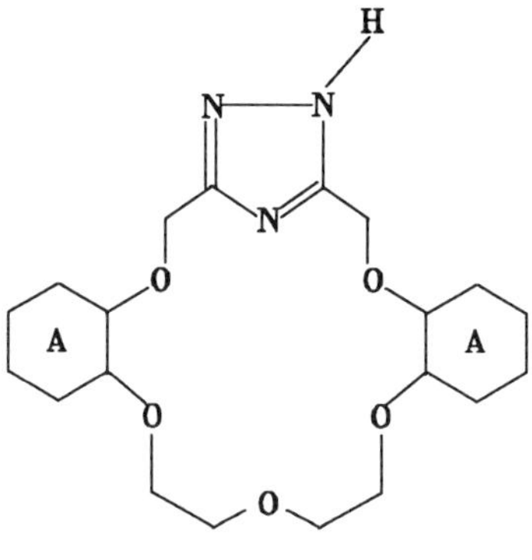

**31**

cyclohexyl). With the pyridone types, co-anion transport is involved. For the triazoles, however, charge balance is maintained by the counter transport of a proton.[113] A series of pyridine-diamide-diester receptors have been investigated for their binding properties to silver(I). Three designs **32-34** have been chosen for targeting selectivity for silver(I). Of these different compounds, **33** shows excellent selectivity for silver(I). In this macrocycle

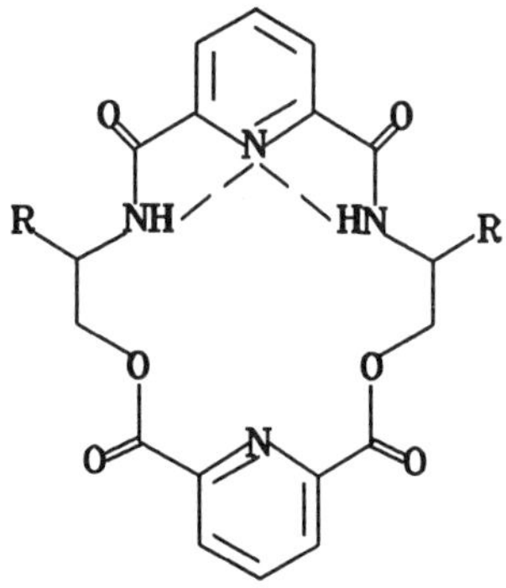

**32** 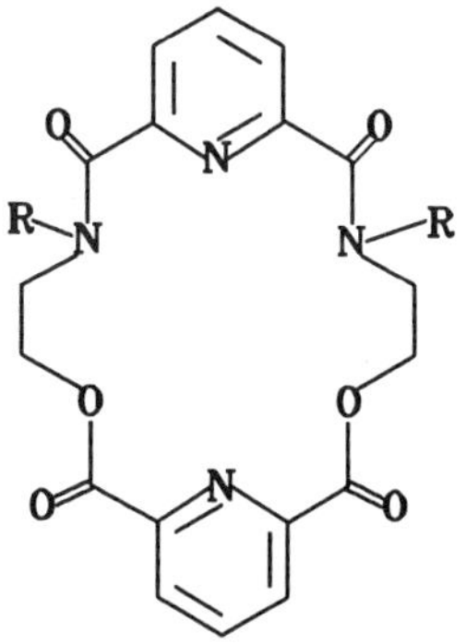 **33**

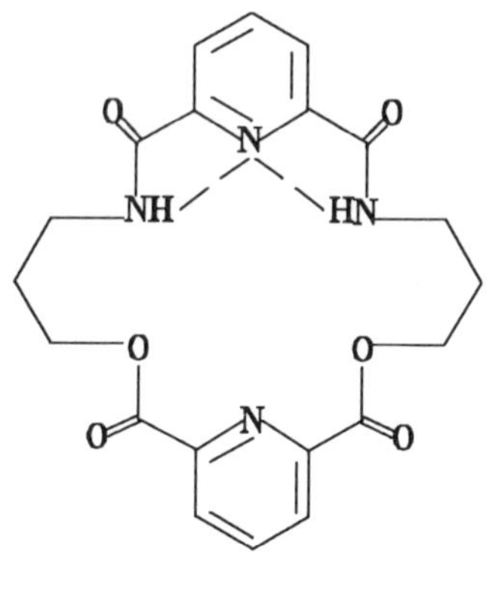

**34**

the amide groups are oriented such that the nitrogens are directed toward the cavity, thereby allowing for the two pyridine nitrogens and the two amide oxygens to define a preorganized binding site.[114] A pyridine-based macrocycle (**35**, R = $CO_2Et$) is highly selective for the extraction of silver(I), and discriminates against copper (both I and II states) and mercury(II). The

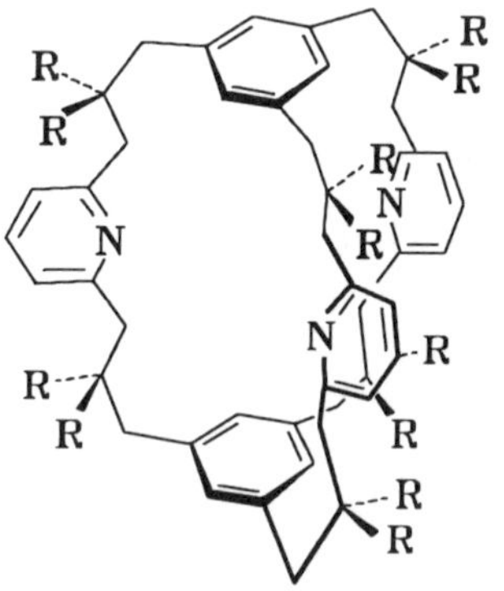

**35**

cavitand has two pyridine nitrogens ideally placed to accommodate the linear coordination geometry of silver.[115] Solid supports for silver adsorption can be obtained by binding the macrocycle to silica.[116]

### 2.3.2. Thiolariat Ethers

Thiolariat ethers with pendant arms are highly selective for silver(I) transport, especially those having a 15-crown-5 ring.[117] The thiacrown **36** (X = S) and selenacrown (X = Se) extract silver(I) better than they do

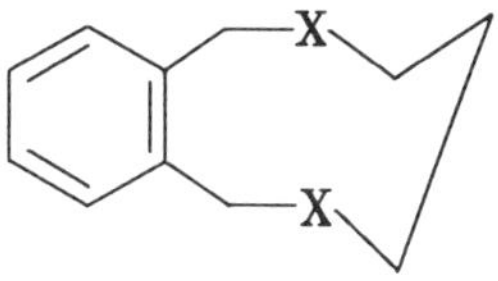

**36**

mercury(II), methylmercury(II) or copper(II).[118] The extraction of silver(I) into chloroform by dibenzo-18-crown-6 as a picrate ion-pair is enhanced by trioctylphosphine oxide due to its coordination to the silver(I) crown complex.[119]

### 2.3.3. Hydrocarbons

A somewhat different approach to designing an extractant for silver(I) uses the host-guest properties of a spherical hydrocarbon. The best such extractant for silver(I) is a macrotricycle $C_{60}H_{60}$. The rationale behind the success of this approach is likely due to the propensity of silver(I) to form strong complexes with unsaturated hydrocarbons. Although thallium(I) and mercury(II) have similar ionic radii to silver(I), they are not extracted.[120]

## 2.4. Calixarenes

A calix[4]arene with ketone functionalities on the narrow rim is a good extractants for silver(I), even from a large excess of palladium(II).[121] A calix[6]arene with azo groups on the upper rim is also an extractant for silver(I) and mercury (both I and II states).[122] Thioamide substituted calix[4]-, [5]- and [6]-arenes are also effective extractants for silver,[123] as are the *N,N*-dimethyldithiocarbamoyl and methylthioether substituted calix[4]arenes.[124,125] Variations other than changing the heteroatoms are also possible. In a consideration of the conformational aspects of silver(I) extraction with conformationally immobilized calix[4]arenes it has been shown that the 1,3-alternate conformer is the most favorable. The silver binds to two benzene rings in the proximal phenyls in addition to a phenolic oxygen.[126] Another approach uses a calix[6]arene that is capped at the three alternate positions of the narrow rim. This cage **37** has a high affinity for

**37**

silver(I), and metal exchange is slow.[127]

# REFERENCES


1. B. G. Stephens, H. L. Felkel, Jr., *Anal. Chem.*, **1975**, *47*, 1676.
2. J. G. V. Lessing, K. F. Fouche, T. T. Retief, *JCS, Dalton Trans.*, **1977**, 2020, 2024.
3. T. Koh, Y. Aoki, Y. Suzuki, *Anal. Chem.*, **1978**, *50*, 881.
4. D. G. Scott, H. Horlings, *Environ. Sci. Technol.*, **1975**, *9*, 849.
5. T. Qiang, S. Xiao-Quan, Q. Jin, N. Zhe-Ming, *Anal. Chem.*, **1994**, *66*, 3562.
6. C. Tondre, S. G. Son, M. Hebrant, P. Tecilla, *Langmuir*, **1993**, *9*, 950.
7. M. Hebrant, P. Tecilla, P. Scrimin, C. Tondre, *Langmuir*, **1997**, *13*, 5539.
8. D. S. Flett, *Accts. Chem. Res.*, **1977**, *10*, 99.
9. E. Papp, J. Inczédy, *J. Chromatogr.*, **1974**, *102*, 225.
10. S. Ichinoki, N. Hongo, M. Yamazaki, *Anal. Chem.*, **1988**, *60*, 2099.
11. A. Apelblat, R. Levin, *JCS, Dalton Trans.*, **1974**, 1476.
12. R. S. Juang, Y. T. Chang, *Ind. Eng. Chem. Res.*, **1993**, *32*, 207.

13. T. Hayashita, H. Sawano, T. Higuchi, M. Indo, K. Hiratani, Z-Y. Zhang, R. A., Bartsch, *Anal. Chem.*, **1999**, *71*, 791.
14. R. A. Jacob, L. M. Klevay, *Anal. Chem.*, **1975**, *47*, 741.
15. W. H. Chan, S. Y. Lam-Leung, W. S. Fong, F. W. Kwan, *J. Appl. Polym. Sci.*, **1992**, *46*, 921.
16. M. Y. Khuhawar, S. N. Lanjwani, *J. Chromatogr. A*, **1996**, *740*, 296.
17. S. Miyazaki, H. Mukai, S. Umetani, S. Kihara, M. Matsui, *Inorg. Chem.*, **1989**, *28*, 3014.
18. A. W. Ashbrook, *J. Chromatogr.*, **1975**, *105*, 151.
19. M. B. Bogacki, A. Jakubiak, G. Cote, J. Szymanowski, *Ind. Eng. Chem. Res.*, **1997**, *36*, 838.
20. A. W. Ashbrook, *J. Chromatogr.*, **1975**, *105*, 141.
21. D. Reiner, D. P. Poe, *Anal. Chem.*, **1977**, *49*, 889.
22. R. D. Eliason, *Proc. Symp. Solv. Extr. Exch.*, *AICHE*, Tucson, AZ, May 1973.
23. R. D. Eliason, E. Edmunds, Jr., *CIM Bull.*, **1974**, *67*, 82.
24. C. R. Mingold, R. B. Sudderth, *Proc. Int. Symp on Hydrometallurgy*, D. J. I. Evans, R. S. Shoemaker, Eds., AIME, New York, 1973, pp 552-588.
25. N. M. Rice, M. Nedved, G. M. Ritcey, *Hydrometallurgy*, **1978**, *3*, 55.
26. B. D. Pandey, V. Kumar, D. Bagchi, D. D. Akerkar, *Ind. Eng. Chem. Res.*, **1989**, *28*, 1664.
27. G. Kyuchoukov, M. B. Bogacki, J. Szymanowski, *Ind. Eng. Chem. Res.*, **1998**, *37*, 4084.
28. Z. -F. Yang, A. K. Guha, K. K. Sirkar, *Ind. Eng. Chem. Res.*, **1996**, *35*, 1383.
29. J. R. Parrish, *Anal. Chem.*, **1977**, *49*, 1189.
30. S. J. Kirchner, Q. Fernando, *Anal. Chem.*, **1977**, *49*, 1636.
31. S. P. Carter, H. Freiser, *Anal. Chem.*, **1980**, *52*, 511.
32. W. J. Albery, R. A. Choudhery, *J. Phys. Chem.*, **1988**, *52*, 1142.
33. W. J. Albery, R. A. Choudhery, P. R. Fisk, *Farad. Disc.*, **1984**, *77*, 53.
34. H. Freiser, *Acc. Chem. Res.*, **1984**, *17*, 126.
35. W. J. Albery, R. A. Choudhery, *J. Phys. Chem.*, **1988**, *92*, 1151.
36. J. O. Morley, *JCS, Perkin Trans II*, **1987**, 1243.
37. M. Jin, F. C. Michel, Jr., R. D. Noble, *Ind. Eng. Chem Res.*, **1989**, *28*, 193 .
38. J. Piotrowicz, M. D. Mariusz, S. Wasylkiewicz, J. Szymanowski, *Ind. Eng. Chem. Res.*, **1989**, *28*, 284.
39. M. D. Bogacki, J. Szymanowski, *Ind. Eng. Chem. Res.*, **1990**, *29*, 601.
40. W. Richmond, C. Tondre, E. Krzyzanowska, J. Szymanowski, *JCS, Farad. Trans.* , **1995**, *91*, 657.
41. R. Cierpiszewski, M. Hebrant, J. Szymanowski, C. Tondre, *JCS Farad Trans.*, **1996**, *92*, 249.
42. M. A. Hughes, P. K. Kuipa, *Ind. Eng. Chem. Res.*, **1996**, *35*, 1976.
43. B. G. Stephens, H. L. Felkel, Jr., W. M. Spinelli, *Anal. Chem.*, **1974**, *46*, 692.
44. D. Jeffries, J. Fresco, *Can. J. Chem.*, **1981**, *51*, 1497.
45. N. Kishii, K. Araki, S. Shiraishi, *JCS, Dalton Trans.*, **1985**, 373.
46. M. Gallego, M. Silva, M. Valcàrcel, *Anal. Chem.*, **1986**, *58*, 2265.
47. K. Hiratani, K. Taguchi, K. Ohhashi, H. Nakayama, *Chem. Lett.*, **1989**, 2073.
48. T. Hirose, J. Hiratoni, K. Kasuga, K. Saito, T. Koike, E. Kimura, Y. Nagawa, H.

Nakanishi, *JCS, Dalton Trans.*, **1992**, 2679.
49. S. G. Son, M Hebrant, P. Tecilla, P. Scrimin, C. Tondre, *J. Phys. Chem.*, **1992**, *96*, 11072.
50. C. Tondre, M. Hebrant, *J. Phys. Chem.*, **1992**, *96*, 11079.
51. H. Sakamoto, J. Ishikawa, H. Nakagami, Y. Ito, K. Ogawa, K. Doi, M. Otomo, *Chem. Lett.*,**1992**, 481.
52. T. Takeuchi, R. Horikawa, T. Tanimura, *Anal. Chem.*, **1984**, *56*, 1152.
53. T. N. Tweeten, J. W. Knoeck, *Anal. Chem.*, **1976**, *48*, 64.
54. S. Bajo, A. Wyttenbach, *Anal. Chem.*, **1976**, *48*, 902.
55. J. D. Kinrade, J. C. Van Loon, *Anal. Chem.*, **1974**, *46*, 1894.
56. S. Ichinoki, M. Yamazaki, *Anal. Chem.*, **1985**, *57*, 2219.
57. A. F. Ellis, M. J. Hudson, A. A. G. Tomlinson, *JCS, Dalton Trans.*, **1985**, 1655.
58. T. Honjo, Y. Fujioka, H. Itoh, T. Kiba, *Anal. Chem.*, **1977**, *49*, 2241.
59. T. Honjo, H. Freiser, *Anal. Chem.*, **1981**, *53*, 1258.
60. H. Matsunaga, T. M. Suzuki, *Chem. Lett.*, **1986**, 225.
61. R. Ma, F. Adams, *Spectrochim. Acta. Part B*, **1996**, *51B*, 1917.
62. H. Sakamoto, S. Ito, M. Otomo, *Chem. Lett.*, **1995**, 37.
63. V. Louvet, P. Appriou, H. Handel, *Tetrahedron Lett.*, **1982**, *23*, 2445.
64. F. R. Muller, H. Handel, *Tetrahedron Lett.*, **1982**, *23*, 2769.
65. M. Di Casa, L. Fabbrizzi, A. Perotti, A. Poggi, R. Riscassi, *Inorg. Chem.*, **1986**, *25*, 3984.
66. J. M. Lloris, R. Martinez Mañez, T. Pardo, J. Soto, M. E. Padilla-Tosta, *JCS, Dalton Trans*, **1998**, 2635.
67. B. A. Moyer, G. N. Case, S. D. Alexandratos, A. A. Kriger, *Anal. Chem.*, **1993**, *65*, 3389.
68. R. D. Adams, R. Layland, K. McBride, *Organometallics*, **1996**, *15*, 5425.
69. L. Tomaszewski, G. Lagger, H. H. Girault, *Anal. Chem.*, **1999**, *71*, 837.
70. I. Yoshida, S. Fujii, K. Ueno, S. Shinkai, T. Matsuda, *Chem. Lett.*, **1989**, 1535.
71. F. Arnaud-Neu, G. Barrett, D. Corry, S. Cremin, G. Ferguson, J. F. Gallagher, S. J. Harris, M. A. McKervey, M. -J. Schwing-Weill, *JCS, Perkin Trans*, **1977**, *2*, 575.
72. M. Tabata, M. Kumamoto, J. Nishimoto, *Anal. Chem.*, **1996**, *68*, 758.
73. M. Kumamoto, J. Nishimoto, T. Takamuku, M. Tabata, *Pure Appl. Chem.*, **1998**, *70*, 1925.
74. K. E. Laintz, C. M. Wai, C. R. Yonker, R. D. Smith, *Anal. Chem.*, **1992**, *64*, 2875.
75. C. M. Wai, S. Wang, J. -J. Yu, *Anal. Chem.*, **1996**, *68*, 3516.
76. J. Li, E. J. Beckman, *Ind. Eng. Chem. Res.*, **1998**, *37*, 4768.
77. J. M. Murphy, C. Erkey, *Environ. Sci. Technol.*, **1997**, *31*, 1674.
78. J. M. Murphy, C. Erkey, *Ind. Eng. Chem. Res.*, **1997**, *36*, 5371.
79. J. M. Lalancette, L. Roy, and J. LaFontaine, *Can. J. Chem.*, **1976**, *54*, 2505.
80. W. B. Kerfoot, R. F. Vaccaro, *Limnol. Oceanogr.*, **1973**, *18*, 689.
81. L. L. Edwards, B. Oregioni, *Anal. Chem.*, **1975**, *47*, 2315.
82. C. Brombacher, R. Bachofen, H. Brand L, *Appl. Environ. Microbiol.*, **1998**, *64*, 1237.
83. J. S. Fritz, R. K. Gillette, H. E. Mishmash, *Anal. Chem.*, **1966**, *38*, 1869.

84. S. Tanaka, T-Kaneta, H. Yoshida, *J. Chromatogr.*, **1988**, *447*, 383.
85. L. Giroux, D. J. Barkley, *Can. J. Chem.*, **1994**, *72*, 269.
86. W. Buchberger, P. R. Haddad, *J. Chromatogr.*, **1994**, *687*, 343.
87. B. Grigorova, S. A. Wright, M. Josephson, *J. Chromatogr.*, **1987**, *410*, 419.
88. P. Fagan, B. Paull, P. R. Haddad, R. Dunne, H., Kamar, *J. Chromatogr., A*, **1997**, *770*, 175.
89. V. F. Hodge, T. R. Folsom, *Anal. Chem.*, **1972**, *44*, 381.
90. S. De Moraes, A. Abrao, *Anal. Chem.*, **1974**, *46*, 1812.
91. K. S. Patel, K. H. Lieser, *Anal. Chem.*, **1986**, *58*, 1547.
92. C. Pohlandt, J. S. Fritz, *J. Chromatogr.*, **1979**, *176*, 189.
93. D. T. Gjerde, J. S. Fritz, *J. Chromatogr.*, **1980**, *188*, 391.
94. K. H. Koenig, R. P. H. Garten, *J. Chromatogr.*, **1975**, *103*, 193.
95. M. G. B. Drew, L. R. Glaves, M. J. Hudson, *JCS Dalton Trans.*, **1985**, 771.
96. E. Borgarelb, N. Serpone, G. Emo, R. Harris, E. Pelizzetti, C. Minero, *Inorg. Chem.*, **1986**, *25*, 4499.
97. Imdadullah, T. Fujiwara, T. Kumamaru, *Anal. Chem.*, **1993**, *65*, 421.
98. X. H. Wang, *Proc-Electrochem Soc.* 92-17 (*Proc. Int. Symp. Electrochem. Miner. Met. Process. III, 3rd, 1992*), **1992**, 452.
99. H. Tsukube, J.-i. Venishi, N. Kojima, O. Yonemitsu, *Tetrahedron Lett.*, **1995**, *36*, 2257.
100. W. Lin, Y. Lu, H. Zeng, *J. Appl. Polym. Sci.*, **1993**, *49*, 1635.
101. H. Sakamoto, J. Ishikawa, T. Mizuno, K. Doi, M. Otomo, *Chem. Lett.*,**1993**, 609.
102. J. D. Kinrade, J. C. Van Loon, *Anal. Chem.*, **1974**, *46*, 1894.
103. L. Bromberg, G. Levin, *J. Chromatogr.*, **1993**, *634*, 183.
104. S. S. Sandhu, P. Nelson, *Environ. Sci. Technol.*, **1979**, *13*, 476.
105. S. Bajo, A. Wyttenbach, *Anal. Chem.*, **1977**, *49*, 1771.
106. J.-M. Lo, J.-D. Lee, *Anal. Chem.*, **1994**, *66*, 1242.
107. J. S. Fritz, R. K. Gillette, H. E. Mishmash, *Anal. Chem.*, **1966**, *38*, 1869.
108. E. M. Moyers, J. S. Fritz, *Anal. Chem.*, **1976**, *48*, 1117.
109. D. A. Choudhury, S. Kamata, *Chem. Lett.*, **1994**, 589.
110. T. Kai, T. Hagiwara, H. Haseba, T. Takahashi, *Ind. Eng. Chem. Res.*, **1997**, *36*, 2757.
111. C. Y. Liu, M. J. Chen, T. J. Chai, *J. Chromatogr.*, **1991**, *555*, 291.
112. S. S. Lee, J. M. Pak, D. Y. Kim, J. H. Jung, M. H. Cho, *Chem. Lett.*, **1995**, 1009.
113. R. M. Izatt, G. C. Lind, H. R. L. Bruening, P. Huszthy, C. W. McDaniel, J. S. Bradshaw, J. J. Christensen, *Anal. Chem.*, **1988**, *60*, 1694.
114. S. Kumar, M. S. Hundal, N. Kaur, R. Singh, H. Singh, *J. Org. Chem.*, **1996**, 61, 7819.
115. F. Vogtle, A. Ibach, M. Nieger, C. Chartroux, T. Kruger, H. Stephan, K. Gloe, *JCS, Chem. Comm.*, **1997**, 1809.
116. R. M. Izatt, R. L. Bruening, M. L. Bruening, B. J. Tarbet, K. E. Krakowiak, J. S. Bradshaw, J. J. Christensen, *Anal. Chem.*, **1988**, 60, 1825.
117. T. Nabeshima, N. Tsukada, K. Nishijima, H. Ohshiro, Y. Yano, J. *Org. Chem.*, **1996**, *61*, 4342.
118. T. Kumagai, S. Akabori, *Chem. Lett.*, **1989**, 1667.

119. Y. Hasegawa, K. Suzuki, T. Sekine, *Chem. Lett.*,**1981**, 1075.
120. J. Gross, G. Harder, F. Vogtle, H. Stephan, K. Gloe, *Angew. Chem., Int. Ed. Engl.,* **1995**, *34*, 481.
121. K. Ohto, E. Murakami, K. Shiratsuchi, K. Inoue, M. Iwasaki, *Chem Lett.*, **1996**, 173.
122. E. Nomura, H. Taniguchi, S. Tamura, *Chem. Lett.*, **1989**, 1125.
123. F. Arnaud-Neu, G. Barrett, D. Corry, S. Cremin, G. Ferguson, J. Gallagher, J. S. Harris, M. A. McKervey, M.-J. Schwing-Weill, *JCS., Perkin Trans.*, **1997**, *2*, 575.
124. A. T. Yordanov, O. M. Falana, H. F. Koch, D. M. Roundhill, *Inorg. Chem.*, **1997**, *36*, 6468.
125. A. T. Yordanov, D. M. Roundhill, *Inorg. Chim. Acta.*, **1998**, *270*, 216.
126. A. Ikeda, S. Shinkai, *J. Am. Chem. Soc.*, **1994**, *116*, 3102.
127. H. Otsuka, Y. Suzuki, A. Ikeda, K. Araki, S. Shinkai, *Tetrahedron,* **1998**, *54*, 423.

# 9

# EXTRACTION OF ACTINIDES AND LANTHANIDES

## 1. ACTINIDES

### 1.1. Introduction

Since the actinide ions are classified as hard acids they associate most strongly with hard base complexants. Among these complexants, those with oxygen donor groups have been the most widely used. By contrast with transition metal and post-transition metal ions, actinides can frequently achieve higher coordination numbers than six. Therefore multidentates that can accommodate to high coordination numbers are good candidates for being extractants for actinides. Another feature of the actinides is their occurrence as oxycations. The most studied one is the linear uranyl ($UO_2^{2+}$) ion, but others such as $PuO_2^{2+}$ and $PuO_2^{+}$ are stable in aqueous solution. These cations present a different challenge in designing complexants since there is the possibility of introducing functional groups that can associate with the oxide groups as well as the metal center of the oxycation.

### 1.2. Monodentates

Simple monodentate complexants have been used as extractants for actinides, with phosphates and phosphine oxides being among the most commonly used. Several reviews have been written on the liquid-liquid extraction of actinide ions from aqueous solutions with such compounds.[1-3] The transuranium elements can be separated by gas chromatography as their chlorides in an analogous manner to that used for the lanthanides. The technique involves using helium carrier gas containing aluminum chloride, which converts the transuraniums into their chlorides. Plutonium forms a tetravalent chloride that is eluted after those of curium(III) and americium(III).[4]

### 1.2.1. Carboxylic Acids

Carboxylic acids have been used for the extraction of uranium. Among those that have been used are α-hydroxyisobutyric acid,[5] malic acid,[6] tartaric acid[7] and citric acid.[7] Citric acid is widely used to mobilize both sorbed and precipitated uranium from both soils and nuclear reactor facilities. An advantage of its use is that its uranium(VI) complexes are biodegradable. In solutions of pH 8-9, over 99% of the citrate is biodegraded. Furthermore, at pH 8-9 insignificant amounts of uranium became associated with biomass. The ready biodegradation of citrate at pH 8-9 is believed to be due to its liberation in the free state from its complexed form (equation 1). The free citrate ion then readily biodegrades to carbon dioxide.[8]

$$3UO_2(\text{citrate})_2^{2-} + 7OH^- = (UO_2)_3(OH)_7^- + 6\text{citrate}^{2-} \quad (1)$$

### 1.2.2. Phosphates and Phosphine Oxides

The most commonly used extractants for the actinides are phosphates and phosphine oxides. The preferred phosphate is tributyl phosphate (TBP, **1**), and the preferred phosphine oxide is trioctylphosphine oxide (TOPO, **2**). Extractions using TBP can be carried out both in the

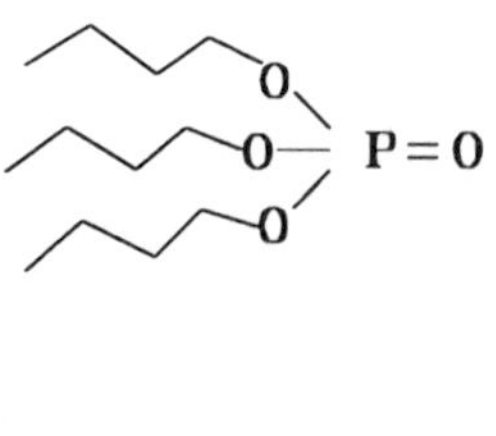

O
O— P=O
O

P=O

**1** 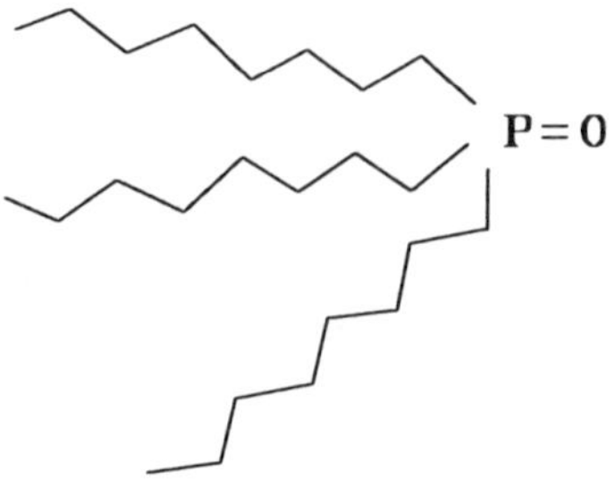 **2**

presence and absence of a solvent. The interaction of TBP with uranium(VI) has a favorable enthalpy and free energy. The stoichiometry of the adduct is $UO_2(NO_3)_2(TBP)_2$.[9] Among the solvents that have been used as the organic phase are dodecane[9, 10] and liquid carbon dioxide.[11-13] The solubility of uranium(VI) in the carbon dioxide phase correlates with the density of the fluid, which is influenced by the operating pressure of the extraction. Experimental data indicate that the selectivity between uranium(VI), plutonium(IV) and hydrogen ion also depends on this operating pressure. Thus one can optimize pressure and/or temperature to increase selectivity in supercritical fluid extractions and metal separations.[12] The stoichiometry of the uranium(VI)-TBP complexes formed, and the kinetics of extraction into supercritical carbon dioxide, are similar to those found for conventional solvents. Supercritical carbon dioxide can also be used to extract uranium(VI) and thorium(IV) by using a binary mixture of TBP and a fluorinated β-diketone.[14]

Tributyl phosphate can also be incorporated onto resin beads to produce a material that extracts uranium(VI) and other actinide ions. A resin bead impregnated with TBP has been used to separate nanogram quantities of uranium and thorium from oxide fuels.[15] A polystyrene bead impregnated with TBP has been used for the extraction and separation of actinides from solutions of dissolved fuels. The resin extracts uranium and plutonium while allowing americium and curium to elute through.[16]

Di-2-ethylhexyl phosphoric acid (HDEHP, **3**) has been used for the liquid-liquid extraction of uranium(VI). From a dilute nitric acid solution

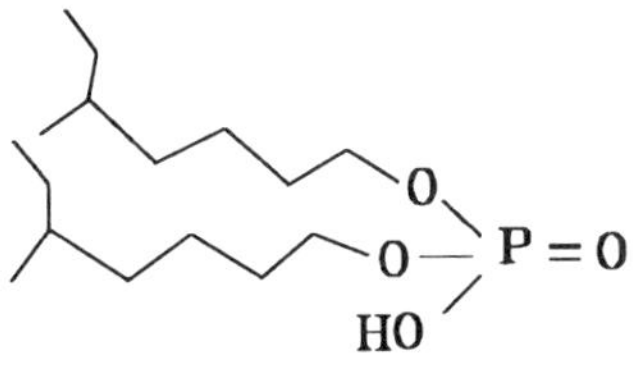

**3**

the extracted species is $UO_2(DEHP)_2$, but at higher acid concentrations, $UO_2(NO_3)(DEHP).HDEHP$ becomes the predominant complex.[17] The compound HDEHP can be used to extract uranium(VI) into kerosine,[18] and it has also been used for the extraction of trace amounts of uranium. The efficiency of the process can be followed by determining the U-239 by neutron activation analysis.[19] The compound HDEHP has also been used as the stationary phase in high-efficiency liquid-liquid chromatography.[20]

Two resins that are commercially available from Eichom Industries Inc. are U/TEVA-spec and TEVA-spec. These resins have high selectivities for the removal of tetravalent actinides and uranium from solutions having somewhat lower concentrations of nitric acid or hydrochloric acid.[21, 22] Chelate resins comprised of diamyl amylphosphonate sorbed onto an inert polymeric support can be used to separate uranium and thorium from silicate rock samples.[23]

Phosphine oxides are also effective extractants for actinides because they can also bind to the metal ion through an oxygen donor atom. The equilibrium constant for binding TOPO with uranium(VI) in nitric acid solution in the presence of dichlorobenzene, cyclohexane or carbon tetrachloride is in the $10^4$ - $10^8$ range at ambient temperature. The adduct has the stoichiometry $UO_2(NO_3)_2(TOPO)_2$.[24] In sulfuric acid solution the stoichiometry of the adduct is $UO_2(SO_4)(H_2O)(TOPO)_2$.[25] TOPO modified octadecyl-bonded silica membrane disks have been used for the solid-phase extraction of ultratrace quantities of uranium(VI). The uranium can then be determined spectrophotometrically with dibenzoylmethane.[26] The phosphine oxide TOPO, tributylphosphine oxide and triphenylphosphine oxide in supercritical carbon dioxide have high extraction efficiencies for both uranium(VI) and thorium(IV).[27] The compound TOPO is particularly effective because the presence of a long octyl chain in the molecule causes it to have good detergent-like properties as an extractant molecule.

Cathodic stripping voltammetry can be used to detect uranium(VI) in raffinate following liquid-liquid extraction.[28] Both ion chromatography and capillary electrophoresis have been used to separate uranium(VI), with arzenazo III being used for absorbance detection.[29, 30]

### 1.2.3. Amines and Ketones

Tri-*n*-octylamine in xylene has been used to separate uranium(VI) from its uranium(IV) state by extraction from a mixture of phosphoric acid and hydrochloric acid. Impurities such as vanadium, molybdenum and chromium do not interfere.[31] Uranium(VI) and thorium(IV) have been separated by chromatography using a diethylaminoethyl functionalized cellulose material. The eluent is either sulfuric or acetic acid.[32, 33] Ketones are also effective for the extraction of actinides. An early example involves the use of a polymer bound hexaketone **4** for the extraction of uranium(VI) from seawater. The hexaketone is attached to polystyrene. The uranium(VI) can be subsequently removed from the polymer by treatment with 1M hydrochloric acid.[34] Dibenzoylmethane has also been used for

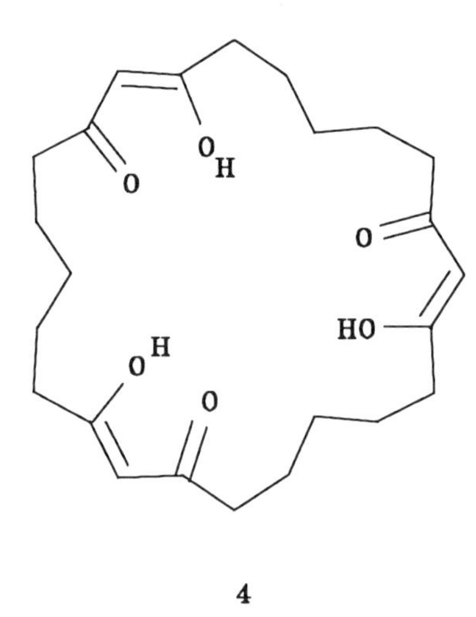


4

the liquid-liquid extraction of uranium(VI), americium(III), thorium(IV), and neptunium(V). Each actinide can be recovered selectively in high purity.[35] 2-Thenoyltrifluoroacetone has been used for the separation and extraction of americium(III). The procedure involves extracting the americium into xylene.[36]

## 1.3. Multidentates

As with other metal ions, multidentates and chelates have been found to be very useful in the extraction of actinides and lanthanides from both acidic and basic aqueous solutions.

### 1.3.1. N, O-Chelates

Chelates with *N,O*-ligating groups have been used for the extraction of actinides. Among these are EDTA and related ligand types. Uranium(IV) forms complexes with both EDTA and DTPA (diethylenetriaminepentaacetate) at pH 2-3. The complexes have the stoichiometries $[U(EDTA)(H_2O)_2]$, $[U(HOEDTA)]^+$ and $[U(DTPA)]^-$.

Raising the pH of the solution allows the uranium(IV) to be extracted.[37] A solid phase chelate prepared by the polymerization of 4- and 5-acrylamidosalicylic acids are selective extractants for uranium(VI), copper(II) and iron(III) at pH 4.5 - 5.5.[38]

### 1.3.2. Malonamides

Malonamides are excellent extractants for removing uranium(VI) and lanthanide(III) ions from aqueous solution. Two particular malonamides that have been used are DMDBO3NPDA **5** and DMDBMA **6**, with former being the better extractant.[39] Other malonamides have also

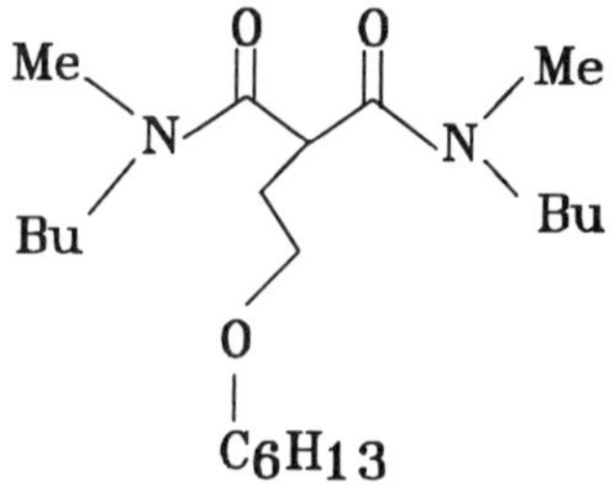

Me O O Me N N Bu Bu O $C_6H_{13}$

5

Me O O Me N N Bu Bu

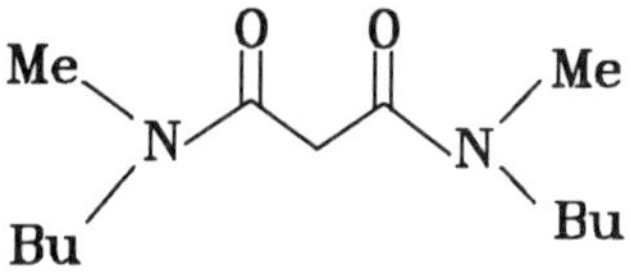

6

been effectively used for the liquid-liquid extraction of actinides. The use of malonamides as extractants for actinides has generated considerable interest because it is desirable that combustion of the extracted metal complexes leaves no residue; a situation that is not found with phosphorus-containing extractants. Malonamides have the general structure **7**, and they

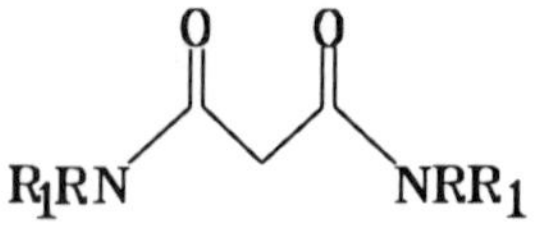

O O $R_1RN$ $NRR_1$

7

usually bind to actinide ions through their oxygens. A further advantage of malonamide extractants is that they can be used to remove actinides from concentrated nitric acid solutions. Distribution ratios for americium(III) between benzene and nitric acid solutions for a series of different functional groups R and $R_1$ on **7** are shown in Table 1.[40] These observations correlate with earlier findings that both dialkyl *N,N*-dialkyl

**Table 1** **Distribution Ratios (D) for Americium(III) between Benzene and Nitric Acid for $R_1RNC(O)CH_2C(O)NRR_1$**

| R | $R_1$ | D |
|---|---|---|
| $C_6H_{11}$ | $C_2H_5$ | 0.11 |
| $C_4H_9$ | $C_4H_9$ | 0.18 |
| $C_3H_7$ | $C_8H_{17}$ | 0.45 |
| $CH_3$ | $C_4H_9$ | 0.55 |
| $CH_3$ | $C_8H_{17}$ | 1.18 |

carbamoylmethylenephosphonates $(RO)_2P(O)CH_2C(O)NR_2$ and tetraalkyl methylene diphosphonates $(RO)_2P(O)CH_2P(O)(OR)_2$ are also good extractants for americium(III) from aqueous nitric acid solutions.[41] Amides are also chosen for the liquid-liquid extraction of actinide ions because they form complexes with uranium(VI) that are soluble in alkanes.[42] Another feature that makes such compounds useful is that they are resistant to radiolysis and hydrolysis in the presence of nitric acid.[43] These properties, along with the desire to use complexants that can be completely incinerated to volatile products, had led to investigations on a broad range of amide derivatives. Thus, in addition to malonamides, a broad range of *N,N*-dialkylamides have been found to be useful for the liquid-liquid extraction of actinide ions. Higher homologues of malonamide are also useful liquid-liquid extractants for actinides. In a study of the extraction of uranium(VI) and plutonium(IV) from aqueous nitrate solutions by *N,N*-tetrabutylglutaramide (TBGA), it is found that at low acidities the complexes formed are $UO_2(NO_3)_2$.TBGA and $Pu(NO_3)_4$.TBGA. At high acidities the anionic complexes [$HTBGA^+$] [$UO_2(NO_3)_3^-$] and [$HTBGA^+$] [$HPu(NO_3)_6^-$] are formed, although there is some uncertainty about the formulation of the latter complex.[44] Homologues with alkyl substituents on the methylenic carbon of the malonamide have also been used as extractants.[45] As a result the homologous series of diamides $R_1RNC(O)CHR_2C(O)NRR_1$ have been used as extractants for actinides. In particular, these compounds have been used to remove americium(III) and curium(III) from aqueous nitric acid solution. The distribution ratios D for the extraction of americium(III) into *tert*-butylbenzene with different functional groups on the diamide are shown in Table 2.[46] These compounds

**Table 2 Distribution Ratios (D) for Americium(III) between *tert*-butylbenzene and Nitric Acid for $R_1RNC(O)CHR_2C(O)NRR_1$**

| R | $R_1$ | $R_1$ | D |
|---|---|---|---|
| $C_4H_9$ | $CH_3$ | $C_6H_{13}$ | 1.16 |
| $C_4H_9$ | $CH_3$ | $(CH_2)_2OC_2H_5$ | 3.28 |
| $C_4H_9$ | $CH_3$ | $(CH_2)_2OC_6H_{13}$ | 7.55 |
| $C_4H_9$ | $CH_3$ | $(CH_2)_2O(CH_2)_2OC_6H_{13}$ | 11.48 |

generally meet the requirements for use as liquid-liquid extractants in the nuclear industry.[47] Other modifications have been made to these malonamide-type extractants, and their chemistry with actinide ions studied. These studies are focused on developing a better understanding of the nature of the complexes that are formed in the different phases. Diamides can be used as actinide extractants from solutions that are high in chloride ion concentration. Another finding is that the addition of oxalic acid allows for the separation of iron and zirconium from both americium and plutonium.[48] Malonamides can also be tailored to induce higher selectivity. A comparison between the *n*-butyl and *iso*-butyl malonamides shows that the former is the better extractant from nitric acid solutions, but that it has a poor plutonium(IV) to uranium (VI) selectivity. In each case the extraction is favored by enthalpy, whereas the entropy is counteracting the extraction of uranium(VI) and favoring the extraction of plutonium(IV). The actinides can be released by the use of oxalic acid or a dilute solution of uranium(IV) for plutonium(IV), and a dilute sodium carbonate solution for uranium(VI).[49] The extraction constants for these actinides, in addition to that of americium(II), increase with increasing nitric acid concentration in the aqueous layer.[50]

Further studies on the extraction of uranium(VI) and plutonium(IV) by 1,3-diamides indicate that they are involved in both inner and outer sphere interactions with these metal ions.[51] Comparisons have been made between the extraction of both lanthanides and actinides with diamides. With one group of extractants a high separation factor of 8.64 for europium(III) to americium(III) has been achived,[52] but the loading capacity of *N,N*-dimethyl-*N,N*-dibutylmalonamide for lanthanide(III) ions is insufficient for practical application.[53] Both *N,N,N,N*-tetrabutylsuccinylamide and the analogous adipamide have been used for the extraction of uranium(VI) and thorium(IV).[54, 55] For malonamides there is restricted third-phase formation, and long alkyl groups and different substituents on the amide nitrogen limit the extent of this.

In seeking generalizations for the extraction of actinide(III) ions by diamides and picolinamides, it is observed that the latter show promise for the separation of actinide(III) ions from lanthanide(III) ions.[56] Tetrahexylmalonamides are, however, good extractants for europium(III). Three tetrahexylmalonamides THMA **8**, MeTHMA **9**, and DiMeTHMA **10**, that have been compared differ by the presence of hydrogens or methyl groups on the methylene carbons. In each case the malonamide to europium(III) concentration ratio is 3 : 1. The extraction constant decreases by seven-fold on going from **8** to **9**, and a large decrease on changing to **10**.[57] These tetrahexylmalonamides are also extractants and complexants

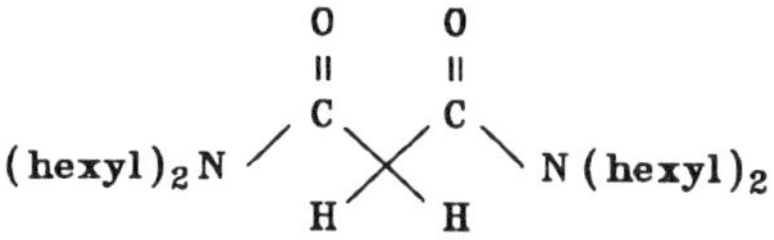

(hexyl)$_2$N–C(=O)–C(H)(H)–C(=O)–N(hexyl)$_2$

**8**

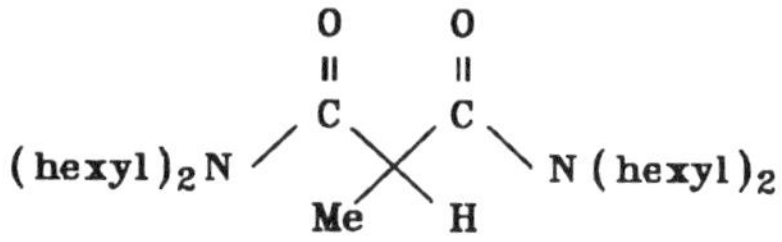

(hexyl)$_2$N–C(=O)–C(Me)(H)–C(=O)–N(hexyl)$_2$

**9**

(hexyl)$_2$N–C(=O)–C(Me)(Me)–C(=O)–N(hexyl)$_2$

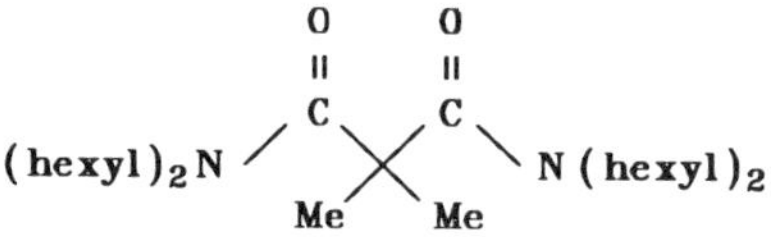

**10**

for uranium(VI). For these three extractants L, the extracted species from 1M sodium nitrate solution are $UO_2(NO_3)_2L_2$ and $UO_2(NO_3)_2L_3$. The $UO_2(NO_3)_2L_2$ complex is dominant for **10**, with very little formation of $UO_2(NO_3)_2L_3$. By contrast, for **9** the complex $UO_2(NO_3)_2L_3$ predominates. The complexant THMA is intermediate between these extremes. The greater propensity of MeTHMA over THMA to bind in a 3 : 1 ligand :

uranium (VI) ratio may be due to the higher basicity of the carbonyl oxygens in the former. The observed 2 : 1 ratio for DiMeTHMA is ascribed to steric reasons. These ligand : uranium (VI) ratios are dependent on the nitrate ion concentration.

Fourier transform infrared spectroscopy has been used to determine the structures of the plutonium(IV) and uranium(VI) nitrate complexes with diamides.[58] The nitrates coordinate in a bidentate mode.[60] Similar structures for the uranium(VI) complexes have been determined by a combination of $^{1}H$, $^{13}C$, and $^{14}N$ nuclear magnetic resonance spectroscopy.[60, 61] The complex $UO_2(NO_3)_2$(TMMA), where TMMA is *N,N,N,N*-tetramethylmalonamide, has been structurally characterized by single-crystal x-ray diffraction. The TMMA is bidentate as are both of the nitrate ions.[62] A spectroscopic investigation of complexes of thorium(IV) and uranium(VI) with malonamides, glutaramides, and succinamides in aqueous nitric acids shows the presence of nitrate ion in the metal coordination sphere in either a monodentate or a bidentate coordination mode.[63]

An amide resin obtained by introducing a -$CH_2N$(Me)C(O)Me group onto a polystyrene-divinylbenzene polymer selectively extracts uranium and thorium from aqueous solution by flotation of the precipitate that it forms with 8-quinolinol. [64] The flotation can be carried out both in the presence or absence of surfactants such as CTAB or sodium lauryl sulfate.[65]

### 1.3.3. Chemical Detection and Analysis

Although radiotracer methods are commonly used, uranium(VI) can be determined as its chelate complex with 1,3-dimethyl-4-acetyl-2-pyrazolin-5-one (DMAP) by direct injection onto a liquid chromatography column.[66] Similarly, the uranium complexes of *N,N*-ethylene-*bis*-(salicylaldiimine) **11** have been extracted into chloroform .[67]

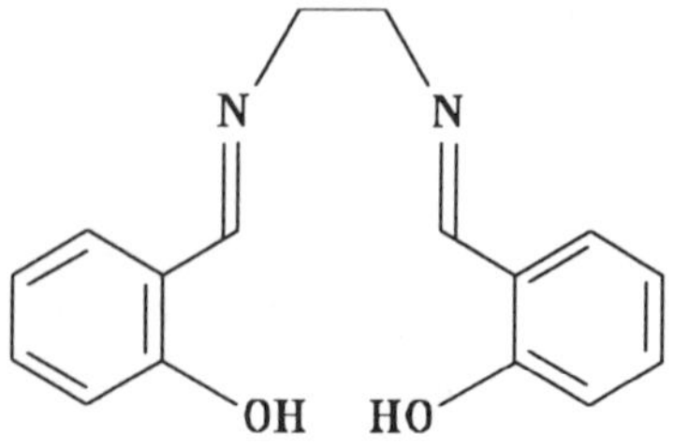

N N OH HO

**11**

and subsequently analyzed by HPLC

### 1.3.4. Stereognostic Coordination

A novel approach to addressing the molecular recognition of oxometal cations such as uranium(VI) involves stereognostic coordination chemistry. This approach employs a strategy where the complexant binds not only to the metal center, but that it provides at least one hydrogen bond donor group that can to the oxide center. A series of complexants that have been prepared for the selective extraction of uranium(VI) are the hydrochlorides of NEB **12**, NPB **13**, NPN **14**, and NpodB **15**. Distribution coefficients of the order of $10^{11}$ are observed for the transfer of uranium(VI) from a neutral aqueous solution into an

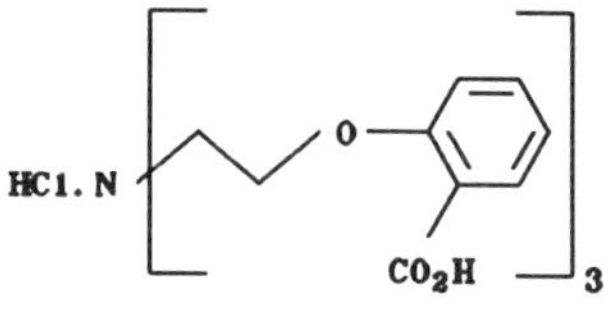

HCl.N [ O, $CO_2H$ ]$_3$

**12**

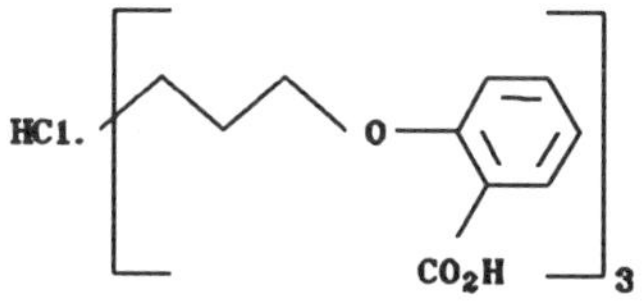

HCl. [ O, $CO_2H$ ]$_3$

**13**

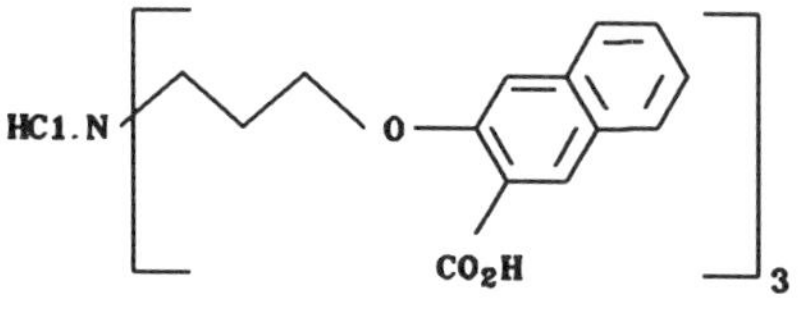

HCl.N [ O, $CO_2H$ ]$_3$

**14**

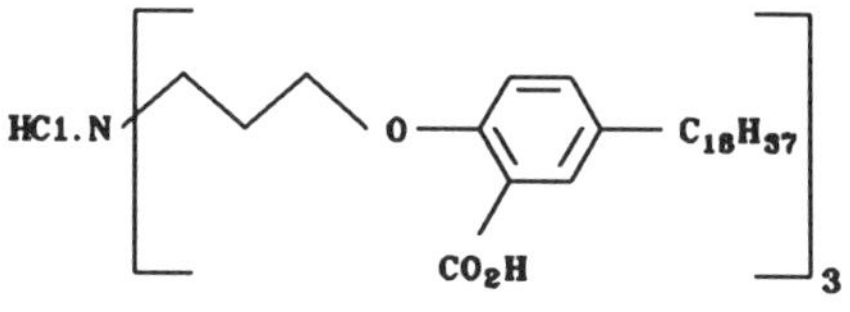

HCl.N [ O, $C_{18}H_{37}$, $CO_2H$ ]$_3$

**15**

organic phase with these compounds. A schematic representation of the cooperative complexant binding to the metal center and hydrogen bonding to an oxo ligand is shown in **16**.[68]

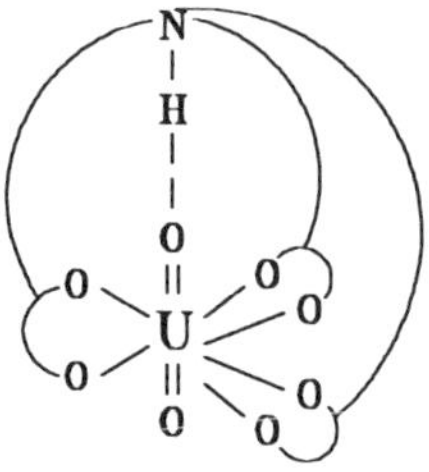

N–H···O=U=O

**16**

### 1.3.5. Carbamoylphosphonates

Extractants of the carbamoylphosphonate type have been widely used for the actinides. Both tri-(2-ethylhexyl) phosphate (TEHP) and di-*n*-hexyl-*N,N*-diethylcarbamoyl methylenephosphonate (DHDECMP) coated trifluorochloroethylene polymers have been used in the purification of uranium. Subsequent extraction of the polymers with organophosphorus compounds recovers the uranium, but leaves behind the rare earth element products of nuclear fuel cycle materials. These remaining elements can, however, be eluted by nitric acid.[69] Chelate resins containing dihydroxyphosphino, phosphono or phosphonomethyl groups have been used as adsorbents for uranium. These materials are effective in 0.25-0.50M $H_2SO_4$ solution, with the uranium being subsequently eluted as its carbonate using 0.25 M sodium carbonate as eluent.[70, 71] These types of materials are useful adsorbent in addition to the conventional ion exchange resins such as Dowex AD50W-X8, Purolite 940,[72] Dowex 1-X8,[73] and Diaion SA10 anion exchange resin.[74] Alternatively a colloid flotation method using ferric hydroxide particles can be used for the separation of uranium from seawater.[75]

### 1.3.6. Crown Ethers and Cavitands

The dicyclohexano-24-crown-8 in conjunction with a mixture of 1, 2-dichloroethane and aqueous hydrochloric acid has been used for the extraction of uranium(VI). The extraction constant increases with decreasing temperature, and the uranium is present in solution as the $UO_2Cl_4^{2-}$ anion. The counterions are $H_5O_2^+$, in contrast to $H_3O^+$ which is observed for the dicyclohexano-24-crown-6 complex.[76] Specific cavitands have been designed for use as extractants for actinides such as uranium(VI). These can be based on amines or CMP(O)-type derivatives **17** which are known to be excellent complexants for actinides. Two such

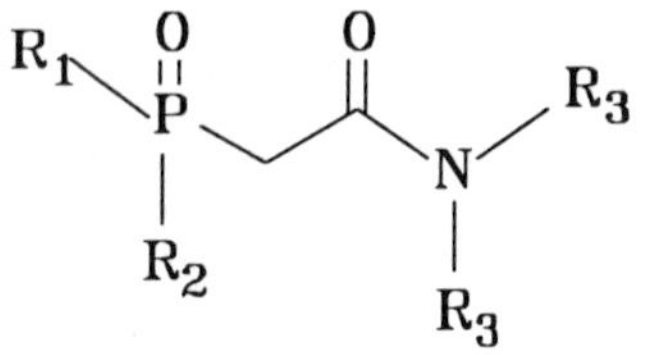

R1 O O R3
P N
R2 R3

17

groups of cavitands are the tetrafunctionalized resorcinarenes **18** and **19**, or the CMP(O) derivative **20** (R = H, L = Ph, ethoxy; R = *n*-propyl, L = Ph, ethoxy). The steric preorganization of the four CMP(O) moieties on

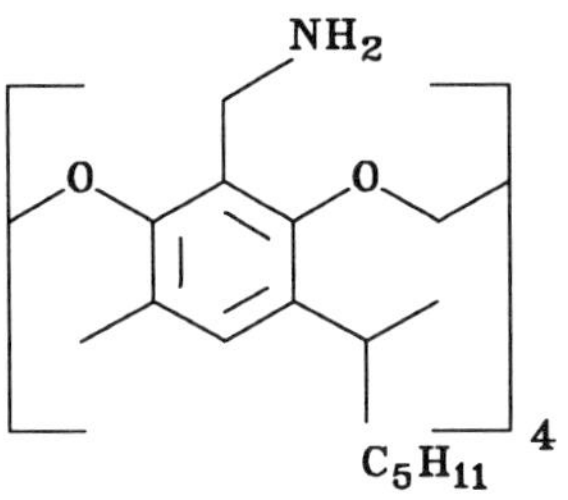

$NH_2$ $C_5H_{11}$ 4

18

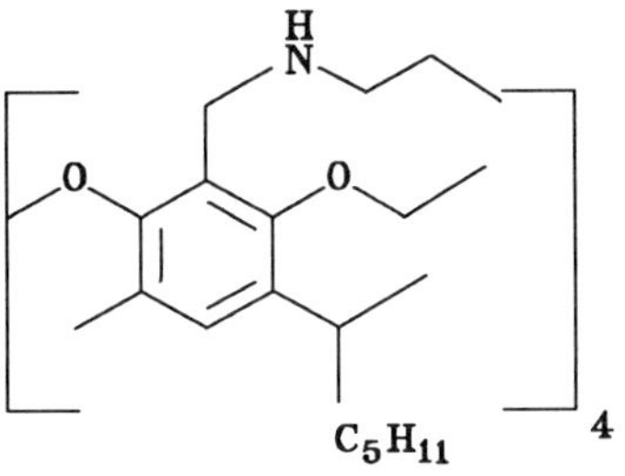

H N $C_5H_{11}$ 4

19

L L O=P O= N−R $C_5H_{11}$ 4

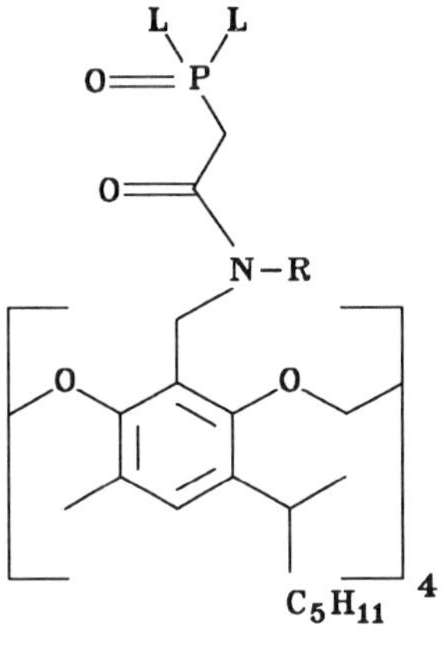

20

the resorcinarene cavitand **20** increases both the extraction constants and the selectivities over their simple analogs. As an example, the selectivity for europium(III) over both uranium(VI) and iron(III) is increased.[77]

### 1.3.7. Calixarenes

The use of calixarenes as extractants for actinides and lanthanides has been recently reviewed.[78] Calix[4]arenes having polyamino carboxylic acid **21** or CMPO-type **22** substituents on the narrow rim have been used for the liquid-liquid extraction of actinide ions. Both the iminocarboxylic acid and the CMPO-type derivatives

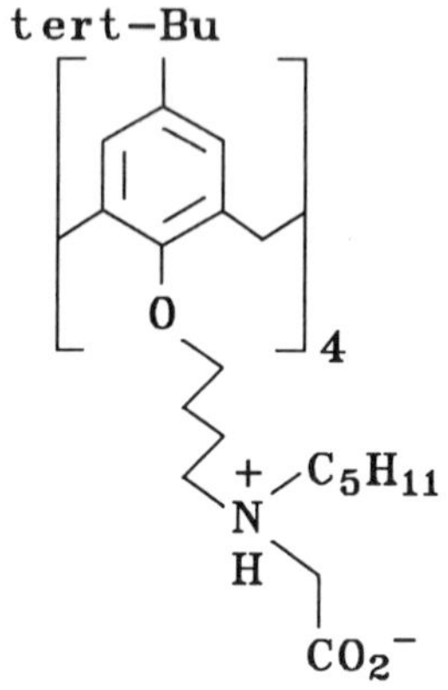

21 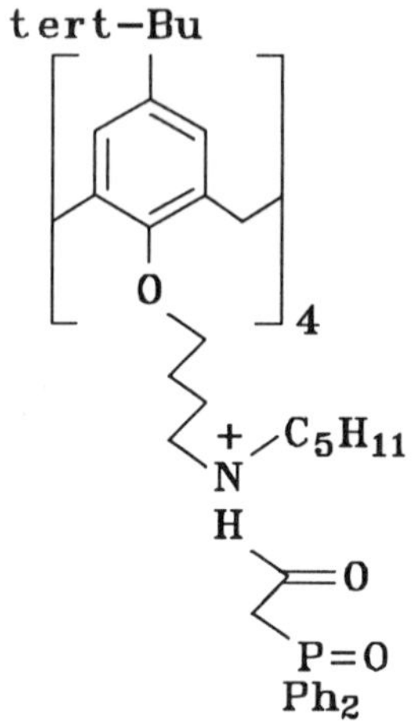 22

preferentially extract thorium(IV) over europium(III) into a chloroform phase from an aqueous solution containing sodium nitrate and nitric acid at a pH of 0.33.[79]

## 2. LANTHANIDES

The extraction and separation of the fourteen lanthanides presents a challenge because they all have closely similar chemistries. The stable form in each case is the trivalent(III) ion, and they differ from the transition and post-transition metal ions in that they commonly adopt coordination numbers greater than six. In terms of their complexation chemistry they act as a hard acid, with oxygen donor ligands being the ones of preference. Because of the lanthanide contraction across the row, the ions can be separated by conventional cation exchange resins. For liquid-liquid extraction, europium is the focus of considerable interest because it is a product of nuclear fission, and therefore is an element that needs to be extracted from mixtures of other metals. A unique feature of europium is that it is stable in solution in reduced form as europium(II), and this property can potentially lead to strategies for its selective removal from other trivalent metal ions.

## 2.1. Monodentates

The common simple ligands that are used for the extraction of lanthanides are carboxylates, phosphates and ketones. These complexants are being classified in this section as monodentates, although in certain cases it is clear that they can function in either a monodentate or bidentate mode.

### 2.2.1. Carboxylates

Hydroxybutyric acids have been used for the extraction of both lanthanides and actinides, the most common of which is 1-hydroxyisobutyric acid **23**. These acids have been used in conjunction

Me $CO_2H$

Me OH

23

with capillary tube isotachophoresis to separate the fourteen lanthanide ions. The separation and analysis can be completed in 20 minutes.[80] By using a mixture of tartaric acid and 1-hydroxyisobutyric acid in a solution of ammonia it is also possible to separate yttrium(III) from the lanthanide ions with this technique.[81] Other acids have been added to improve the separation, among which are malonic acid, malic acid, tartronic acid and glycolic acid. Malonic acid is the most effective in this group because of the lower stability constants of yttrium(III) malonate complexes as compared with those of the lanthanide(III) ions.[82] In addition to 1-hydroxyisobutyric acid, 1-hydroxy-1-methylbutyric acid **24**

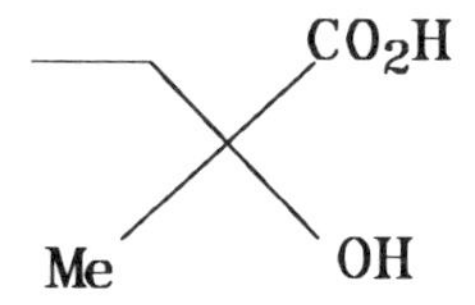


24

has been used to separate lanthanides. Compound **24** is useful for the lighter lanthanides, with 1-hydroxyisobutyric acid being preferable for the heavier lanthanides.[83] In addition to its use in isotachophoresis, 1-hydroxyisobutyric acid has been used in conjunction with capillary zone electrophoresis for the separation of yttrium and the lanthanide ions.[84] Other carboxylic acids have been used for the extraction of lanthanides. Among these are oxalic acid,[85] mandelic acid,[86] and acids having both nitrogen and carboxylic acid functionalities. These acids can be used as eluents, as is the case with *N*-(2-hydroxyethyl) iminodiacetic acid. Ion-exchange chromatography of lanthanide oxides on Dowex 50W-X2 or IX4 resins with either *N*-(2-hydroxyethyl) iminodiacetic acid or EDTA results in their separation.[87, 88] Similarly, a column packed with iminodiacetic acid bonded to silica has been used as the solid phase for the separation of lanthanide ions.[88] Solvent extraction methods for measuring both the stability constants and the thermodynamic parameters for the complexation of lanthanides and actinides by glycine leads to the conclusion that the enthalpy of complexation is endothermic with the complexes being stabilized by a positive entropy change.[90]

A novel approach to designing solid resins involves the preparation of polymeric materials for lanthanide ion extraction using the ring-opening metathesis polymerization (ROMP) route. One of these polymeric resins has been prepared from norborn-2-ene-5,6-dicarboxylic acid anhydride with a molybdenum carbene as the ROMP catalyst. This particular resin has a hydrophobic interior and a hydrophilic exterior surface, and it is proposed that the lanthanide ions are complexed at the surface as in **25** (X = nitrate; n = 5, 6).[91] Block copolymers of norborn-2-ene and 7-oxanorborn-2-ene-5, 6-dicarboxylic acid have also been prepared by ROMP techniques. These materials have been coated onto silica and used for the selective extraction of lanthanides. The optimum pH range for lanthanide ion extraction is in the 3.5-5.5 range. Actinides such as uranium(VI) and thorium(IV) are quantitatively co-extracted by this resin.[92] Materials of 200-400 mesh are used to facilitate handling in terms of back pressure. Breakthrough curves, the loading characteristics at different pH values, as well as recoveries for both Pm-147 and Eu-152 have been determined by $\beta$ liquid scintillation counting. The retention of potentially interfering metals such as magnesium(II), calcium(II), barium(II), manganese(II), cobalt(II), nickel(II), aluminum(III), iron(III), and zinc(II) is less than 5% in all cases. Since the material is stable within a pH range of 0-12 it can be rapidly recycled and reconditioned. This is particularly important for radioactive materials because of the high cost of their disposal.

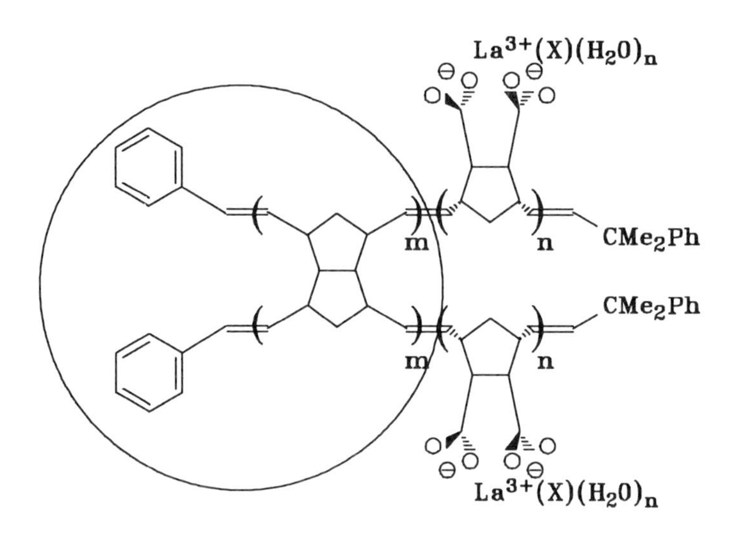


25

### 2.2.2. Phosphates

Phosphates are also widely used for the extraction of lanthanides. One method used in the extraction of trivalent lanthanides and actinides is the cleanex solvent extraction process which uses di-(2-ethylhexyl) phosphoric acid. The process is applicable to situations that contain a broad range of contaminants.[93] A variety of experimental conditions have been employed with this reagent. One of these involves adsorbing the di-(2-ethylhexyl) phosphoric acid onto Teflon powder, and then using this material to separate lanthanides from actinides. The lanthanide ions are adsorbed on the column while the actinide ions are eluted.[94] The lanthanides can be subsequently eluted with 6M hydrochloric acid.[95] Another approach involves a solid-liquid extraction procedure using paraffin wax as the solvent.[96] Naphtha has also been used as the organic phase,[97] For the extraction of either gadolinium(III) or yttrium(III) by di-(2-ethylhexyl) phosphoric acid, the procedure is not affected by the presence of the other ion.[98]

Lanthanide ions have been separated with di-(2-ethylhexyl)

phosphoric acid in conjunction with high-speed countercurrent chromatography. The system uses the di-(2-ethylhexyl) phosphoric acid in chloroform as the lower phase, and a hydrochloric acid solution as the upper phase.[99] Toluene or heptane can also be used as the organic phase.[100, 101] Di-(2-ethylhexyl) phosphoric acid has also been used in conjunction with countercurrent chromatography and reversed micelles for the extraction and separation of lanthanide ions [102]

Tributyl phosphate has also been used for the liquid extraction of lanthanides with supercritical carbon dioxide as the solvent. Thus lanthanides can be extracted from a solution of 6M nitric acid and 3M lithium nitrate using tributyl phosphate and carbon dioxide. The extraction is enhanced if the system also contains fluorinated ketones.[103-105] Other phosphates that have been used for the extraction of lanthanides include sodium metaphosphate with a Dowex I-X4 anion exchange column,[106] and β-ketophosphonates. The β-ketophosphonates that have been used are dibutyl phenacylphosphonate **26** and two of its derivatives. Chelation of the lanthanide *via* both the ketonic and phosphonate oxygens is observed.[107]

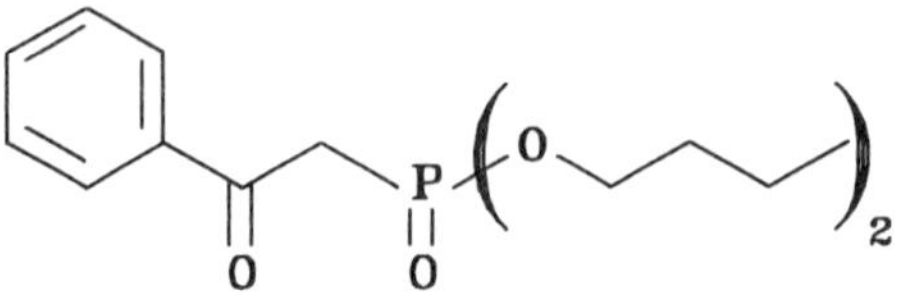

**26**

## 2.2.3. Ketones and Phosphine Oxides

Ketones and phosphine oxides have also been used for the liquid-liquid extraction of lanthanides. In a study of the extraction of lanthanum(III), samarium(III), terbium(III), thulium(III), and lutetium(III) by the ketones hexfluoroacetylacetone(Hhfa), pivaloyltrifluoroacetone(Hpta), benzoylacetone(Hba), and acetylacetone(Hacac), the extraction constants follow the order: Hacac < Hpta < Hhfa. This order correlates with their acid dissociation constants. The extraction is enhanced by the presence of bipyridine due to its complexation with the lanthanide ion along with the complexed ketone.[108] Further studies with Hacac using a series of organic solvents show that there is a clear correlation between the partition coefficient of the lanthanide(III) chelates and their successive formation constants.[109] Lanthanide ions have also been extracted into organic solvents using 2-thenoyltrifluoroacetone. With the larger lanthanides the hydration number of the extracted complex is about 3, but decreases to about 2.4 for the smaller ones.[110] Introducing different substituents onto the β-diketone changes their acidity, and hence their function as extractants for lanthanides. As example, the $pK_a$ of the methyl derivative $PhC(O)CHMeC(O)CF_3$ is 10.36, which is larger than that of $PhC(O)CH_2C(O)CF_3$, which is 8.10. As a result lanthanides can be extracted from a lower pH region with the non-methylated compound because of its higher acidity.[111] A charge-transfer mechanism has been proposed for the extraction of trivalent lanthanides and actinides with β-diketones.[112] Another probe of the study of effects that govern the extraction of lanthanide ions by diketones has involved comparing the three cyclic ketones, trifluoroacetylcyclopentanone, trifluoroacetylcyclohexanone and trifluoroacetylcycloheptanone. The O-O distance in the enol form is larger for the smaller five-membered ring, which correlates with both a weaker intramolecular hydrogen bond strength and a higher acidity. The larger ring compounds give higher extraction constants for lanthanum(III), praesodymium(III), europium(III), holmium(III) and ytterbium(III).[113] Fluoroketones can also be used to extract lanthanides into supercritical carbon dioxide.[114]

Chromatography has been used with ketones for lanthanide separations. Techniques can involve using fluorinated diketones as extractants, followed by chromatography of the complexes,[115] or by direct chromatography using polyacryloylacetone as the stationary phase.[116]

Phosphine oxides have been used for the liquid-liquid extraction of lanthanides. In one application a combination of β-diketones and phosphine oxides has been employed. The direct separation of lanthanide β-diketonates cannot be achieved by supercritical fluid chromatography due to the hydration of the complexes. Nevertheless, adduct formation of lanthanide β-diketonates with triphenylphosphine oxide or trioctylphosphine oxide alters their hydration properties and allows for their separation.[117] Trioctylphosphine oxide (TOPO) can be used for the liquid-liquid extraction of lanthanides from chloride containing solutions. The lanthanides extract as the hydroxide complexes having the stoichiometries $Ln(OH)_3$.nTOPO.[118] A combination of ketones or trifluoroacetylcycloalkanones, and TOPO show a synergistic effect whereby addition of TOPO improves the liquid-liquid extraction of lanthanides into a chloroform phase.[119] Similarly, lanthanides can be analyzed by gas chromatography using mixed ligand complexes with hexafluoroacetylacetone and TOPO.[120] Methylene-*bis*-(dialkylphosphine oxides) can be used for the liquid-liquid extraction of lanthanides from aqueous hydrochloric, nitric and perchloric acids into a carbon tetrachloride or 1,2-dichlorobenzene phase. Up to three molecules of the methylene-*bis*-(dialkylphosphine oxide) is bound to the lanthanide ion.[121]

### 2.2.4. Crown Ethers

Crown ethers have been used for the extraction of lanthanides. This class of compound has been used both as complexants themselves for lanthanide ions, and also as synergistic components with other complexants in the organic phase. The simple unsubstituted crown ethers have been used as extractants for lanthanides. Two such crowns that have been used are 15-crown-5 and 18-crowns-6. Lanthanide(III) picrates are quantitatively extracted by these two crowns at low ionic strength. The peak extraction constants for 15-crown-5 are found for samarium, and for 18-crown-6 at cerium and praesodymium. The cerium complex has a 1 : 1 stoichiometry, and the praseodymium complex a 1 : 2 stoichiometry.[122] Thermal lensing detection of lanthanide ions by solvent extraction has also been carried out using crown ethers. The thermal lens signal intensity of the extracted ions in the organic phase can be enhanced 24-fold. This enhancement is due to the thermal lens signal being dependent on the thermo-optical properties of the solvent, and water is a poor thermo-optical solvent. The selectivity of these crowns for the lanthanide ions makes this method useful.[123] Synergistic effects of crown ethers in the extraction of lanthanides has been observed. In the extraction of lanthanides into cyclohexane with di-(2-ethylhexyl) phosphoric acid, the addition of 18-crown-6 results in the extraction of those metal ions with the larger ionic radii being shifted to the higher pH region. As a consequence the separation among the ions is enhanced.[124] The extraction constant of lanthanum(III) and gadolinium(III) has been increased two-fold by using $D_2O$ instead of $H_2O$ in the aqueous phase. The extractant is 18-crown-6, and the organic phase is dichloromethane. By contrast, alkali and alkaline earth ions show no such effect. This lanthanide-specific solvent isotope effect is attributed to hydrogen-bond breaking between waters of hydration in the first hydration sphere and bulk water upon the extraction of partially dehydrated lanthanide ions into the dichloromethane phase.[125]

Substituted crowns have also been effectively used in the liquid-

liquid extraction of lanthanide ions. The compound 4-*tert*-butylcyclohexyl-15-crown-5 has been used for its synergistic effects. Didodecylnaphthalenesulfonic acid is a useful liquid cation extractant for lanthanide ions, and its high acidity allows for it to be used in low pH conditions. A limitation of its general use is its low selectivity, but the addition of 4-*tert*-butylcyclohexyl-15-crown-5 to the toluene phase results in higher selectivities in the extraction of the light lanthanides.[126] Substituted crowns have been used along with thenoyltrifluoroacetone(TTA) for the extraction of lanthanum(III) and neodymium(III). The crown ethers that have been used are 1,10-diaza-4,7,13,16-tetraoxacyclooctadecane-*N,N*-diacetic acid (DACDA, **27**) and 1,7-diaza-4,10,13-trioxacyclopentadecane-*N,N*-diacetic acid (DAPDA, **28**). These macrocycles form strong monocationic complexes with the lighter lanthanides. The extracted species is pH dependent with $Ln(TTA)_3$ being the predominant species at pH $\leq$ 5.0, and Ln(DAPDA/DACDA)TTA being prevalent at pH ~ 7.5. The maximum distribution coefficient into the organic phase is obtained in the pH 7.5-

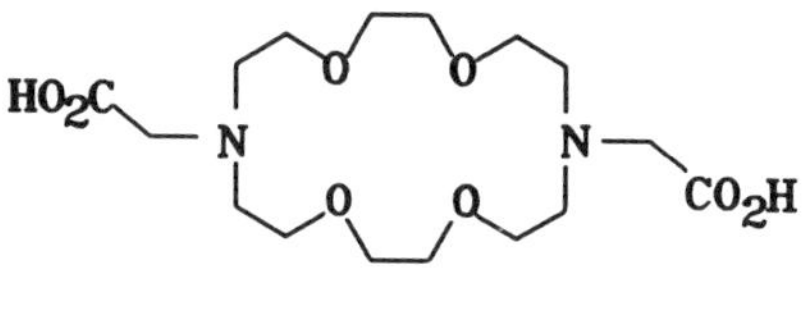

**27** 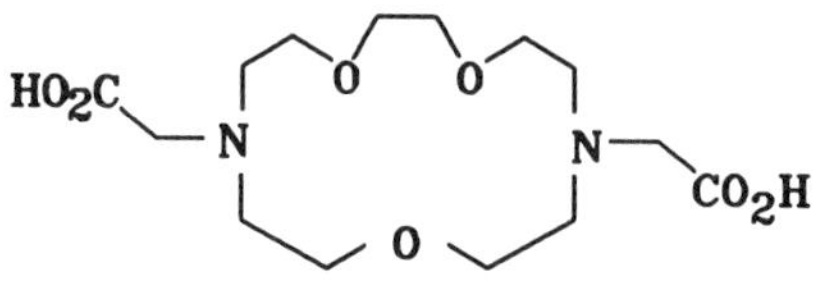 **28**

8.0 range.[127] These extraction studies with DAPADA and DACDA and TTA have been extended to europium(III), ytterbium(III) and lutetium(III). For the lighter lanthanides the formation of the ternary complex Ln(DAPDA/DACDA)TTA is the rate-determining step, whereas for the heavier lanthanides it is the dissociation of the $Ln(DAPDA/DACDA)^+$ complex that governs the rate of extraction.[128] Other crown carboxylates selectively extract lanthanide(III) ions into an organic phase comprised of chloroform and heptanol. The extraction constant of lutetium(III) is two orders of magnitude greater than that of sodium(I), and a selectivity ratio of six is found for lutetium(III) over lanthanum(III).[129] Linked-crowns have also been used for the liquid-liquid extraction of lanthanide(III) ions. The linked crown 1,8-dioxooctamethylene-*bis*-(4-benzo-15-crown-5) **29** that has been prepared from benzo-15-crown-5 and suberic acid extracts light lanthanides. This particular *bis*-(crown) gives much higher extractabilities of lanthanides than does the analogous *mono*-(crown), 4-acetylbenzo-15-crown-5, **30**.[130]

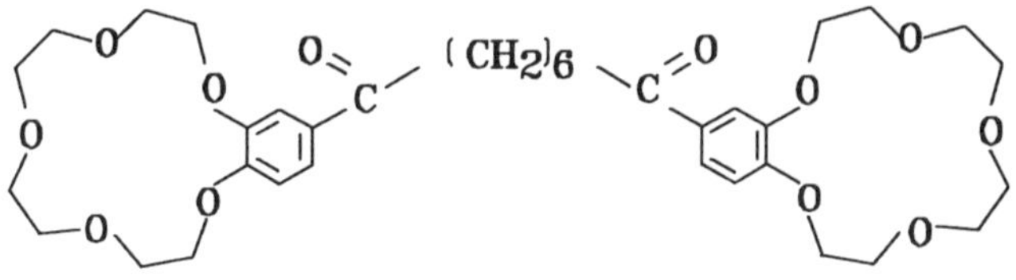

29

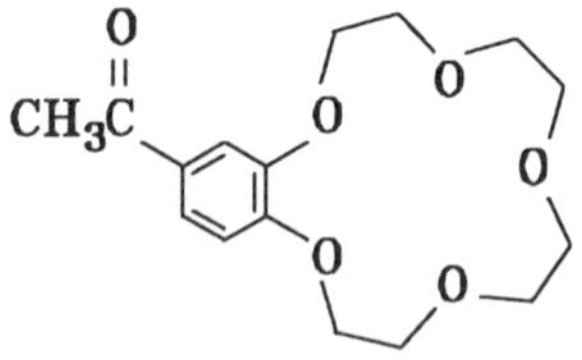

30

### 2.2.5. Nitrogen Donors

Nitrogen donor complexants are used as extractants for lanthanide(III) ions. These can be amines, amides, Schiff bases, and EDTA-type multidentate ligands. Like oxygen donors, nitrogen donors are hard bases which are expected to complex strongly with hard acid metals such as lanthanides. Amides are particularly favorable for lanthanides because they have both hard oxygen and nitrogen donor groups available for complexation. Two groups of amides that have been used for the extraction of lanthanide ions are malonamides and amide substituted calixarenes. In studying the tailoring of systems with malonamides to maximize the extraction constants, a mixed solvent comprising *n*-dodecane and an aliphatic alcohol is favored. In general the larger the carbon chain and the higher the concentration of the aliphatic alcohol, the greater the improvement in the phase compatibility. For neodymium(III) the extraction constant increases with increasing chain length of the aliphatic alcohol, but it is favored by lower concentrations of the alcohol.[131] Vibrational spectroscopy has been used to investigate the complexes of neodymium(III), thorium(IV) and uranium(VI) formed by extraction by malonamides, glutaramides and succinamides from aqueous nitrate solutions. The nitrate groups in the complexes are present in both monodentate and bidentate coordination modes.[132] Two hydrophobic malonamides butyl-*N,N*-dimethyl-*N,N*-diphenylmalonamide **31** and ethoxyethyl-*N,N*-dimethyl-*N,N*-diphenyl malonamide **32** are good extractants for lanthanide ions. The lanthanide

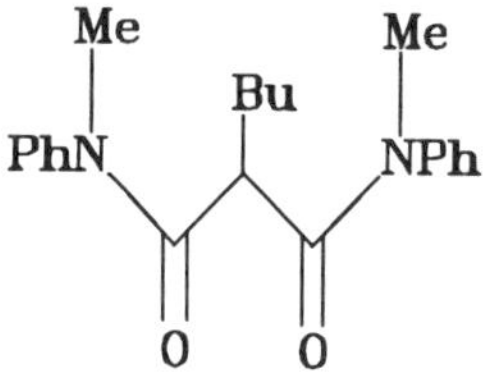

31 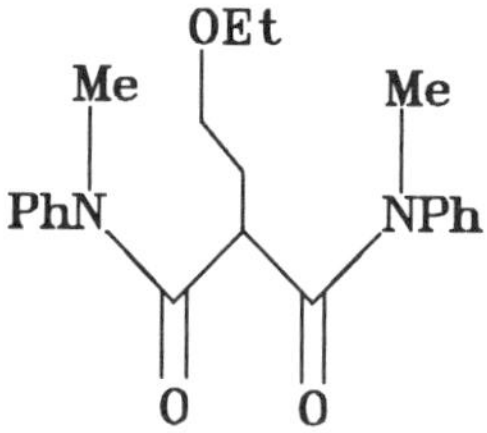 32

complexes that are extracted from aqueous nitrate solutions with these malonamides are ten-coordinate, with bonds to four oxygens of two malonamides, and six oxygens of three nitrates. Although malonamides with alkoxy groups rather than alkyl groups are better extractants for lanthanide(III) and actinide(III) ions, this structural evidence does not support the premise that this increase in the extraction constant is not a consequence of the binding of the alkoxy oxygens to the trivalent metal ions.[133]

Calixarenes with amide functionalities appended have been used as extractants for lanthanide(III) ions. One such calixarene **33** with diethylamide functionalities on the 1,3-positions of the narrow rim has high extraction constants for lanthanides into a dichloromethane phase.

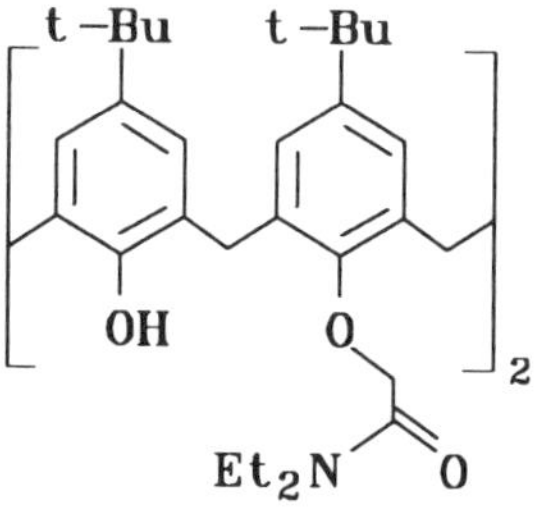

33

The lanthanide(III) ions are extracted as their picrate salts. The structures of the lanthanum(III), samarium(III), ytterbium(III) and lutetium(III) complexes show the lanthanide cation encapsulated in an eight-coordinate oxygen environment consisting of six oxygens from the calixarene, a water molecule, and a picrate ion.[134] A calix[4]arene **34** has

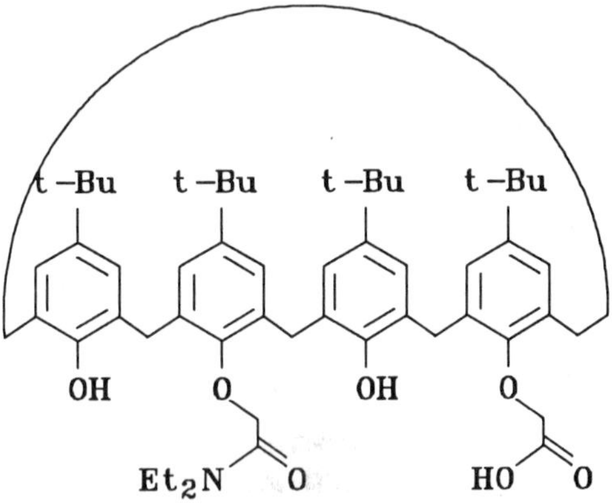

t-Bu t-Bu t-Bu t-Bu

OH O OH O

$Et_2N$ O HO O

34

been prepared with a diethylamide and a carboxylic acid group in the 1,3-positions of the narrow rim. For this compound the extraction of lanthanides into dichloromethane follows the sequence lanthanum(III) > europium(III) > thulium(III) > lutetium(III), which correlates with a preference for the larger lanthanides.[135] This mixed diethylamide-carboxylate calix[4]arene selectively extracts uranium(VI), and to a lesser extent, yttrium(III). A similar amide calix[4]arene bound to a polymeric support has been used as a selective adsorbent for uranium(VI).[136] These calixarene-derivatized resins can show the metal ion selectivities of the calixarene cavity, but with the advantages of a solid-phase support.

Schiff bases have been used as extractants. Three examples that have been used as extractants for lanthanides are *N,N-bis*-(5-nitrosalicylidene)ethylenediamine ($H_2$Nsalen, **35**), *N,N-bis*-(5-nitrosalicylidene)-*o*-phenylenediamine ($H_2$Nsaloph, **36**) and *N,N-bis*-(5-nitrosalicylidene) propylenediamine ($H_2$Nsalpn, **37**). These Schiff bases form anionic 1 : 2 (M : L) complexes with lanthanide(III) ions which can be extracted into nitrobenzene. The extraction order is **36** > **35** >> **37**,

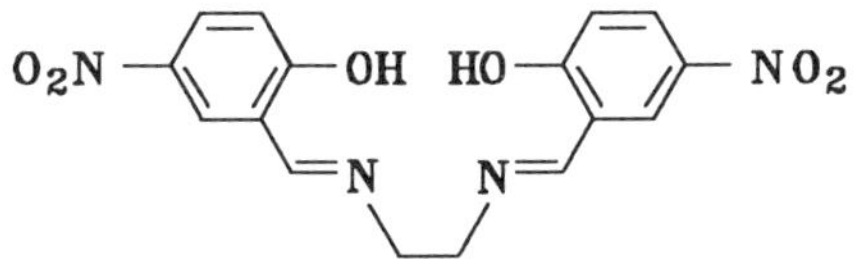

**35**

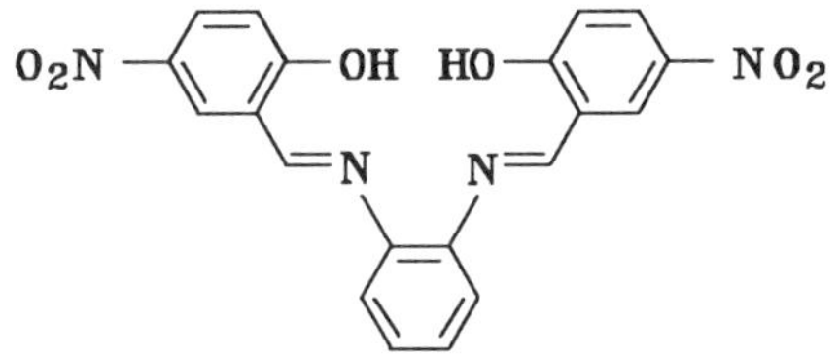

**36**

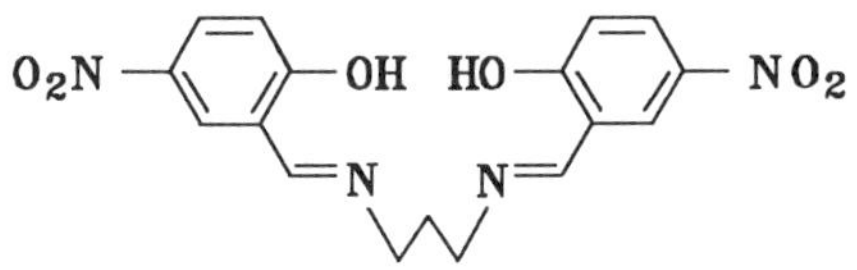

**37**

but the selectivity order is **35** > **36**. The extractabilities are also influenced by the cation. With decreased cation size the selectivity is improved, although the extraction constant is decreased.[145]

### 2.2.6. EDTA and Analogues

Chelates are liquid-liquid extractants for lanthanides. As with other metals, multidentate ligands of the EDTA type can be used effectively as both complexants and extractants. Altematively, EDTA can be used to elute lanthanide ions that have been bound to the protonated form of Dowex 50W-X8 ion exchange resin. Such a technique results in improvements in the separation of the erbium-holmium, holmium-dysprosium, dysprosium-terbium, and terbium-gadolinium pairs.[138] Similarly diethylenetriaminepentaacetic acid

(DTPA) has been used as the eluent with a cation exchanger. The main application for this system is the separation of yttrium(III) from erbium(III) and other lanthanides. Polyaminoacetic acids are generally preferred for this because of their high stability constants with the lanthanide ions, and also because of the large differences in the stability constants between adjacent lanthanides over a wide pH range. As a result, they are preferred over complexants such as citrate that were previously used. Although EDTA has an advantage over DTPA in being more soluble, the limited solubility of some of its complexes with the heavier lanthanides makes it less preferable than DTPA. When the lanthanides ($M^{3+}$) are eluted with a solution of ammonium DTPA, the solution reaction at the rear boundary of the absorbed band is given by equation 2:

$$M^{3+}_{(ads)} + nNH_4^+ + H_{5-n}DTPA^{n-} = 3NH_4^+{}_{(ads)} + MDTPA^{2-} + (5-n)H^+ + (n-3)H^+ \qquad (2)$$

whereas in the solution phase of the band the reactions are given by equations 3-6.

$$MDTPA^{2-} + H^+ = MHDTPA^- \qquad (3)$$

$$MHDTPA^- + H^+ = MH_2DTPA \qquad (4)$$

$$MH_2DTPA = M^{3+} + H_2DTPA^{3-} \qquad (5)$$

$$H_{5-n}DTPA^{n-} + nH^+ = H_5DTPA \qquad (6)$$

At the front edge $H_5DTPA$ is adsorbed by the resin (equation 7):

$$2H^+_{ads} + H_5DTPA = H_7DTPA^{2+}_{ads} \qquad (7)$$

and the reactions at the front edge are given by equation 8 and 9.

$$3H_7DTPA_{ads} + 2M^{3+} + 2H_2DTPA^{3-} = 2M^{3+}_{ads} + 5H_5DTPA \qquad (8)$$

$$H_7DTPA_{ads} + 2NH_4^+ = 2NH_4^+ + H_5DTPA + 2H^+ \qquad (9)$$

The quantity of yttrium(III) and erbium(III) in the resin band decreases with decreasing pH of the eluent.[139]

The combination of EDTA and a cation exchanger has been used for the separation of lanthanides by high performance liquid chromatography. The resolution is limited, however, by the slow decomplexation of the lanthanide(III) chelate complexes with EDTA.[140] Capillary tube isotachophoresis with EDTA as terminating electrolyte has also been used for the separation of lanthanides.[141]

### 2.2.7. Aromatic Chelates

Quinolinols have been used as extractants for lanthanides. With 8-quinolinol (HQ) in the presence of tetra-*n*-heptylammonium chloride ($R_4NCl$), lanthanum(III) extracts into chloroform as the simple ion-pair complex $[R_4N][LnQ_4]$. For praesodymium(III) and ytterbium(III), the extracted complex has the formula $[R_4N][LnQ_4].HQ$, where Ln is Pr or Yb.[142] An alternate procedure involves the use of 5,7-dibromo-8-quinolinol (HQ′). When this extractant is used alone, praesodymium(III), europium(III) and ytterbium(III) extract as the neutral complex LnQ′.HQ′ (Ln = Pr, Eu, Yb). If, however, 1,10-phenanthroline (phen) is present, the extracted complex has the formula LnQ′.phen. Alternatively, if tetra-*n*-heptylammonium chloride is present, the species is $[R_4N][LnQ_4']$. Although HQ′ forms complexes with lanthanides that are less stable than those with HQ, the higher acid dissociation constant of HQ′ allows it to be used in the lower pH range.[143] The reagent Kelex 100, 7-(1-vinyl-3,3,6,-tetramethylhexyl)-8-quinolinol (DDQ) behaves similarly in the extraction of lanthanides into chloroform. The extracted complexes are either $Ln(DDQ)_3$ or $Ln(DDQ)_3$.phen. The addition of phenanthroline or 8-quinolinol improves both the extraction constants and the selectivities.[144]

The compounds 1,8-bipyridine, 1,10-phenanthroline, and 2,2:6,2-terpyridine **38** are themselves good liquid-liquid extractants for lanthanide(III) ions, as are pyridines and pyrazolones. Lanthanides have been separated in a chromatographic process involving their sorption on an amphoteric resin comprised of vinylpyridine and iminodiacetic acid. The diffusion process in the resin can be improved by the use of ultrasound.[145] The chromatographic separation of lanthanides has also been achieved by precolumn chelating reversed-phase ion-pair high performance liquid chromatography with 2-(2-arsenophenylazo)-1,8-dihydroxy-7-(4-chloro-2,6-dibromophenylazo) naphthalene-3,6-disulfonic acid as the complexant. The mobile phase is the tetrabutylammonium ion in methanol.[146] Phenyl substituted derivatives of 2,2:6,2-terpyridine (**39-42**) have comparable extraction constants and separation factors for americium(III) and europium(III) as does 2,2:6,2-terpyridine itself. Extraction from a nitric acid solution into either *tert*-butylbenzene or hydrogenated tetrapropene gives $D_{Am}/D_{Eu}$ separation

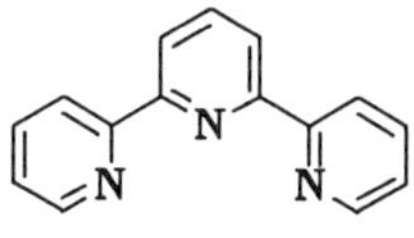

38

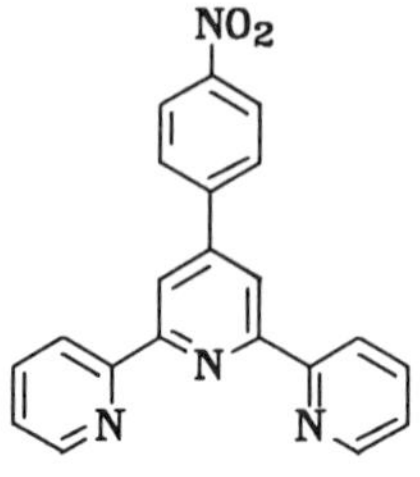

39

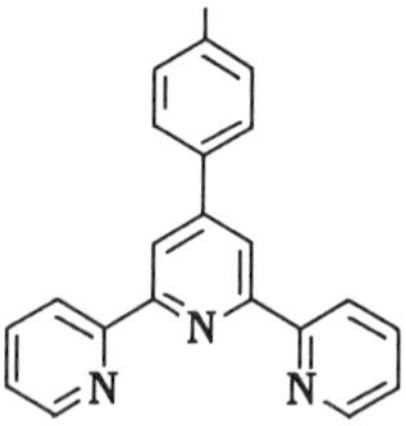

40

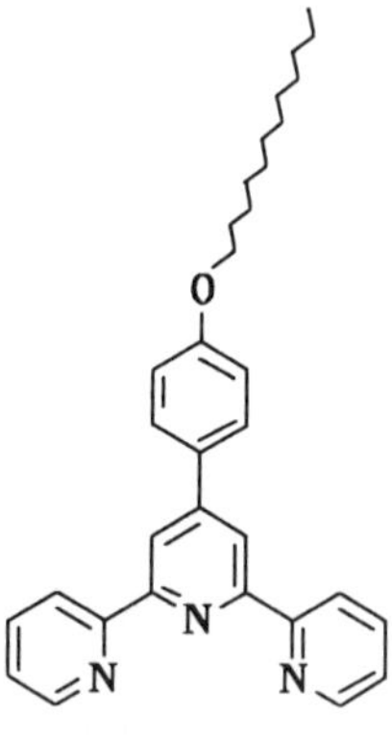

41

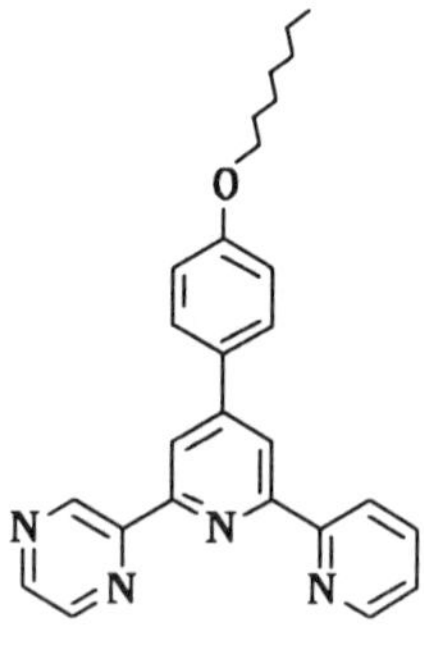

42

factors in the 7-9 range. These hydrophobic extractants have little or no solubility in the aqueous phase. For the lanthanides, six structural types have been identified for these complexants L.. These are: $Nd(NO_3)_3L(H_2O)$, $[M(NO_3)_2L_2]^+[M(NO_3)_4L]^-$ (M = Nd, Sm, Tb, Dy, Ho), $[M(NO_3)_3L(H_2O)].L$ (M = Ho, Er, Tm, Yb), $Tm(NO_3)L(H_2O)$ (9-coordinate), $Yb(NO_3)_3L$, and $[(H_2L)(NO_3)]^+[H_2L]^{2+}[La(NO_3)_6]^{3-}$.[147]

Pyrazolones have been used as liquid-liquid extractants for lanthanides. The acylpyrazolones are a particularly attractive series of compounds to target as liquid-liquid extractants into chloroform because they are stronger acids than are the 8-quinolinols. The compounds exist

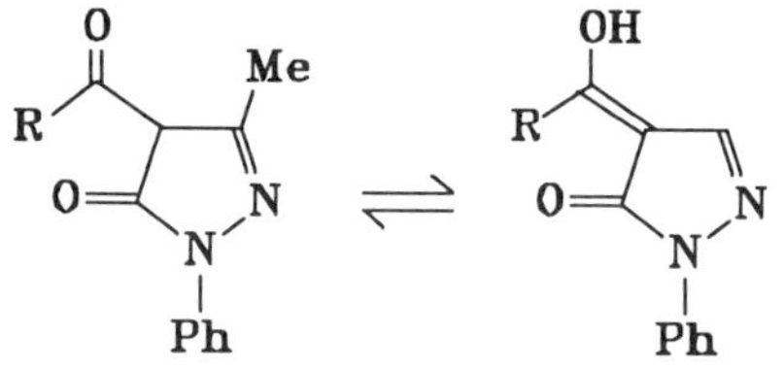

**43**

in a keto-enol equilibrium as shown for 1-phenyl-2-methyl-4-acyl-5-pyrazolone (HP, **43**). The lanthanide(III) ions of lanthanum(III), praesodymium(III), europium(III) and ytterbium(III) are extracted as the compounds $LnP_3$. In the presence of 1,10-phenanthroline or triooctylphosphine oxide, the extracted species are $LnP_3$.phen and $LnP_3$.TOPO respectively.[148] Further studies show that when the trifluoroacetyl pyrazolone (R = $CF_3$) is used, the complex that extracts from 0.1M sodium perchlorate solution has the stoichiometry $LnP_3.(TOPO)_2$. An advantage of using this particular pyrazolone is its high acidity ($pK_a$ = 3.12), which is caused by the strong electron-withdrawing nature of the trifluoromethyl group.[149] The extraction processes with these pyrazolones have been investigated by polarography and chronopotentiometry in a water-dichloroethane (DCE) two-phase mixture. The limiting current is directly proportional to the concentration of pyrazolone in the DCE, and is independent of the pH of the aqueous phase. The results correspond to the transfer of a singly charged species, and for a process where there is no adsorption or chemical reaction in the transfer process within the time scale of the electrochemical experiment. The data strongly indicate that the extractable metal species $LnP_3$ or $LnP_3$.TOPO are formed in the aqueous phase.[150] Centrifugal partition chromatography (CPC) has also been used in the separation of lanthanide(III) ions with pyrazolone extractants and toluene as the organic phase. The CPC efficiencies are mainly limited by the slow dissociation of the complexes $LnP_3$ which occurs exclusively at the toluene-water interface. These systems represent the first examples of a separation being achieved by multistage countercurrent distribution where the efficiencies are mainly limited by interfacial processes analogous to conventional liquid chromatography. Major improvements in the efficiencies of these separations is obtained by addition of the neutral surfactant Triton X-100 to the toluene phase, and the metallochromic indicator Arsenazo III (AZ) to the aqueous phase. The enhancement from Triton X-100 is due to the increased interfacial areas resulting from the adsorption of the surfactant, and that of AZ is due to the increased interfacial area and interfacial catalysis of the formation and dissociation of the $MP_3$ complexes. The addition of AZ provides much higher efficiencies than does the addition of Triton X-100. Thus the addition of Triton X-100 does not significantly alter the selectivities of the ligands for the lanthanide ions, but the addition of AZ results in a lowering of the selectivites[151, 152]

Lanthanides can also be extracted using 1,7-diaza-4,10,13-trioxacyclopentadecane-*N,N*-diacetic acid ($H_2L$) as the chelate, and a mixture of thenoyltrifluoroacetone (TTA) and benzene as the organic phase. The use of this ionizable macrocycle allows for the thermodynamically unfavored LaL(TTA) complex to be extracted faster than the favored complexes $Yb(TTA)_3$ or $Eu(TTA)_3$.[153]

### 2.2.8. Phophoramides

Phosphoramides are used as extractants for both lanthanides and actinides. The liquid-liquid extraction of nuclear wastes is currently carried out industrially using the TRUEX process.[154] This uses *N, N*-(diisobutylcarbamoylmethyl) octylphenylphosphine oxide (CMPO, **44**) as extractant. Both the C=O and P=O groups act as ligating functions to

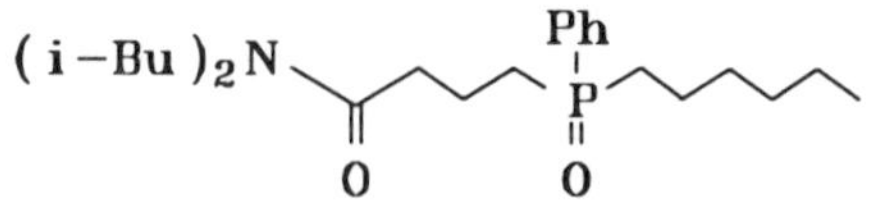

(i-Bu)$_2$N ... Ph P O O

44

the metal center, and the extracted complex contains three molecules of CMPO per cation. Since calixarenes are good phase transfer agents, the phosphoramide moiety has also been appended to both the narrow and wide rims of a calixarene, and the products used as liquid-liquid extractants for lanthanides and actinides. These calix[4]arenes can have alkylether (OR) groups on the narrow rim, and CMPO-like functionalities (-NH-C(O)-$CH_2$-P(O)$Ph_2$) on the wide rim (**45**, **46**). The

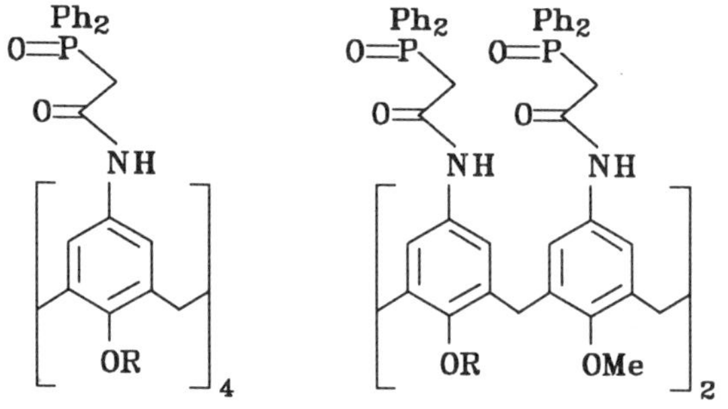

45 46

R groups have chain lengths from C-5 to 18. For comparison, the acyclic analogs **47** (R = $C_{18}H_{37}$) and **48** (R = $C_{18}H_{37}$, n = 0, 1) have been synthesized. All of these compounds extract europium(III), thorium(III), neptunium(III), plutonium(III) and americium(III) from aqueous nitric

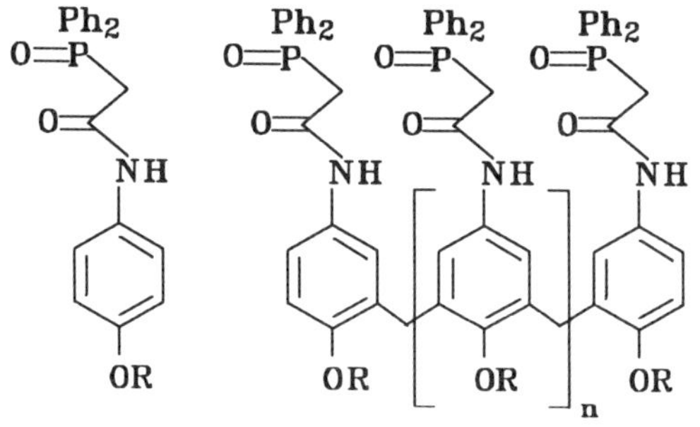

47 48

acid into methylene chloride or *o*-nitrophenyl hexyl ether, and are considerably better than is CMPO itself. This improvement is the greatest with the cyclic calixarene derivatives, which lends support to their being particularly good compounds as liquid-liquid extractants.[155] Calix[4]arenes substituted by acetamidophosphine oxide groups at the wide rim and pentoxy groups on the narrow rim **49** show high extractabilities for trivalent cations,and have good selectivity for the light lanthanides and actinides. Since this selectivity can reach nearly three

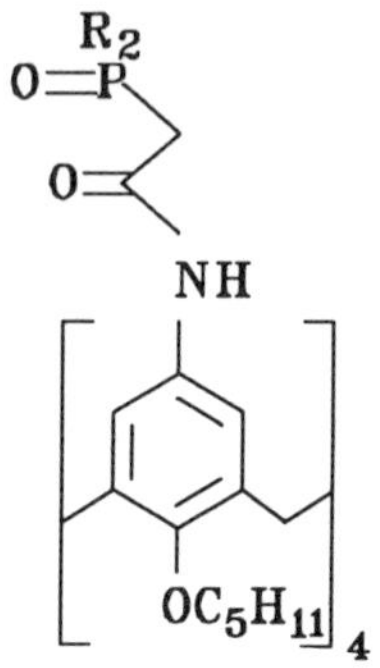

49

orders of magnitude, these compounds are potentially useful for the extraction of actinides or light lanthanides from heavy lanthanides.[156] Calix[4]arenes with CMPO-type functionalities on the narrow rim **50** have also been used as liquid-liquid extractants for lanthanides. These

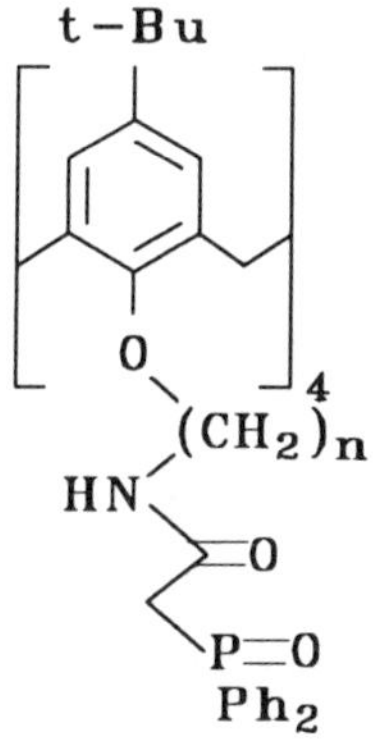

50

molecules are designed with a methylene group spacer $(CH_2)_n$ at the narrow rim in order that the effect of the spacer length n on the extraction properties of the calix[4]arene can be investigated. For the extraction of lanthanum(III), europium(III), ytterbium(III) and thorium(IV) into dichloromethane from 1M $HNO_3$, the extraction constant increases as the spacer length is increased. By contrast, with the derivative having the CMPO-type derivative on the wide rim, extraction into *o*-nitrophenyl hexyl ether shows no decrease in the distribution ratio for higher nitric acid concentrations.[157]

### 2.2.9. Miscellaneous Methods

Several other miscellaneous methods and techniques have been used in the extraction and separation of lanthanides. One approach used a photoredox method for separating either cerium(III) or europium(III) from lanthanide mixtures in aqueous solution. For europium(III) in aqueous isopropanol solution, photoreduction to europium(II) occurs. Since the solution also contains potassium sulfate, the europium(II) precipitates as $EuSO_4$.[158] By contrast, cerium(III) can be photooxidized to cerium(IV), and then precipitated with iodate. Separation factors of over a hundred are achieved, with product yields in excess of 95%. Quantum yields in the region of 0.1 are observed.[159]

In addition to the separation methods discussed in this chapter, other methods have been applied to the extraction and separation of lanthanides. Among the chromatographic methods used are the reversed-phase separation on silica gel impregnated with Amberlite LA-2,[160] and HPLC on bonded-phase strong-acid ion exchangers.[161] Capillary electrophoresis has also been used for the separation of lanthanide ions.[162] This technique is used in conjunction with arsenazo III as complexant because of the strong absorption of the complexes.[163]

A metallochromic indicator method has been used to simultaneously obtain back extraction kinetics and extraction equilibrium data. The method is illustrated in the use of Arsenazo III to characterize *bis*-(2,4,4-trimethylpentyl) phosphinic acid lanthanide(III) complexes in surfactant micelles.[164]

## REFERENCES

1. G. W. Mason, H. E. Griffin, *ACS Sympos. Ser.*, **1980**, *117*, 89.
2. U. Wenzel, C. L. Branquinho, D. Herz, G. Ritter, *ACS Sympos. Ser.*,**1980**, *117*, 533.
3. W. J. McDowell, B. L. McDowell, *ACS Sympos. Ser.*, **1992**, *509*, 206.
4. T. S. Zvarova, I. Zvara, *J. Chromatogr.*, **1970**, *49*, 290.

5. D. Dubuquoy, S. Gusmini, D. Poupard, M. Verry, *J. Chromatogr.*,**1971**, *57*, 455.
6. T. Oi, Y. Sakuma, M. Okamoto, M. Maeda, *J. Chromatogr.*, **1982**, *248*, 281.
7. A. Nakagawa, Y. Sakuma, M. Okamoto, M. Maeda, *J. Chromatogr.*, **1983**, *256*, 231.
8. F. Y. C. Huang, P. V. Bradley, E. Lindgren, P. Guerra, *Environ. Sci. Technol.*, **1998**, *32*, 379.
9. Y. Marcus, Z. Kolarik, *J. Chem. Eng. Data,* **1973**, *18*, 155.
10. V. Friehmelt, C. Frydrych, R. Gauglitz, G. Marx, *Inorg. Chim. Acta.*, **1987**, *140*, 265.
11. S. Iso, Y. Meguro, Z. Yoshida, *Chem. Lett.*, **1995**, 365.
12. Y. Meguro, S. Iso, Z. Yoshida. *Anal. Chem.*, **1998**, *70*, 1262.
13. C. M. Wai, Y. Lin, M. Ji, K. L. Toews, N. G. Smart. ACS *Sympos. Ser.* **1999**, *716*, 390.
14. Y. Lin, C. M. Wai, F. M. Jean, R. D. Braver, *Environ. Sci. Technol.*, **1994**, *28*, 1190.
15. L. W. Green, N. L. Elliot, T. H. Longhurst *Anal. Chem.*, **1983**, *55*, 2394.
16. C. Apostolidis, H. Bokelund, A. Moens, M. Ougier, *Inorg. Chim. Acta.*, **1987**, *140*, 253.
17. A. E. Lemire, A. F. Janzen, K. Marat, *Inorg. Chim. Acta.*, **1985**, *110*, 237.
18. T. C. Huang, C. T. Huang, *Ind. Eng. Chem. Res.*, **1988**, *27*, 1675.
19. D. A. Becker, P. D. LaFleur, *Anal. Chem.*, **1972**, *44*, 1508.
20. E. P. Horwitz, W. H. Delphin, C. A. A. Bloomquist, G. F. Vandergrift, *J. Chromatogr.*, **1976**, *125*, 203.
21. E. P. Horwitz, M. L. Dietz, R. Chiarizin, H. Diamond, *Anal. Chim. Acta.*, **1992**, *266*, 25.
22. E. P. Horwitz, M. L. Dietz, R. Chiarizin, H. Diamond, *Anal. Chim. Acta.*, **1995**, *310*, 63.
23. T. Yokoyama, A. Makishima, E. Nakamura, *Anal. Chem.*, **1999**, *71*, 135.
24. D. M. Petkovic, A. Lj. Ruvarac, J. M. Konstantinovic, V. K. Trujic, *JCS, Dalton Trans.* **1973**, 1649.
25. T. Sato, *ACS Sympos. Ser.* **1980**, *117*, 117.
26. M. Shamispur, A. R. Ghiasvand, Y. Yamini, *Anal, Chem.*, **1999**, *71*, 4892.
27. Y. Lin, N. G. Smart, C. M. Wai, *Environ. Sci. Technol.*, **1995**, *29*, 2706.
28. J. T. van Elteren, C. M. G. van den Berg, H. Zhang, T. D. Martin, E. P. Achterberg, *Anal. Chem.*, **1995**, *67*, 3903.
29. M. P. Harrold, A. Siriraks, J. Riviello, *J. Chromatogr.*, **1992**, *602*, 119.
30. N. Macka, P. Nesterenko, P. Andersson, P. R. Haddad, *J. Chromatogr.*, **1998**, *803*, 279.
31. I. Hodara, I. Balovka, *Anal. Chem.*, **1971**, *43*, 1213.
32. R. Kuroda, K. Oguma, H. Watanabe, *J. Chromatogr.*, **1973**, *86*, 167.
33. T. Shimizu, A. Muto, *J. Chromatogr.*, **1974**, *88*, 351.
34. I. Tabushi, Y. Kobuke, T. Nishiya, *Nature,* **1979**, *280*, 665.
35. A. Saito, G. R., Choppin, *Anal. Chem.*, **1983**, *55*, 2454.
36. J. R. Stokely, Jr., F. L. Moore, *Anal. Chem.*,**1967**, *39*, 994.
37. F. H. Al-Dabbagh, J. I. Bullock, *Inorg. Chim. Acta.*, **1985**, *110*, 69.
38. J. F. Kennedy, S. A. Baker, A. W. Nicol, A. Hawkins, *JCS Dalton Trans.*, **1973**, 1129.
39. T. Nakamura, Y. Matsuda, C. Miyake, *Mater. Res. Symp. Proc.* 353 (Scientific Basis for Nuclear Waste Management XVIII, Pt. 2) 1293 (1995).

40. C. Musikas, H. Hubert, *Solv. Extr. Ion. Exch.*, **1987**, *5*, 877.
41. T. H. Siddall III, *J. Inorg. Nucl. Chem.*, **1963**, *25*, 883.
42. C. Musikas, *Inorg. Chim. Acta.*, **1987**, *140*, 197.
43. C. Musikas, *Sep. Sci. Technol.*, **1988**, *23*, 1211.
44. M. C. Charbonnel, C. Musikas, *Solv. Extr. Ion Exch.*, **1989**, *7*, 1007.
45. G. Thiollet, C. Musikas, *Solv. Extr. Ion Exch.*, **1989**, 7, 813.
46. C. Cuillerdier, C. Musikas, P. Hoel, L. Nigond, X. Vitart, *Sep. Sci. Technol.*, **1991**, *26*, 1229.
47. C. Musikas, N. Condamines, C. Cuillerdier, *Anal. Sci.*, **1991**, *7*, 11.
48. C. Cuillerdier, C. Musikas, L. Nigond, *Sep. Sci. Technol.*, **1993**, *28*, 155.
49. G. M. Nair, D. R Prabhu, G. R. Mahajan, J. P. Shukla, *Solv. Extr. Ion Exch.*, **1993**, *11*, 831.
50. G. M. Nair, D. R. Prabhu, G. R. Mahajan, *J. Radioanal. Nucl. Chem.*, **1994**, *186*, 47.
51. L. Nigond, C. Musikas, C. Cuillerdier, *Solv. Extr. Ion Exchange*, **1994**, *12*, 287.
52. Y. Sasaki, G. R. Choppin, *Anal. Sci.*, **1996**, *12*, 225.
53. T. Nakamura, C. Miyake, *Solv. Extr. Ion Exch.*, **1995**, *13*, 253.
54. Y.-S. Wang, C.-H. Shen, Y-h. Yang, J.-K. Zhu, B.-R. Bao, *J. Radioanal. Nucl. Chem.*, **1996**, *213*, 199.
55. Y.-S. Wang, G.-X. Sun, D.-F. Xie, B.-R. Bao, W.-G. Cao, *J. Radioanal. Nucl. Chem.*, **1996**, *214*, 67.
56. L. Nigond, N. Condamines, P. Y. Cordier, J. Livet, C. Madic, C. Cuillerdier, C. Musikas, *Sep. Sci. Technol.*, **1995**, *30*, 2075.
57. B. K. McNamara, G. J. Lumetta, B. M. Rapko, *Solvent Extr. Ion Exch.* **1999**, *17*, 1403.
58. G. J. Lumetta, B. K. McNamara, B. M. Rapko,, J. W. Hutchison, *Inorg. Chim. Acta*, **1999**, *293*, 195.
59. P. B. Ruikar, M. S. Nagar, *Polyhedron*, **1995**, *14*, 3125.
60. T. Nakamura, C. Miyake, *J. Alloys Compd.*, **1996**, *233*, 1.
61. T. H. Siddall III, M. L. Good, *J. Inorg. Nucl. Chem.*, **1967**, *29*, 149.
62. G. J. Lumetta, B. K. McNamara, B. M. Rapko, R. L. Sell, R. D. Rogers, G. Broker, J. W. Hutchison, *Inorg. Chim. Acta*, **2000**, *309*, 103.
63. H. G. M. Edwards, E. Hicknott, M.A. Hughes, *Spectrochim. Acta, Part A*, **1997**, *53A*, 43.
64. C. Pohlandt, J. S. Fritz, *J. Chromatogr.*, **1979**, *176*, 189.
65. K. Shakir, K. Benyamin, M. Aziz, *Can. J. Chem.*, **1984**, *62*, 51.
66. M. D. Plamieri, J. S. Fritz, *Anal. Chem.*, **1988**, *60*, 2244.
67. M. Y. Khuhawar, S. N. Lanjwani, *J. Chromatogr.*, **1996**, *740*, 296.
68. T. S. Franczyk, K. R. Czerwinski, K. N. Raymond, *J. Am. Chem. Soc.*, **1992**, *114*, 8138.
69. E. A. Huff, *Spectrochim Acta, Part B*, **1987**, *42B*, 275.
70. H. Egawa, T. Nonaka, M. Nakayama, *Ind. Eng. Chem. Res.*, **1990**, *29*, 2273.
71. M. C. Duff, C. Amrhein, *J. Chromatogr. A.*, **1996**, *743*, 335.
72. I. Ortiz, A. I. Alonso, A. M. Urtiaga, M. Demircioglu, N. Kocacik, N. Kabay, *Ind. Eng. Chem. Res.*, **1999**, *38*, 2450.
73. O. A. Vita, C. R. Walker, C. F. Trivisonno, R. W. Sparks, *Anal. Chem.*, **1970**, *42*, 465.
74. K. Gonda, A. Ohnishi, D. Narita, T. Murase, *J. Chromatogr.*, **1971**, *55*, 395.
75. Y. S. Kim, H. Zeithlin, *Anal. Chem.*, **1971**, *43*, 1390.
76. W. Wang, J. Lin, A. Wang, P. Z. Zheng, M. Wang, B. Wang, *Inorg. Chim.*

*Acta.*, **1988**, *149*, 151.
77. H. Boerrigter, W. Verboom, B. N. Reinhoudt, *J. Org. Chem.*, **1997**, *62*, 7148.
78. "CALIX 2001", Vicens, Böhmer and Harrowfield, eds., Chapter 22: Phase Transfer Extraction of Heavy Metals, D. M. Roundhill and J. Y. Shen, Kluwer, 2001 in press.
79. T. N. Lambert, G. D. Jarvinen, A. S. Gopalan, *Tet. Lett.*,**1999**, *40*, 1613.
80. I. Nukatsuka, M. Taga, H. Yoshida, *J. Chromatogr.*, **1981**, *205*, 95.
81. T. Hirokawa, W. Xia, Y. Kiso, *J. Chromatogr.*, **1995**, *689*, 149.
82. T. Hirokawa, Y. Hashimoto, *J. Chromatogr.* A, **1997**, *772*, 357.
83. M. Vobecky, *J. Chromatogr.*, **1989**, *478*, 446.
84. Q. Mao, Y. Hashimoto, Y. Manabe, N. Ikuta, F. Nishiyama, T. Hirokawa, *J. Chromatogr.* A, **1998**, *802,* 203 .
85. E. A. Jones, H. S. Bezuidenhout, J. F. Van Stoden, *J. Chromatogr.*, **1991**, *537*, 277.
86. S. Elchuk, K. I. Burns, R. M. Cassidy, C. A. Lucy, *J. Chromatogr.*, **1991**, *558*, 197.
87. J. P. Quinche, *Helv. Chim. Acta.*, **1973**, *56*, 1073.
88. F. Schoebrechts, E. Merciny, G. Duyckaerts, *J. Chromatogr.* **1973**, *79*, 293.
89. P. N. Nesterenko, P. Jones, *J. Chromatogr.* **A**, **1998**, *804*, 223.
90. S. P. Tanner, G. R. Choppin, *Inorg. Chem.*, **1968**, *7*, 2046.
91. M. R. Buchmeiser, R. Tessadri, G. Seeber, G. K. Bonn, *Anal. Chem.*, **1998**, *70*, 2130.
92. G. Seeber, P. Brunner, M. R. Buchmeiser, G. K. Bonn, *J. Chromatogr.* **A**, **1999**, *848*, 193.
93 J. E. Bigelow, E. D. Collins, L. J. King, *ACS Sympos. Ser.* **1980**, *117*, 147.
94. D. B. Martin, D. G. Pope, *Anal. Chem.*, **1982**, *54*, 2552.
95. T. D. Filer, *Anal. Chem.*, **1974**, *54*, 608.
96 L. Cheng, Y. Yang, M. Luo, D. Zhang, *Inorg. Chim. Acta.*, **1987**, *130*, 119.
97. T. Harada, M. Smutz, R. G. Bautista, *J. Chem. Eng. Data,* **1972**, *17*, 203.
98. G. Brunisholz, W. Hirsbrunner, *Helv. Chim. Acta,* **1974**, *57*, 2483.
99. Z. Ma, L. Zhang, S. Han, *J. Chromatogr. A,* **1997**, *766*, 282.
100. S. Nakamura, H. Hashimoto, K. Akiba, *J. Chromatogr.*, **1997**, *789*, 381.
101. E. Kitazume, M. Bhatnagar, Y. Ito, *J. Chromatogr.*, **1991**, *538*, 133.
102. A. Berthod, J. Xiang, S. Alex, C. Gonnet-Collet, *Can. J. Chem.*, **1996**, *74*, 277.
103. K. E. Laintz, E. Tachikawa, *Anal. Chem.*, **1994**, *66*, 2190.
104. Y. Lin, C. M. Wai, *Anal. Chem.*, **1994**, *66*, 1971.
105. C. M. Wai, Y. Lin, M. Ji, K. L. Toews, N. G. Smart, *ACS Sympos. Ser.* **1999**, *716*, 390
106. S. Kutun, A. Akseli, *J. Chromatogr.*, **1999**, *847*, 261.
107. M. Burgard, B. Ceccaroll, *J. Phys. Chem.*, **1982**, *86*, 4817.
108. S. Nakamura, N. Suzuki, *Inorg. Chim. Acta.*, **1986**, *114*, 101.
109. N. Suzuki, J. Kodera, H. Imura, *Inorg. Chim. Acta.,* **1987**, *128*, 261.
110. Y. Hasegawa, E. Ishiwata, T. Ohnishi, G. R. Choppin, *Anal. Chem.*, **1999**, *71*, 5060.
111. T. H. LeQuyen, S. Umetani, M. Suzuki, M. Matsui, *JCS, Dalton Trans,* **1997**, 643.
112. J. N. Mathur, P. K. Khopkar, *Inorg. Chim. Acta.*, **1985**, *109*, L19.
113. S. Umetani, Y. Kawase, T. H. LeQuen, M. Matsui, *Inorg. Chim. Acta.,* **1998**, *267*, 201.
114. Y. Lin, R. D. Brauer, K. E. Laintz, C. M. Wai, *Anal. Chem.*, **1993**, *65*, 2549.

115. C. A. Burgett, J. S. Fritz, *Anal. Chem.*, **1972**, *44*, 1738.
116. T. Tomida, K. Inagawi, S. Masudo, *Chem. Lett.*, **1991**, 1253.
117. H. Wu, Y. Lin, N. G. Smart, C. M. Wai, *Anal. Chem.*, **1996**, *68*, 4072.
118. T. Cecconie, H. Freiser; *Anal. Chem.*, **1990**, *62*, 622.
119. S. Umetani, Y. Kawase, T. H. LeQuyen, M. Matsui, *Chem. Lett.*,**1997**, 771.
120. K. S. R. Murthy, R. J. Krupadam, Y. Anjaneyulu, *J. Chromatogr Sci.*, **1998**, *36*, 595.
121. J. W. O'Laughlin, D. F. Jensen, *Anal. Chem.*, **1969**, *41*, 2010.
122. K. Nakagawa, S. Okada, Y. Inoue, A. Tai, T. Hakushi, *Anal. Chem.*,**1988**, *60*, 2527 .
123. C. D. Tran, W. Zhang, *Anal. Chem.*, **1990**, *62*, 830.
124. S. Tsurubou, M. Mizutani, Y. Kadota, T. Yamamoto, S. Umetani, T. Saski, T. H. LeQuyen, M. Matsui, *Anal. Chem.*, **1995**, *67*, 1465.
125. Y. Inoue, K. Nakagawa, T. Hakushi, *JCS, Dalton Trans.*, **1993**, 1333.
126. D. D. Ensor, G. R. McDonald, C. G. Pippin, *Anal. Chem.*, **1986**, *58*, 1814.
127. V. K. Manchanda, C. A. Chang, *Anal. Chem.*, **1986**, *58*, 2269.
128. V. K. Manchanda, C. A. Chang, *Anal. Chem.*, **1987**, *59*, 813.
129. J. Tang, C. M. Wai, *Anal. Chem.*, **1986**, *58*, 3233.
130. Y. Inoue, K. Nakagawa, T. Hakushi, *JCS, Dalton Trans.*, **1993**, 2279.
131. T. Nakamura, Y. Matsuda, C. Miyake, *Mater. Res. Soc. Symp. Proc.* 353 (Scientific Basins for Nuclear Waste Management XVIII, Pt. 2) 1293 (1995).
132. H. G. M. Edwards, E. Hickmott, M. A. Hughes, *Spectrochim. Acta. Part A*, **1997**, *53A*, 43.
133. P. B. Iveson, M. G. B. Drew, M. J. Hudson, C. Madic, *JCS, Dalton Trans.*, **1999**, 3605.
134. P. D. Beer, M. G. B. Drew, M. Kan, P. B. Leeson, M. I. Odgen, G. Williams, *Inorg. Chem.*, **1996**, *35*, 2202 .
135. P. D. Beer, M. G. B. Drew, A. Grieve, M. Kan, P. B. Leeson, G. Nicholson, M. I. Oden, G. Williams, *JCS, Chem. Comm.*, **1996**, 1117.
136. P. D. Beer, M. G. B. Drew, D. Hesek, M. Kan, G. Nicholson, P. Schmitt, P. D. Sheen, G. Williams, *JCS, Chem. Comm.*, **1998**, 2783.
137. N. Hirayama, I. Takeuchi, T. Honjo, K. Kubono, H. Kokusen, *Anal. Chem.*, **1997**, *69*, 4814.
138. J. E. Powell, H. R. Burkholder, *J. Chromatogr.*, **1968**, *36*, 99.
139. L. Kogan, R. Ratner, *J. Chromatogr.*, **1971**, *62*, 449.
140. F. Schoebrechts, E. Merciny, G. Duyckaerts, *J. Chromatogr.*, **1979**, *179*, 63.
141. S. Tanaka, T. Kaneta, K. Nishima, H. Yoshida, *J. Chromatogr.*, **1991**, *540*, 475.
142. M. Kawashima, H. Freiser, *Anal. Chem.*, **1981**, *53*, 284.
143. O. Tochiyama, H. Freiser, *Anal. Chem.*, **1981**, *53*, 874.
144. E. Yamada, H. Freiser, *Anal. Chem.*, **1981**, *53*, 2115.
145. N. N. Matorina, L. V. Shepetyuk, T. I. Bakaeva, L. V. Karlina, O. V. Kryuchkova, *J. Chromatogr.*, **1986**, *365*, 89.
146. X. Zhang, M. Wang, J. Cheng, *J. Chromatogr. Sci.*, **1988**, *26*, 517.
147. M. G. B. Drew, P. B. Iveson, M. J. Hudson, J. O. Liljenzin, L. Spjuth, P.-Y. Cordier, A. Enarsson, C. Hill, C. Madic, *JCS, Dalton Trans.*, **2000**, 821.
148. Y. Sasaki, H. Freiser, *Inorg. Chem.*, **1983**, *22*, 2289.
149. S. Umetani, H. Freiser, *Inorg. Chem.*, **1987**, *26*, 3179.
150. L. Sinru, H. Freiser, *Anal. Chem.*, **1987**, *59*, 2834.
151. G. Ma, H. Freiser, S. Muralidharan, *Anal. Chem.*, **1997**, *69*, 2835.

152. G. Ma, H. Freiser, S. Muralidharan, *Anal. Chem.*, **1997**, *69*, 2827.
153. C. A. Chang, V. K. Manchanda, J. Peng, *Inorg. Chim. Acta.*, **1987**, *130*, 117.
154. E. P. Horwitz, D. G. Kalina, H. Diamond, D. G. Vandergrift, W. W. Schultz, *Solv. Extr. Ion Exch.*, **1985**, *3*, 75.
155. F. Arnaud-Neu, V. Böhmer, J.-F. Dozol, C. Grüttner, R. A. Jakobi, D. Kraft, O. Mauprivez, H. Rouquette, M.-J. Schwing-Weill, N. Simon, W. Vogt, *JCS, Perkin Trans.2*, **1996,** 1175.
156. L. H. Delmau, N. Simon, M.-J. Schwing-Weill, F. Arnaud-Neu, J.-F. Dozol, S. Eymard, B. Tournois, V. Böhmer, C. Grüttner, C. Musigmann, A. Tunayar, *JCS, Chem. Comm.*, **1998**, 1627.
157. S. Barboso, A. Garcia Carrera, S. E. Matthews, F. Arnaud-Neu, V. Böhmer, J.-F. Dozol, H. Rouquette, M.-J. Schwing-Weill, *JCS. Perkin Trans.*, **1999**, *2*, 719.
158. T. Donohue, *J. Chem. Phys.*, **1977**, *67*, 5402.
159. T. Donohue, *Chem. Phys. Lett.*, **1979**, *61*, 601.
160. T. Shimizu, *J. Chromatogr.*, **1974**, *96*, 262.
161. S. Elchuk, R. M. Cassidy, *Anal. Chem.*, **1979**, *51*, 1434.
162. M. Chen, R. M. Cassidy, *J. Chromatogr.*, **1993**, *640*, 425.
163. M. Macka, P. Nesterenko, P. Andersson, P. R. Haddad, *J. Chromatogr. A*, **1998**, *803*, 279.
164. K. Inaba, S. Muralidharan, H. Freiser, *Anal. Chem.*, **1993**, *65*, 1510.

# 10

## EXTRACTION OF ANIONS AND OXYANIONS

### 1. INTRODUCTION

The synthesis of complexants and hosts for specific metals is an important goal for high value toxic metals. For cations, the metal can be directly bound to a ligating group on the host, but for anions or uncharged molecules it may not be possible to get a direct interaction between the metal center and the atoms in the host. Although numerous molecules act as hosts and complexants for cations, fewer molecules function as hosts for anions.[1-4] Recently, however, a number of chemically modified calixarenes have been synthesized that can be used as hosts for simple anions, and this has helped to stimulate renewed interest in developing hosts for anions.[5,6] From an environmental viewpoint, a series of anions for which selective hosts would be useful are the oxyanions. Three such oxyanions are chromate, selenate and phosphate. Under basic solution conditions each of these anions has a *pseudo*-tetrahedral structure, therefore conceptually they can be viewed as a group where hosts may be interchangeable. Chromium(VI) is somewhat of an exception, however, because of the reversible formation of a dimer at low acidity. Furthermore, since remediation of chromate frequently involves its reduction to chromium(III), it is necessary to be aware of methods for extracting this reduced state of chromium when considering the removal of chromium(VI). The selenate, chromate, and dichromate oxyions are all environmentally important, both because of their high toxicity,[7-12] and because of their localized presence in soils and waters.[13]

Selenium is a toxic element that is commonly found in its anionic selenium(IV) or selenium(VI) form. Because of similarities with the chemistry of sulfur, selenium is frequently present in sulfur-containing minerals. From an environmental viewpoint, therefore, the element becomes problematic in operations such as the burning of high sulfur coal, or in the desulfurization of oil products. The similar chemistries of selenium and sulfur can also result in toxicological problems because biochemical systems often fail to discriminate between the two elements. Also in developing a criterion maximum concentration (CMC) for selenium, the U.S. Environmental Protection Agency has not discriminated between the different chemical forms of selenium since selenate(VI), selenite(IV), and organoselenium compounds can all interconvert.[14]

In oxygenated alkaline water, selenate ($SeO_4^{2-}$) is thermodynamically favored, although both selenite ($SeO_3^{2-}$) and organoselenium compounds can also be present. It is also believed that the acute toxicities of the different forms of selenium are additive; that is, these forms are more toxic together than they are separately. The conversion between selenate and selenite simply involves redox chemistry between the selenium(VI) and (IV) forms. For conversion to organic forms the pathway usually involves formation of the selenide ($Se^{2-}$) ion, which is subsequently alkylated to form compounds of the organoselenide ($RSe^-$) type. The metabolism of selenite to dimethyl selenide involves the sequential reaction with glutathione, reduction to selenide, and methylation of selenide.[15] The final methylation step can be enzymatically catalyzed. For example, cell extracts from the ciliate *Tetrahymena thermophila* catalyzes the *S*-adenosylmethionine-dependent methylation of selenide. Methyltransferase activity is also observed with methaneselenol to give dimethyl selenide. As a result of the toxic effects of selenium, and its ready conversion to lipophilic organoselenium compounds, there is a need to discover extractants for both the selenate and selenite oxyanions.

Phosphate is another anion for which receptors are needed. The major problem with phosphate is related to its action as a fertilizer, and since the release of phosphate into waterways causes excess plant growth, it is necessary that this be avoided. Although there are many restrictions on the use of phosphates in household products such as detergents, there are still uses of phosphorus compounds in agriculture and in fire retardant materials. Under aerobic conditions, phosphorus compounds are oxidized to phosphate ($PO_4^{3-}$) and its protonated forms $HPO_4^{2-}$, $H_2PO_4^-$ and $H_3PO_4$. These phosphorus(V) compounds are the thermodynamically stable form of phosphorus, and are therefore widely distributed under aerobic conditions.

Pertechnetate ($TcO_4^-$) is yet another anion which is an environmental problem because it is formed as a radioactive by-product of nuclear fission. This technetium(VII) oxyanion has a high geochemical mobility and bioavailability. However, the form changes through a combination of factors such as redox conditions and microbial activity in soils.[16] Indeed, the activity of microorganisms influence technetium adsorption onto the soil. There is therefore an urgent need to remove pertechnetate before it becomes dispersed in soils. Again, like chromate, selenate and phosphate, pertechnetate has a *pseudo*-tetrahedral structure, although it differs in that it is only monoanionic. However, because all isotopes of technetium are radioactive, perrhenate ($ReO_4^-$) is frequently used as a surrogate in experiments focused on seeking pertechnetate receptors.

Although this chapter covers anions and oxyanions, the main focus is on the extraction of chromium(VI).

## 2. TOXICOLOGICAL EFFECTS

Chromium(VI) is a carcinogen in humans and animals, and chromate compounds are mutagenic and genotoxic. Chromium(VI) requires intracellular reduction for activation, and this *in vivo* reduction can produce several reactive intermediates such as chromium(V) and chromium(IV) that can target and damage DNA.[12] Many epidemiological studies have established that chromium(VI) compounds are human carcinogens.[17,18] They are strong clastogens,[19, 20] and are mutagenic in both bacterial and mammalian test systems.[21-23] The principal chemical form of chromium(VI) in solution at physiological pH is chromate. Chromate enters cells through the general anion channel, resulting in a rapid accumulation of high concentrations of intracellular chromium.[24] By contrast, water soluble chromium(III) compounds are not considered to be carcinogenic, possibly because they do not cross plasma membranes.[23, 25] Nevertheless, the final intracellular reduction product of chromium(VI) is chromium(III), which forms amino acid nucleotide complexes. The mutagenic potential of these complexes is not fully known.

Mussels, which have been widely used as biological monitors of coastal contamination, assimilate chromium.[26] Only chromium(VI) from the dissolved phase and chromium(III) from ingested food contribute to chromium accumulation in marine mussels. The uptake of the toxic chromium(VI) is especially important from solution at lower salinities.

The toxicity of technetium(VII) results from the radioactivity of Tc-99 as a $\beta$ emittor. The half-life of this radionuclide is $\mathbf{10^5}$ years, which causes the isotope to be a long term environmental health problem.

# 3. ANALYTICAL TECHNIQUES

The quantity of dissolved chromium(VI) in waters can be determined by Ion Chromatography.[27] The method detection limits are in the 0.3-0.4 μg/L range. Other methods that have been used include co-precipitation, colorimetry, chelation extraction in combination with atomic absorption spectroscopy (AAS) or inductively coupled plasma-atomic emission spectroscopy (ICP-AES), and differential pulse polarography.[28]

A simple, fast, sensitive field method has been developed for the determination of chromium(VI). The method uses ultrasonic extraction in combination with a strong anion-exchange solid-phase extraction (SAE-SPE) technique. The chromium is assayed by ultrasonification in basic ammonium buffer solution to extract the chromium(VI), followed by SAE-SPE to separate chromium(VI) from chromium(III), elution and complexation of the chromium with 1,5-diphenylcarboxide, and spectrophotometric determination of the chromium-diphenylcarbazone complex. The chromium detection limit is 0.08 microgram. [29]

Phosphate can be determined by $^{31}P$ NMR spectroscopy, or by assays based on its biological activity. The former method is completely specific, although its sensitivity is low. The ICP-AES method also has a rather low sensitivity for phosphate. Technetium-99 can be detected and determined by its radioactive signature.

# 4. EXTRACTION METHODS

Several approaches have been used for the environmental removal of chromium(VI). Conventional methods are based on the reduction of chromium(VI) to chromium(III), followed by $Cr(OH)_3$ precipitation at a pH in the 8-10 range.[30-32] One approach is to reduce chromium(VI) with $\alpha$-hydroxyl carboxylic acids such as mandelic acid in the presence of oxide surfaces.[33] Alternatively oxalic acid or substituted phenols can be used.[34] Conventional pump-and-treat techniques for chromium(VI) removal frequently result in lengthy application times because of the decrease in its removal at the later stages of the process. The addition of sulfate to the extraction water decreases by almost two orders of magnitude the number of pore volumes required to achieve a targeted level of chromium(VI) reduction. This effect is proposed to be a consequence of sulfate releasing chromate from the insoluble barium chromate according to equation 1.[35] The reduction

$$BaCrO_4 + SO_4^{2-} \rightarrow BaSO_4 + CrO_4^{2-} \quad (1)$$

of chromium(VI) by ferrous sulfate is a method that has found use for the remediation of chromate. In industrial wastewater treatment the removal of chromium(VI) involves a two-step process; the reduction of chromium(VI) under acidic conditions (usual pH 2-3), and the hydroxyl precipitation of chromium(III) at pH 8-10. Commonly used reducing agents are sulfur dioxide, sodium sulfite, sodium bisulfite, and ferrous sulfate. In addition, iron(III) has been found to be effective in precipitating the resulting chromium(III) ions.[36-38] The solid chromium hydroxide can then be removed by flotation.[39, 40] An electrochemical precipitation (ECP) process has also been used to remove chromium from an electroplating waste-water. The chromium removal efficiencies are greater than 99%, and the residual chromium concentrations are less than 0.5 mg/1. The ECP process uses a current of 0.5-5.0 amp and an initial pH of 4.5.[41] An alternate approach couples the microbial reduction of chromium(VI) with the anerobic degradation of benzoate.[42] Hexavalent chromium can also be extracted directly from soils by use of a mixture of sodium carbonate and sodium hydroxide at 90-95 °C.[43] An effective reliable method for extracting both soluble and insoluble forms of chromium(VI) from soils without inducing chromium(III) oxidation or chromium(VI) reduction is required.[44] In comparing methods it is found that extraction procedure using 0.28 M sodium carbonate and 0.5 M sodium hydroxide with continuous swirling and heating at 90-95 °C for 90 min results in maximal dissolution of chromium(VI), while minimizing oxidation and reduction.[43] Ambient temperature mixed solutions containing sodium carbonate and sodium hydroxide, pure water, phosphate buffer at pH 7.0, and sonication under alkaline conditions all extract less chromium(VI). Other methods such as ion exchange and adsorption that use solid phase materials have also been successfully used to remove chromium(VI).[45] Examples are the use of anthracite[46] and activated carbon.[47-51] In addition, minerals can also be used as adsorbents for chromium(VI). One such mineral is goethite ($\alpha$-FeOOH).

[52, 53] By EXAFS it has been concluded that the three different surface complexes of chromate on goethite are a monodentate complex **1**, and both a mononuclear **2** and a binuclear **3** bidentate complex. Pressure jump

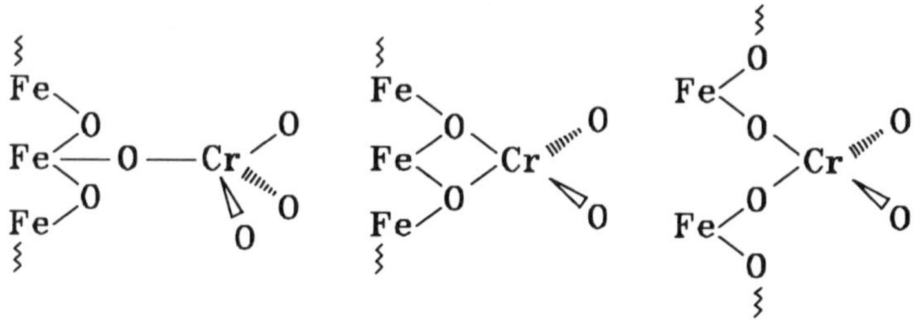

**1** **2** **3**

relaxation measurements conclude that the adsorption of chromate on goethite is a two-step process resulting in the formation of an inner-sphere bidentate surface complex. The initial rapid step involves a ligand exchange reaction of aqueous hydrogen chromate with hydroxyl groups of the goethite surface to give an inner-sphere monodentate surface complex. Zeolites have also been used as adsorbents for chromium(VI).[54] Although chromium(VI) can be removed by reduction to chromium(III) followed by precipitation of $Cr(OH)_3$, at very low concentrations of chromium(VI) the reduction is kinetically slow and does not result in the near-zero levels in waters that is often required.

A polymeric ligand exchanger has been developed that removes trace concentrations of chromium(VI) in the presence of sulfate, chloride, or bicarbonate ions. The material consists of a cross-linked polystyrene-divinylbenzene matrix with picolylamine groups appended. To these picolyl groups is chelated copper(II), and the selectivity is considered to result from the preference of chromium(VI) for the divalent copper center.[55] Chromium(III) has been removed from phosphoric acid solutions by chelating resins containing phosphonic or diphosphonic groups. The resin Diphonix that contains both sulfuric and *gem-* diphosphonic acid groups removes chromium(III). The column sorption is influenced by the acid concentration, and a decrease in elution efficiency of Cr(III) from Diphonix is obtained with increases in phosphoric acid concentration during sorption.[56]

# 5. COMPLEXANT DESIGN

To date there has been little effort directed toward designing a selective receptor for chromium(VI). Nevertheless, as the interest in designing anion receptors grows, it is likely that this situation will change.[57] One design theory is to recognize that chromium(VI) in basic solution has the tetrahedral chromate structure, which is analogous to the anions sulfate, selenate and phosphate. In addition to having *pseudo*-tetrahedral geometries these anions also have oxygens that can potentially hydrogen bond to a site on the host molecule. In the design of host for these anions it is advantageous to have a host molecule that has hydrogen bonding sites and can accommodate to the requirements of a tetrahedral guest.

## 5.1 Acyclic Ligating Sites

To date the complexation and extraction of oxyanions has been carried out with acyclic ligands, although there is future potential for macrocyclic hosts to be designed and synthesized. Chromium(VI) has been recovered from waste waters by extraction with methyl isobutyl ketone.[58] Chromium has been extracted from phosphoric acid solutions by *para*-(1,1,3,3-tetramethyl butyl) phenyl phosphoric acid in kerosine solvent, showing the preference for chromium(VI) over phosphorus(V).[59] Chromium(VI) has also been extracted using polyurea microcapsules.[60] Chromium(VI) has been removed from water by ferrous sulphate,[61] and iron(III) hexamethylenedithiocarbamate has been used as a flotation collector for the separation of chromium.[62] A mixture of ferric and chromic hydroxide has been used as an adsorbent for chromium(VI). Almost complete removal of chromium(VI) is found at pH 5-6.[63] Chromium(VI) removal is some 4-6 times better with the complexant diphenyl carbazide in aqueous micellar surfactant than with the surfactant Dowfax 8390 alone.[64]

In phytoremediation, chromium has been removed from aqueous solutions by Water Hyacinth.[65,66] In bioremediation, chromium(VI) has been removed from aqueous solutions by *Zoogloea ramigera*,[67] and simultaneously both chromium and organics by an anaerobic consortium of bacteria.[68] Chromium(VI) has been removed using an *Enterobacter cloacae* strain that reduces chromium(VI) under anaerobic conditions.[69] A two-stage bioreactor has been designed where *E. Coli* cells were grown in a first stage, and then pumped to a second stage plug-flow reactor where anaerobic reduction of Chromium(VI) occurred.[70]

## 5.2. Amines

One approach to designing an extractant for chromium(VI) is to seek compounds such as polyamines as anion binders.[71] This choice is made because the NH functionality can act to hydrogen bond with other moieties, and also because a pH reversible ion exchange system can be designed. Both the macrocycle **4** and and the cavitand **5** act as hosts for anions. These

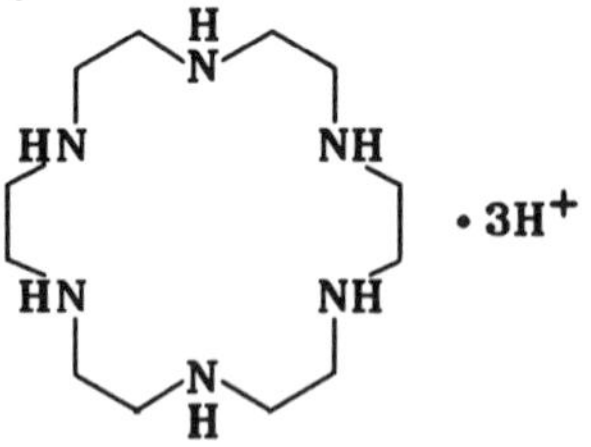

HN NH HN NH N H N H • 3H$^+$

**4**

N N X X X • 6H$^+$

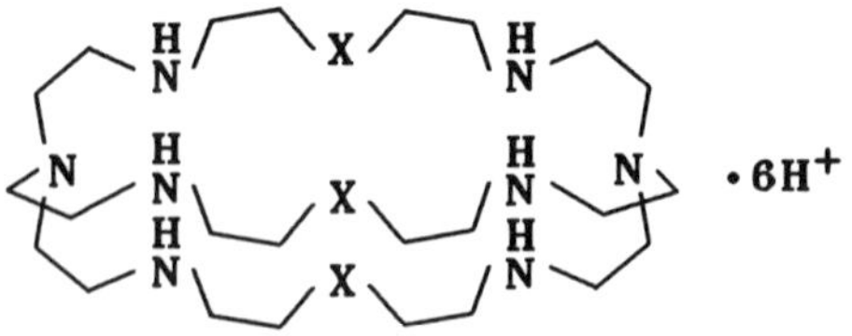

**5**

compounds can act either in a host-guest-mode or in a multiply hydrogen bonded mode to occlude oxyanion guests. Amine extractants can also be used to extract chromium(VI). In acidic solution the chromium(VI) will be primarily present as the monoanionic hydrogen dichromate, and it can be extracted *via* ion-pair formation with a long-chain alkylamines $R_3N$ (equation 2). If, however, the chromium(VI) is present in basic solution it will be

$$HCr_2O_7^- + H^+ + 2R_3N \rightleftharpoons (R_3NH)_2Cr_2O_7 \qquad (2)$$

primarily in the dianionic chromate form. Extraction can then be accomplished with a tetraalkylammonium ion exchanger such as Aliquat 336. Such a procedure for sodium chromate is shown in equation 3. Such metal extractions have been carried out successfully on non-dispersively

$$2\,Na^+ + CrO_4^{2-} + 2\,R_4N^+Cl^- \rightleftharpoons 2Na^+ + 2Cl^- + 2\,(R_4N)_2CrO_4 \qquad (3)$$

using hollow fiber membrane extractors.[72] Simultaneous back-extraction of chromium(VI) into a strongly alkaline stripping solution has been achieved in a two-fiber-set hollow fiber-contained liquid membrane.[73] A hydrophobic microporous hollow fiber membrane-based method has been used to remove anionic chromium(VI) in the presence of a metal cation.[74] The method can be used for the simultaneous extractions of both the cationic metal and the anionic oxyanion. Further development of this process allows for the simultaneous extraction of both the cationic and anionic heavy metal salts into one organic phase containing both an acidic and a basic organic extractant contained in an organic diluent. Alternatively the extraction can be accomplished using three sets of microporous hollow fibers.[75] An important aspect of developing such extractants is that they are kinetically rapid. This is especially important if the complexant will eventually be bound to a polymeric support, because attachment of a chelate to such a support results in a decrease in the rate of metal binding.

Heterocyclic amines have been used as liquid-liquid extractants for oxyanions from nitric acid solutions. One such compound is the symmetrical 4-(5-nonyl) pyridine. This extractant is used because the low solubility of water in it leads to the lower extraction of impurity salts from the aqueous phase. This compound has been used for the separation of pertechnetate from uranium.[76] The compound has also been used for the extraction of chromate from an 0.25M nitric acid solution. At high chromium concentration the metal is extracted as the pyridinium salt of $HCr_2O_7^-$. The chromium can be subsequently transferred back from the organic phase either by lowering the acid concentration or by reducing the extract from its chromium(VI) state to chromium(III) with ascorbic acid, hydrazine or sodium thiosulfate.[77]

### 5.2.1. Ferrocene Derivatives

Oxyanions have also been extracted with a redox-recyclable anion-exchange material. Because of its reversible redox behavior, ferrocene cationic component is the monovalent cation of 1,1,3,3-*tetrakis*-(2-methyl-2-hexyl) ferrocene (HEP, **6**), which is adsorbed onto silica to produce the solid phase material.[78] Although the uncharged HEP does not extract pertechnetate

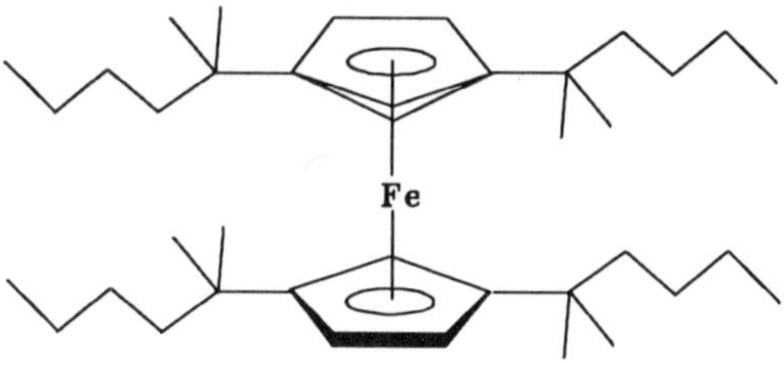

6

or perrhenate ions into an organic phase, the charged $HEP^+$ extracts up to 99% of these anions. The adsorbed perrhenate can be recovered by treating the exchanged materials with aqueous potassium ferrocyanide, which leads to reduction of $HEP^+$ to HEP and release of the anion. The cationic ion-exchange resin can be recovered by the re-oxidation of HEP with aqueous iron(III) nitrate.[79]

## 5.3. Calixarenes

Calixarenes have recently been used for anion recognition. Although much of this research centers on non-metallic anions such as chloride or phosphate, there are examples of metalloanion recognition and extraction. Again the advantage of calixarenes is that they can be designed to have cavities of different sizes and shapes that can be selective for specific oxyanions. In addition, calixarenes have multiple functionalities that can be converted into cationic centers for ion pairing with anions.

### 5.3.1. Calixarene Amines and Amides

A calixcrown **7** has also been used for binding oxyanions. The *bis*-(benzo crown ether) derivative is uncharged and therefore does not itself bind anions, but when an alkali metal ion is bound into each of the two crown ether sites the resulting dication binds oxyanions into the amide cavity. This

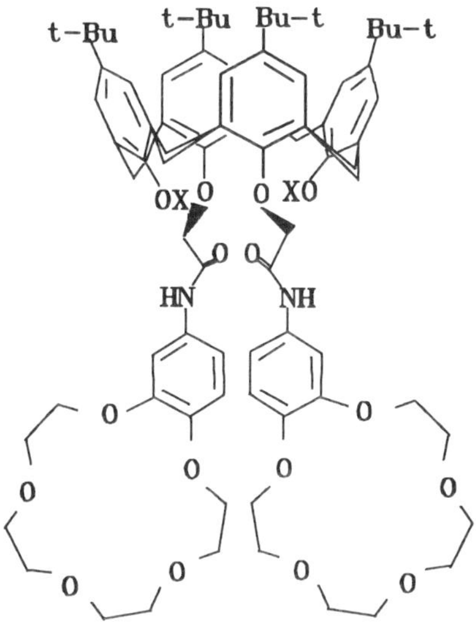

7

receptor forms 1 : 1 adducts with the chloride, nitrate, hydrogen sulfate, and dihydrogen phosphate anions. The highest stability constants are obtained with the latter two anions.[80] Calix[4]arenes with either amine or amide functionalities on their lower rim also act as extractants for high valent oxyanions.[6, 81] Amides have a stability advantage over amines because of their higher aerobic stability, but both have been used as extractants. Another advantage of amides is that they have both ketonic oxygens and amide nitrogens which can hydrogen bond with the periphery of oxyanions. A difference between amides and amines is in the basicities of their nitrogen centers. Amides are much weaker bases than are amines, therefore they do not become protonated until much higher solution acidities are achieved. This discrimination factor between the two compound types can be used to determine which of them is preferable based on the acidity of the aqueous solution from which the oxyanion is to be extracted. Finally, in choosing between amide and amine, the amide may have an advantage because the cation of the salt can bind to the ketonic oxygens. Amines have no such complementary functionality. The amines **8** and **9** and the amides **10** and **11** can extract oxyanions from an

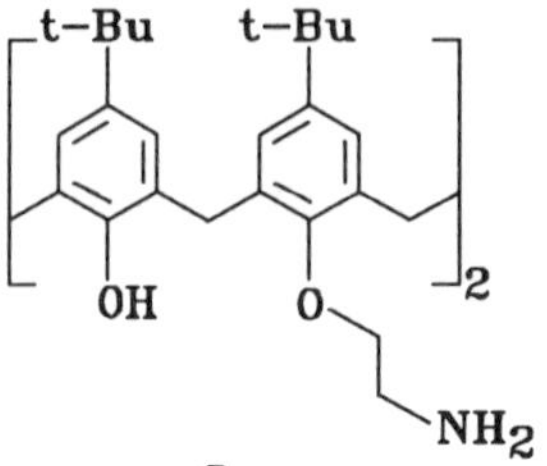

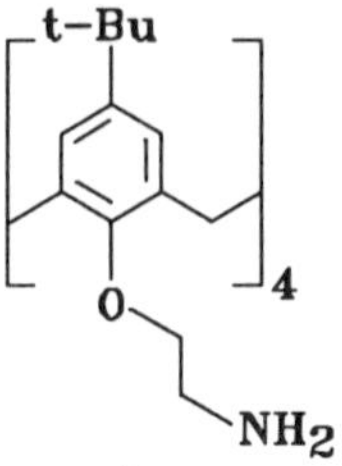

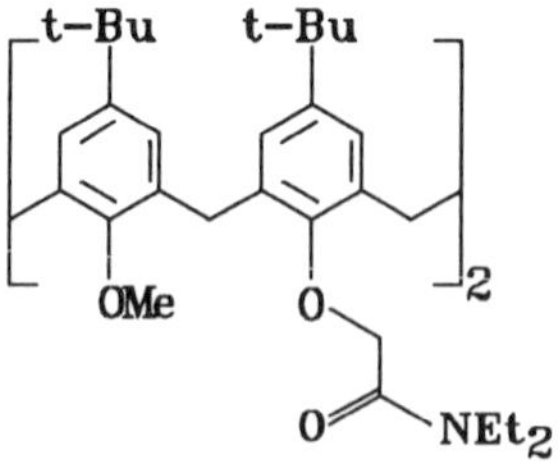

aqueous phase into chloroform. The extraction percentages are given in Table 1.[82] In general the highest extractions are obtained with the amine **9**.

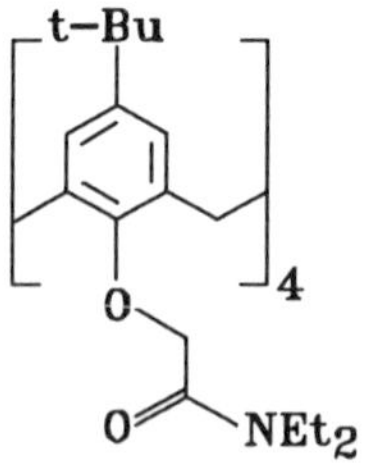

**Table 1. Extraction of Oxyanions by Calix[4]arene Amines and Amides**

| Anion | 8 | 9 | 10 | 11 |
|---|---|---|---|---|
| $HCr_2O_7^-$ | 11 | 16 | - | 1 |
| $CrO_4^{2-}$ | 9 | 88 | 6 | 5 |
| $MoO_4^{2-}$ | 4 | 74 | 3 | 3 |
| $HSeO_4^-$ | 13 | 18 | 14 | 28 |
| $SeO_4^{2-}$ | 3 | 87 | 3 | 6 |

| | | | | |
|---|---|---|---|---|
| $ReO_4^-$ | - | 36 | - | 39 |

This compound has the largest number of NH functionalities for hydrogen bonding to occur with the oxyanion. From Table 1 it is apparent that **9** is an excellent extractant for chromate, molybdate, and selenate. The two amides are poorer extractants than **9**, although they do extract hydrogen selenate and perrhenate. This lower extraction is observed even though the amides have both amide nitrogen and carbonyl oxygen functionalities for hydrogen bonding with the oxyanion. Introducing a more lipophilic *n*-butyl group does not lead to higher extractabilities of oxyanions, but they do extract the oxycation $UO_2^{2+}$.[83] These extractions are all carried out for short time durations in order to gain information about the relative rapidities of the complexant/oxyion pair formation, although. longer contact times cause no significant changes.

## 5.4. Polyethylene Glycol

Polyethylene glycol (PEG) has been used as a medium for the phase transfer recovery of chromium from aqueous salt solutions. Bubbling ammonia through an aqueous solution with the chromium(VI) in the upper PEG layer results in transfer of chromium(VI) to the lower layer. Similarly, passing carbon dioxide through the system in aqueous ammonia with the chromium(VI) in the lower layer results in its transfer to the upper phase. Addition of ferrous sulfate leads to reduction of chromium(VI) to chromium(III), and essentially quantitative transfer of the chromium from the PEG to the salt layer. A combination of electrolysis and the iron(II, III) redox couple can be used to reduce chromium(VI) to chromium(III) because after oxidation of the added iron(II) to iron(III) by chromium(VI), the iron(III) can then be electrolytically reduced back to iron(II). The iron(II) needs therefore to be present in only catalytic quantities.[84]

# 6. EXTRACTION FROM SOILS

The extraction of oxyanions such as chromium(VI) has been covered in earlier chapters. For soil washing and *in situ* stabilization methods the reader should refer to Chapter 3. Electrokinetic extraction is covered in

Chapter 4, and phytoremediation and bioremediation in Chapter 12.

# REFERENCES


1. D. H. Busch, *Chem. Rev.,* **1993**, *93*, 847.
2. L. F. Lindoy, *The Chemistry of Macrocyclic Ligand Complexes,* Cambridge Univ. Press, Cambridge, U.K., 1989.
3. G. Gokel, *Crown Ethers and Cryptands,* Monographs in Supramolecular Chemistry, Fraser Stoddart, J. ed., Royal Society of Chemistry, 1991.
4. F. C. J. M. Van Veggel, W. Verboom, D. N. Reinhoudt, *Chem. Rev.,* **1994**, *94*, 281.
5. H. C. Visser, D. M. Rudkevich, W. Verboom, F. de Jong, D. N. Reinhoudt, *J. Am. Chem. Soc.,* **1994**, *116*, 11554.
6. E. M. Georgiev, N. Wolf, D. M. Roundhill, *Polyhedron,* **1997**, *16*, 1581.
7. D. Burrows, *Chromium: Metabolism and Toxicity,* CRC Press, Boca Raton, FL., 1983.
8. J. A. H. Waterhouse, *Br. J. Cancer,* **1975**, *32*, 262.
9. S. Bonatti, M. Meini, A. Abbondandolo, *Mutat. Res.,* **1976**, *39,* 147.
10. V. Bianchi, A. Zantedeschi, A. Montaldi, J. Majone, *Toxicol. Lett.*, **1984**, *8*, 279.
11. S. De Flora, K. E. Wetterhahn, *Life Chem. Rep.,* **1989**, *7,* 169.
12. D. M. Stearns, L. J. Kennedy, K. D. Courtney, P. H. Giangrande, L. S. Phieffer, K. E. Wetterhahn, *Biochemistry*, **1995**, *34*, 910.
13. P. R. Wittbrodt, C. D. Palmer, *Environ. Sci. Technol.,* **1995**, *29*, 255.
14. Environmental Protection Agency, *Fed. Regist.* **1996**, *61*, (221), 58444.
15. N. Esaki, N. Tanaka, S. Vemura, T. Suzuki, K. Soda, *Biochemistry,***1979**, *18*, 407.
16. K. Tagami, S. Uchida, *Mater. Res. Soc. Symp. Proc. 353, no. Scientific Basis for Nuclear Waste Management XVIII, Pt. 2* (1995), 973.
17. IARC Monograph on the Evaluation of Carcinogenic Risk to Humans. Chromium, Nickel and Welding (1990), Vol. 48 (IARC, Lyon, France), pp. 49-508.
18. S. Langardt, *Am. J. Ind. Med.,* **1990**, *17*, 189.
19. P. Sen, M. Costa, *Carcinogenesis,* **1986**, *7*, 1527.
20. J. P . Wise, D. M. Stearns, K. E. Wetterhahn, S. R. Patierno, *Carcinogenesis,* **1994**, *15*, 2249.
21. J. Chen, W. G. Thilly, *Mutat. Res.,* **1994**, *323*, 21.
22. J.-L. Yang, Y.-C. Hsieh, C.-W. Wu, T.-C. Lee, *Carcinogenesis,* **1995**, *16,* 907.
23. E. Snow, *Pharmacol. Ther.,* **1992**, *53*, 31.
24. B. Buttner, D. Beyersmann, *Xenobiotica,* **1985**, *15*, 735.
25. A. Zhitkovich, V. Voitkun, M. Costa, *Biochemistry,* **1996**, *35*, 7275.
26. W.-X. Wang, S. B. Griscom, N. S. Fisher, *Environ. Sci. Technol.*, **1997**, *31*, 603.
27. E. J. Arar, S. E. Long, J. D. Pfaff, in *Methods for the Determination of Metals in Environmental Samples*, Environmental Monitoring Systems Laboratory, Office of Research & Development, U.S.E.P.A., Cincinnati, OH, C. K. Smoley,, Ed., 1992, CRC Press Inc., Boca Raton, FL, Method 218.6, pp.233-247.

28. W. Mueller, D. L Smith, *Compilation of E.P.A.'s Sampling and Analysis Methods,* L. H. Keith, ed., 1991, Lewis Publishers Inc., CRC Press Inc., Boca Raton, FL.
29. J. Wang, K. Ashley, D. Marlow, E. C. England, G. Carlton, *Anal. Chem.* **1999**, *71*, 1027.
30. J. V. Rouse, *J. Environ. Eng. Div., ASCE,* **1976**, *102*, 5.
31. J. J. Thomas, T. L. Theis, *J. Water Pollut. Control Fed.*, **1976**, *48*, 2032.
32. J. W. Patterson, *Industrial Wastewater Treatment Technology*, Butterworth, Boston, MA, 1985.
33. B. Deng, A. T. Stone, *Environ. Sci. Technol.*, **1996**, *30*, 463.
34. B. Deng, A. T. Stone, *Environ. Sci. Technol.*, **1996**, *30,* 2484.
35. *The Hazard. Waste Consultant*, **1995**, (*Jan/Feb*), L1.
36. D. C. Schroeder, G. F. Lee, *Water, Air and Soil Pollut.*, **1975**, *4*, 355.
37. M. J. Thomas, T. L. Theis, *J. Water Pollut. Control Fed.*, **1976**, *48*, 2032.
38. R. Bewley, R. Jeffries, J. Hellings, *Ground. Eng.*, **1998**, 22
39. L. D. Skrylev, T. L Skryleva, A. N. Purich, *J. Water Chem and Technol.*, **1996**, *18*, 25.
40. Y. Shoubary, N. Speizer, S. Seth, H. Savoia, *Environ. Progr.*, **1998**, *17*, 209.
41. N. Kongsricharoern, C. Polprasert, *Wat. Sci. Tech.*, **1996**, *34*, 109.
42. H. Shen, P. H. Pritchard, G. W. Sewell, *Environ. Sci. Technol.*, **1996**, *30*, 1667.
43. B. R. James, J. C. Petura, R. J. Vitale, G. R. Mussoline, *Environ. Sci. Technol.*, **1995**, *29*, 2377.
44. R. J. Vitale, G. R. Mussoline, J. C. Petura, B. R. James, *Environ. Qual.*, **1994**, *23*, 1249.
45. A. R. Bowers, C. P. Huang, *Prog. Water Technol.*, **1980**, *12*, 629.
46. I. A. Tarkovskaya, A. N. Tomashevskaya, V. E. Goba, O. E. Nagorskaya, L. A. Shurupova, B. S. Povazhnyi, *Sov. Prog. Chem.*, **1991**, *57*, 34.
47. G. N. Manju, T. S. Anirudhan, *Indian J. Environ. Health,* **1997**, *39*, 289.
48. D. C. Sharma, C. F. Forster, *Process Biochem.*, **1996**, *31*, 213.
49. D. G. Gajghate, E. R. Saxena, A. L. Aggarwal, *Water, Air, Soil Pollut.*, **1992**, *65*, 329.
50. S. Dahbi, M. Azzi, M. de la Guardi, *Fresenius J. Anal. Chem.*, **1999**, *363*, 404.
51. S. B. Lalvani, T. Wiltowski, A. Hubner, A. Weston, N. Mandich, *Carbon*, **1998**, *36*, 1219.
52. S. Fendorf, M. J. Eick, P. Grossl, D. L. Sparks, *Environ. Sci. Technol.*, **1997**, *31*, 315.
53. P. Grossl, M. Eick, D. L. Sparks, S. Goldberg, C. C. Ainsworth, *Environ. Sci. Technol.*, **1997**, *31,* 321.
54. M. Pansini, C. Coleela, M. de Gennaro, *Desalintion,* **1991**, *82*, 145.
55. D. Zhao, A. K. SenGupta, L. Stewart, *Ind. Eng. Chem. Res.*, **1998**, *37,* 43
56. N. Kabay, M. Demircioğlu, H. Ekinci, M. Yüksel, M. Sağlam, M. Akçay, M. Streat, *5Ind. Eng. Chem. Res.*, **1998**, *37*, 2541.
57. B. Dietrich, *Pure & Appl. Chem.*, **1993**, *65*, 1457.
58. I. Bojanowska, A. Sniegocka, *Przemysl Chem.*, **1998**, *77*, 262.
59. A. Mellah, D. Bauer, *Hydrometallurgy.*, **1995**, *37*, 117.
60. P. Ni, M. Zhang, N. Yan, *J. Membrane Sci.*, **1994**, *89*, 1.

61. C-J. J. Lin, P. A. Vesilind, *Proc. Mid-Atlantic. Ind. Hazard Waste Conf,*, **1995**, *27*, 568.
62. T. Stafilov, G. Pavlovska, K. Cundeva, *Microchem. J.*, **1998**, *60*, 32.
63. C. Namasivayam, K. Ranganathan, *J. Environ. Pollut.*, **1993**, *82*, 255.
64. N. B. Thirumalai, D. A. Sabatini, B-J. Shiau, J. H. Harwell, *Water Res.*, **1996**, *30*, 511.
65. O. Saltabas, G. Akcin, *Toxicol. Environ Chem.*, **1994**, *43*, 163.
66. O. Saltabas, G. Akcin, *Toxicol. Environ Chem.*, **1994**, *41*, 131.
67. C. Solisio, A. Lodi, A. Converti, M. Del Borgi, *Chem. Biochem. Eng. Quarterly*, **1998**, *12*, 45.
68. Y.-T. Wang, E. M. N. Chirwa, *Proc. Mid-Atlantic. Ind. Hazard Waste Conf.*, **1997**, *29*, 32.
69. K. Komori, A. Rivas, K. Toda, H. Ohtake, *Biotech. Bioeng.*, **1990**, *35*, 951.
70. H. Shen, Y-T. Wang, *J. Environ. Eng.*, **1995**, *121*, 798.
71. J. S. Bradshaw, K. E. Krakowiak, R. M. Izatt, *Tetrahedron*, **1992**, *48*, 4475.
72. C. H. Yun, R. Prasad, A. K. Guha, K. K. Sirkar, *Ind. Eng. Chem. Res.*, **1993**, *32*, 1186.
73. A. K. Guha, C. H. Yun, R. Basu, K. K. Sirkar, *AICHE J*, **1994**, *40*, 1223.
74. Z.-F. Yang, A. K. Guha, K. K. Sirkar, *Ind. Eng. Chem. Res.*, **1996**, *35*, 1383.
75. Z.-F. Yang, A. K. Guha, K. K. Sirkar, *Ind. Eng. Chem. Res.*, **1996**, *35*, 4214.
76. M. Igbal, M. Ejaz, *J. Radioanal. Chem.* **1974**, *23*, 51.
77. M. Igbal, M. Ejaz, *Anal. Chem.* **1975**, *47*, 936.
78. J. F. Clark, D. L. Clark, G. D. Whitener, N. C. Schroeder, S. H. Strauss, *Environ. Sci. Technol.* **1996**, *30*, 3124.
79. C. K. Chambliss, M. A. Odom, C. M. L. Morales, C. R. Martin, S. H. Strauss, *Anal Chem.*, **1996**, *70*, 757.
80. P. D. Beer, M. G. B. Drew, R. J. Knubley, M. I. Ogden, *JCS. Dalton Trans.* **1995**, 3117.
81. D. M. Roundhill, E. Georgiev, A. Yordanov, *J. Incl. Phenom. Mol. Recogn. in Chem.*, **1994**, *19*, 101.
82. N. J. Wolf, E. M. Georgiev, A. T. Yordanov, B. R. Whittlesey, H. F. Koch, D. M. Roundhill, *Polyhedron*, **1999**, *18*, 885.
83. O. M. Falana, H. F. Koch, G. Lumetta, B. Hay, D. M. Roundhill, *JCS, Chem. Comm.*, **1998**, 503.
84. H. F. Koch, J. Shen, D. M. Roundhill, *Sep. Sci. Technol.*, **2000**, *35*, 623.

# 11

# ALKALI AND ALKALINE EARTH METALS

## 1. INTRODUCTION

The selective extraction of alkali metal ions made a dramatic step forward with the discovery of the crown ethers. These oxygen-containing macrocycles are important because they were the first compounds that allowed alkali metal ions to be solubilized in organic solvents. These compounds effectively extract alkali metal ions into an organic solvent from an aqueous phase, and by changing the cavity size within the macrocycle selectivity, between the different alkali metals can be achieved. More recently, chemically modified calixarenes have been found to be good extractants for these ions. The present state of the art in designing extractants for alkali metal ions involves incorporating a crown ether moiety onto a calixarene framework. The resulting calixcrowns can be highly selective complexants and extractants for the individual alkali metal ions. Although early research in seeking hosts focused on discovering extractants for the lighter alkali metals such as sodium, more recently the focus has been on cesium because of its presence as one of the radioactive metals in nuclear wastes.

This chapter introduces the reader to the use of crown ethers as extractants, followed by calixarenes, and then calixcrowns. At the end of the chapter are collected other miscellaneous methods that have been used for the liquid-liquid extraction of alkali metal ions, along with a section on cesium extraction because of the particular requirements for extracting a large radioactive alkali metal ion. This need to find a selective extractant for radioactive cesium-137 from a mixture of other alkali metal ions has led to a renewed interest in synthesizing new alkali metal extractants.

# 2. CROWN ETHERS

This section covers a wide range of crowns, *bis*-crowns, and lariat crowns, and compares and contrasts their ability to effect liquid-liquid extraction of the alkali and alkaline earth metal ions. Crown ethers and their derivatives continue to be important extractants.

## 2.1. Crowns and *Bis*-Crowns

Sodium and potassium halides are extracted from aqueous solution by dibenzo-18-crown-6 and dicyclohexyl-18-crown-6. A protic media is used for the phase transfer because they effectively solvate the anions. Both separate ions and ion pairs are extracted, with the relative extractability order following the sequence of sulfate < chloride < bromide < iodide < nitrate < acetate < fluoride for the dibenzocrown extractant into *m*-cresol as the organic phase. The order of difficulty of removal salts of these anions from the aqueous phase is given by the order of their Gibbs free energies of hydration in kJ $mol^{-1}$ which are: iodide (-283), nitrate (-300), bromide (-321), chloride (-350), fluoride (-406), sulfate (-876). The unexpected positions of fluoride and nitrate in the extraction series is believed to be due to the strong hydrogen bonding properties of the *m*-cresol organic phase. With respect to the choice of organic phase, the best choices are a protic solvent with a low dielectric constant that permits ion pairing, a low Gibbs free energy for the transfer of an ion such as chloride into it from water, a high solubility of water in the liquid, but one that has a low solubility in water in order to avoid its loss.[1]

*Bis*-(benzocrown ether)s linked by a poly-(oxyethylene) chain show better extractabilities and selectivities for larger cations than do the single crown ethers.[2] These linked crowns **1**, **10**, **19**, **20**, **22**, **23** (X = $(CH_2)_2$), **2**, **11**, **21** (X = $O(CH_2)_3$), **3**, **12** (X = $(CH_2)_8$), **4**, **13** (X = $O(CH_2)_6O$), **5**, **14** (X = $O((CH_2)_2O)_2$), **6**, **15** (X = $(O(CH_2)_2O)_3$), **7**, **16** (X = $(O(CH_2)_2O)_4$), **8**, **17** (X = $(O(CH_2)_2O)_5$), **9**, **18** (X = C≡C ), are good extractants for alkali metal ions, and the extraction data, expressed as percentages, are shown in Table 1. The "*bis*-crown" effect is favorably exerted with an oligoethylene glycol linkage rather than a hydrocarbon one, and minimal effect of lipophilic groups or donor oxygens on the side chain of the *mono*-(benzocrown ether)s is observed.[3] Dicyclohexyl-18-crown-6 itself has also been used for the distribution of alkali metal picrates between water and a polyurethane foam. The extraction order follows the sequence: potassium(I) > rubidium(I) > cesium(I) > sodium(I), which generally follows the trend of the stabilities of the alkali metal crown ether complexes.[4]

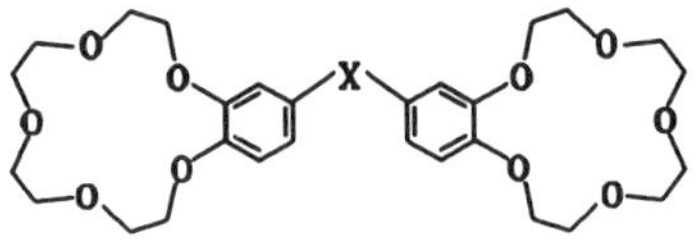

1–9 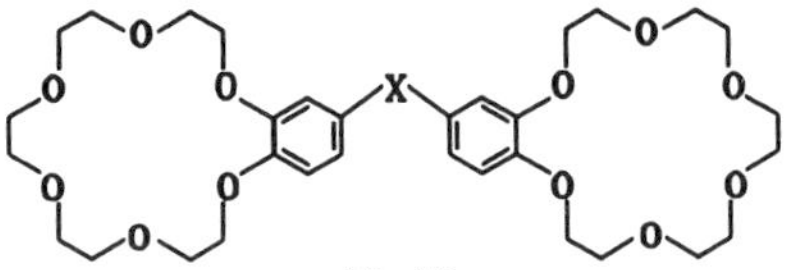 10–18

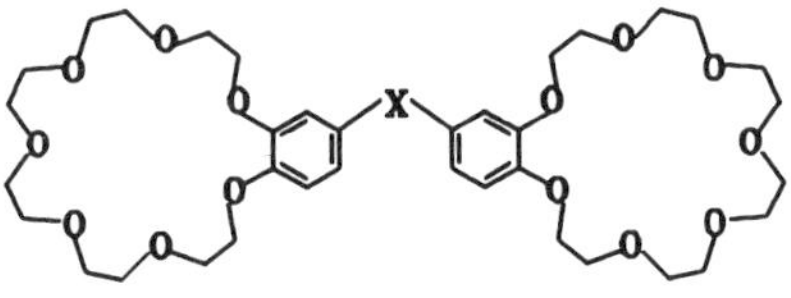

19 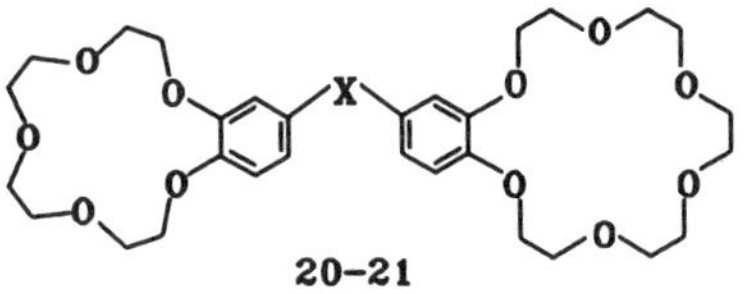 20–21

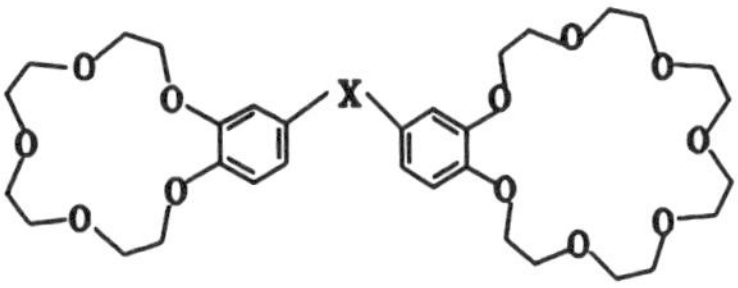

22 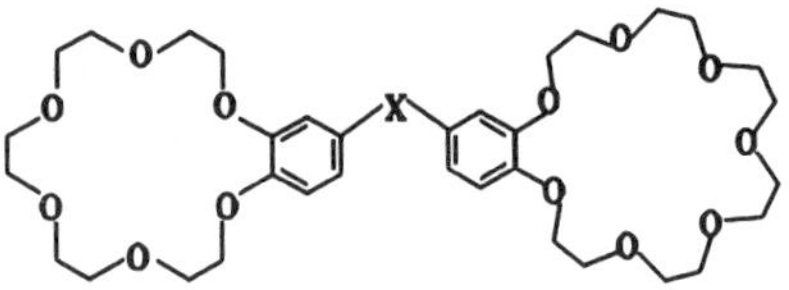 23

**TABLE 1. Solvent extraction of alkali-metal picrates by bis(benzocrown ether)s**

| Extraction (%) | | | | |
|---|---|---|---|---|
| **Crown** | **Sodium(I)** | **Potassium(I)** | **Rubidium(I)** | **Cesium(I)** |
| **1** | 6.5 | 53.2 | 46.8 | 22.2 |
| **2** | 6.3 | 65.4 | 41.4 | 12.5 |
| **3** | 4.9 | 61.8 | 35.1 | 9.8 |
| **4** | 7.5 | 79.2 | 46.9 | 11.6 |
| **5** | 9.1 | 86.5 | 61.5 | 18.2 |

| | | | | |
|---|---|---|---|---|
| **6** | 10.2 | 86.7 | 61.5 | 18.2 |
| **7** | 9.4 | 84.5 | 58.6 | 19.0 |
| **8** | 8.0 | 80.5 | 52.7 | 15.5 |
| **9** | 1.8 | 19.0 | 4.8 | 1.4 |
| **10** | 7.1 | 54.1 | 53.8 | 69.3 |
| **11** | 4.9 | 59.0 | 41.0 | 72.4 |
| **12** | 4.3 | 60.8 | 36.8 | 48.0 |
| **13** | 5.4 | 61.0 | 45.7 | 71.0 |
| **14** | 6.5 | 62.7 | 54.7 | 85.1 |
| **15** | 6.0 | 62.5 | 54.6 | 82.5 |
| **16** | 5.9 | 60.8 | 51.2 | 78.9 |
| **17** | 6.1 | 60.6 | 51.4 | 77.8 |
| **19** | 5.1 | 36.6 | 53.8 | 59.3 |
| The values are based on picrate salts in the aqueous phase. | | | | |

The anion effect on the dicyclohexano-18-crown-6 extraction of alkali metal cations into chloroform follows the sequence perchlorate > iodide, thiocyanate > nitrate > bromide (Table 2). The selectivities for both the potassium : rubidium and potassium : cesium ratios follow the sequence nitrate > thiocyanate > perchlorate = iodide > bromide. Again, the extraction order does not correlate with the hydration enthalpy of the anion.[5]

**TABLE 2. Solvent extraction percentages of alkali metal salts by dicyclohexano-18-crown-6**

| **Anion** | **Potassium(I)** | **Rubidium(I)** | **Cesium(I)** |
|---|---|---|---|
| Perchlorate | 41 | 17.5 | 4.1 |

| | | | |
|---|---|---|---|
| Iodide | 16.5 | 7.0 | 2.1 |
| Thiocyanate | 18.1 | 6.0 | 1.6 |
| Nitrate | 4.0 | 1.1 | 0.25 |
| Bromide | 0.6 | 0.3 | 0.2 |

The stoichiometries of the extracted complexes have been investigated. A study of the extraction equilibrium constants of picrate salts shows that both 1 : 1 and dicationic 2 : 1 complexes are formed between alkali metal ions and crown ethers Dicationic complexes are observed for 18-crown-5 with lithium(I), 24-crown-8 with lithium(I), sodium(I) and potassium(I), and dibenzo-24-crown-8 with sodium(I), potassium(I), and silver(I). This formation of dicationic complexes is attributed to the effective charge-shielding coordination of the bulky, lipophilic picrate anions in the contact ion-pair complex that is extracted.[6] Quantitative solvent extractions have been carried out in the water-dichloromethane system over the temperature range of 10-25°C. These experiments yield the extraction equilibrium constants ($K_{ex}$) and the thermodynamic quantities ($\Delta H^\circ$ and $\Delta S^\circ$) for the ion-pair extraction of aqueous alkali metal picrates with 15-crown-5, benzo-15-crown-5, *cis*-cyclohexano-15-crown-5, 16-crown-5, 18-crown-6, dibenzo-18-crown-6 and dibenzo-24-crown-8. The values of log $K_{ex}$ generally fall in the range of 3-7, with the 25°C values being collected in Table 3. The

**TABLE 3. Extraction equilibrium constants ($K_{ex}$) at 25°C for extraction of picrates into dichloromethane**

| Crown | Cation | Log $K_{ex}$ | Crown | Cation | Log $K_{ex}$ |
|---|---|---|---|---|---|
| 15-Crown-5 | $Li^+$ | 2.78 | Benzo-15-crown-5 | | |
| | $Na^+$ | 4.48 | | $Na^+$ | 3.74 |
| | $K^+$ | 4.33 | | $K^+$ | 3.99 |
| | $Rb^+$ | 4.14 | | $Rb^+$ | 3.70 |

| | | | | | |
|---|---|---|---|---|---|
| | $Cs^+$ | 3.67 | | $Cs^+$ | 3.31 |
| Cyclohexano -15-crown-5 | $Na^+$ | 4.85 | 16-Crown-5 | $Na^+$ | 4.54 |
| | $K^+$ | 4.23 | | $K^+$ | 3.60 |
| | $Rb^+$ | 4.16 | | $Rb^+$ | 3.48 |
| | $Cs^+$ | 3.81 | | -- | -- |
| 18-Crown-6 | $Na^+$ | 4.02 | Dibenzo-18-crown-6 | $Na^+$ | 2.83 |
| | -- | -- | | $K^+$ | 4.80 |
| | $Rb^+$ | 5.89 | | $Rb^+$ | 4.44 |
| | $Cs^+$ | 5.19 | | $Cs^+$ | 4.09 |
| Dibenzo-24-crown-8 | $Rb^+$ | 4.33 | | | |
| | $Cs^+$ | 4.55 | | | |

crown-5 complexants show a preference for sodium(I) and potassium(I), with the crown-6 and crown-8 derivatives showing a more favorable preference for the larger alkali metal ions.[7] The 1 : 2 "sandwich" complexes formed between alkali metal ions and crown ethers have been investigated in more detail. The overall process is stepwise with the formation of a 1 : 1 then a 1 : 2 complex. The slopes of $\Delta H°$ against $T\Delta S°$ plots for 1 : 2 complexation are larger than those found for 1 : 1, and are similar to those found for *bis*-(crown ethers). These results suggest that upon complexation of the second crown, the initially bound crown also undergoes substantial conformational changes.[8] In a further study of this system it has been verified that 1 : 1 : 1 (metal ion : crown ether : picrate ion) complexes are formed.[9] These ion-paired compounds can be extracted into organic liquids. For lithium(I), the 13-membered ring compound benzo-13-crown-4 (**24**) is an effective crown complexant. With this complexant, lithium(I) can be selectively extracted into benzene or methylene chloride. The extraction increases along the anion series in the sequence hydroxide < chloride < perchlorate ≈ thiocyanate

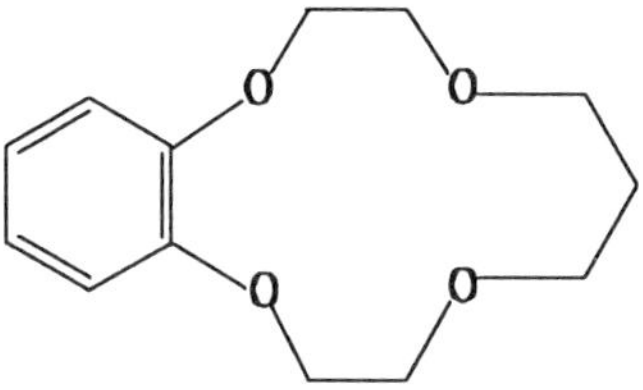

24

< picrate. A 1 : 1 complex predominates in the organic layer.[10]

## 2.2. Capped Cleft Molecules

Both *bis-* and *tetra-*capped clefts **25** (n = 2) and **26** have been used for the selective extraction of alkali metal picrates. In addition, the analogue having only a single crown **27** has been used for comparison. The comparative extraction data collected in Table 4. These data show

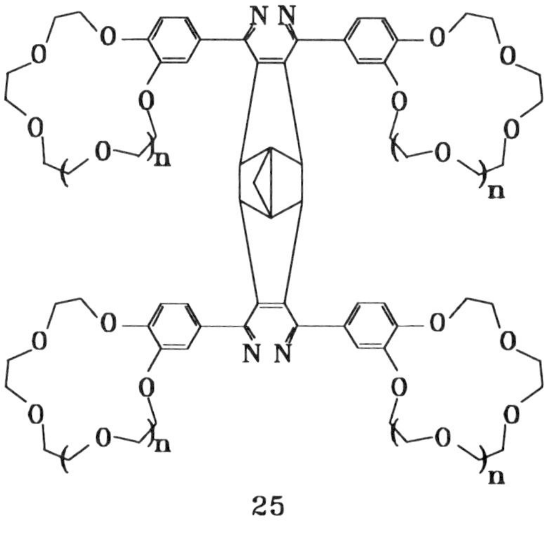

25

26 27

**TABLE 4. Percent Extraction of Monovalent Alkali Metal Picrates**

| Cmpd | Lithium | Sodium | Potassium | Rubidium | Cesium |
|---|---|---|---|---|---|
| **25** | 18.7 | 24.7 | 67.5 | 71.5 | 88.6 |
| **26** | 22.1 | 25.7 | 19.9 | 10.0 | 6.4 |
| **27** | 5.8 | 8.6 | 17.5 | 16.7 | 7.6 |

that cooperativity occurs between the two spatially proximate crown ether units of the host **25** with the guest metal cation.[11]

## 2.3. Biphasic Media

Poly(ethylene glycol) has been used as a biphasic medium for the extraction of these metals. In the presence of nitrate ion, the partitioning of sodium(I) correlates with the log K values for 18-crown-6 complexation in water, but a similar trend is not observed for 15-crown-5.[12]

## 2.4. Transport Phenomena

The transport phenomena that occur in liquid-liquid extractions of alkali metal ions with crowns have been investigated for 18-crown-6 and chloroform. The transport rate can be simplified to equation 1,

$$V = AK_a'L_oa^2 \qquad (1)$$

where $L_o$ is the total concentration of the carrier in the membrane, the term a is the activity of the salt in the aqueous phase, and $K_a'$ is the conditional extraction constant.[13] A plot of the function log V against log a is linear with a slope of 2 in the range of low activities.[14] A study has been carried out on the liquid-liquid extraction of MX salts (M = potassium, rubidium; $X^-$ = thiocyanate, iodide, nitrate, bromide) in this system. The rate of transport of the MX salts confirms the direct relationship that exists between extraction constants and transport rates.[15]

A computational analysis method has been used for the estimation of two-phase extraction constant data from stability constants for the complexation of alkali metal cations by a series of crown ethers and cryptands. This total free energy change for the extraction of alkali metal salts from water to a non-polar organic solvent is evaluated as a sum of free energy terms for complexation, transfers of cation, anion, and complex between solvents, and ion pairing in the non-polar solvent. The method has been used to fit known extraction constants, and can be used predictively.[16]

## 2.5. Carboxylate Crowns

The extraction of alkali metal cations into an organic solvent facilitated by a neutral complexant involves the simultaneous phase transfer of an anion. This phase transfer process for a salt can be aided if the complexant contains both a cation and an anion binding site within the molecule. Such a situation exists with crown ethers that have carboxylic acid substituents appended to their periphery, because deprotonation to give a carboxylate anion provides the necessary charge balance for extraction of the cationic metal. A series of carboxylate crowns **28** ( R = Et), **29** ( R = n–Bu), **30** ( R = *n*-hexyl), and **31** ( R = *n*-octyl) have been used for the extraction of alkali metals. These compounds have a single carboxylic acid functionality which provides the monoanion for charge balance with a single alkali metal cation. The lipophilicities of **29**, **30**, and **31** are sufficiently high that they are entirely in the chloroform phase in extractions of alkali metal cations from strongly alkaline aqueous solutions. By contrast, **28** is distributed between both phases.[17,18] For **31**, the selectivity order for liquid-liquid

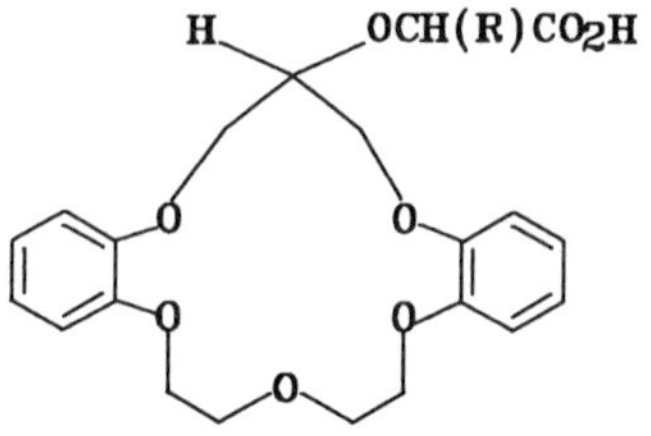
H
OCH(R)CO2H
O
O
O
O
O

28–31

extraction follows the sequence: sodium(I) > potassium(I) ~ lithium(I) > rubidium(I) > cesium(I).[19] For **31** the extraction efficiency for alkali metal cations decreases in the order chloroform > 1,1,1-trichloroethane > tetrahydronaphthalene > benzene > toluene > *para*-xylene. This ordering is explained on the basis that the hydrogen-bonding ability of the organic solvent may plan an important role in determining the extraction selectivity of these lipophilic crown ether carboxylic acids.[20] The effect of structural variations within lipophilic dibenzocrown ether carboxylic acids has also been investigated. For these compounds **32-40** the effect

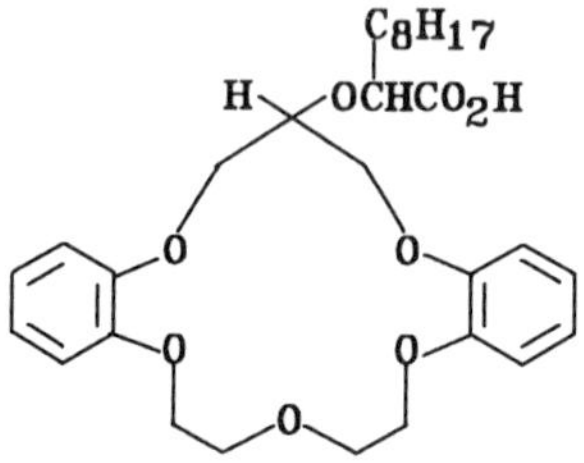
C8H17
H
OCHCO2H
O
O
O
O
O

33

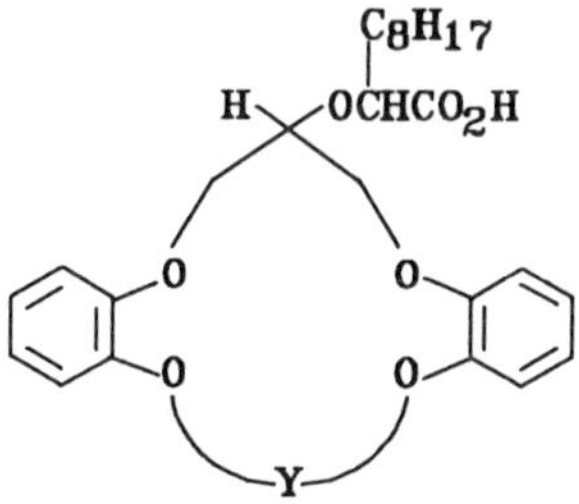
C8H17
H
OCHCO2H
O
O
O
O
Y

32,34–35

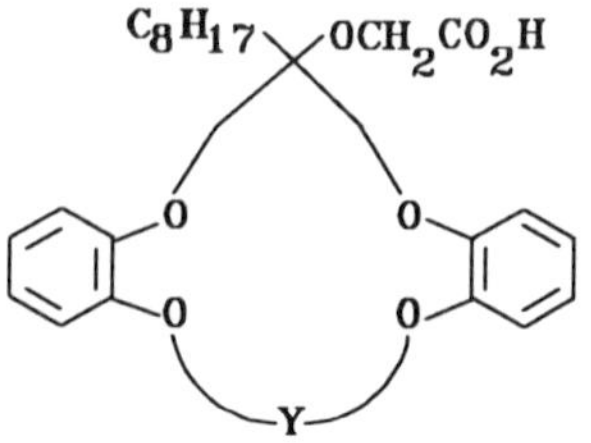
C8H17
OCH2CO2H
O
O
O
O
Y

36–38

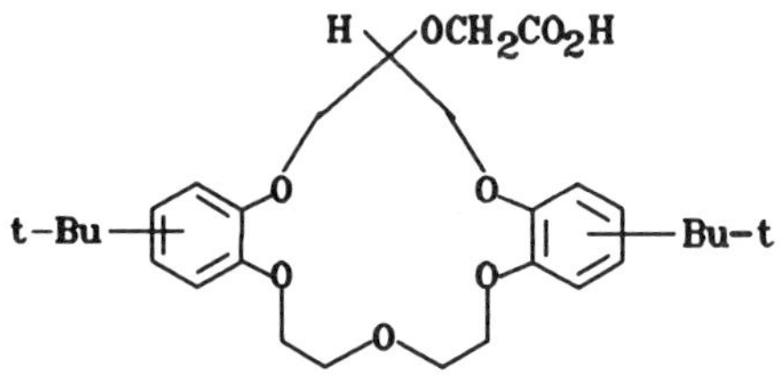

**39** 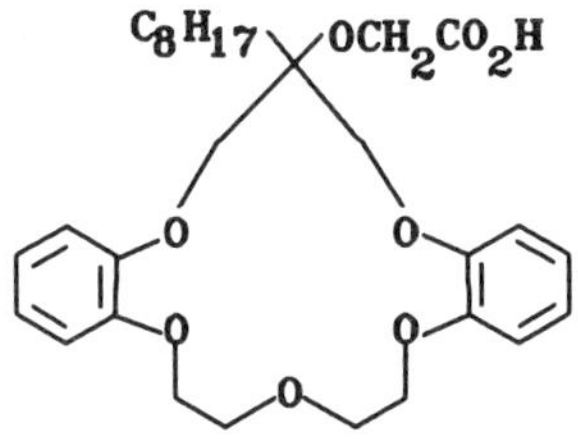 **40**

of ring size variations and lipophilic group attachment site are shown in Tables 5 and 6. These results show that the selectivity and efficiency of extraction are strongly affected by the crown ether ring size (Table 7)

**TABLE 5. Effect of Crown Ether Ring Size Variation on Selectivity and Efficiency of Competitive Alkali-Metal Cation Extraction from Aqueous Solutions into Chloroform by Lipophilic Dibenzocrown Ether Carboxylic Acids.**

| cmpd | ring size | selectivity order and selectivity coefficients[a,b] | | | | | max metal (%) |
|---|---|---|---|---|---|---|---|
| **32** | 14C4 | $Na^+$ > | $Li^+$ > | $K^+$ > | $Rb^+$ > | $Cs^+$ | 100 |
| | | | 2.5 | 5 | 19 | 60 | |
| **33** | 16C5 | $Na^+$ > | $K^+$ > | $Li^+$ > | $Rb^+$ > | $Cs^+$ | 68 |
| | | | 5 | 17 | 36 | 83 | |
| **34** | 19C6 | $K^+$ > | $Na^+$ > | $Rb^+$ > | $Cs^+$ > | $Li^+$ | 100 |
| | | | 4 | 7 | 11 | 28 | |
| **35** | 22C7 | $K^+$ > | $Rb^+$ > | $Li^+$ > | $Cs^+$ > | $Na^+$ | 100 |
| | | | 1.5 | 1.9 | 2.8 | 3.0 | |
| **36** | 14C4 | $Li^+$ > | $Na^+$ > | $K^+$ > | $Rb^+$ > | $Cs^+$ | 100 |

| | | | | | | | |
|---|---|---|---|---|---|---|---|
| | | | 3 | 7 | 12 | 13 | |
| **37** | 16C5 | $Na^+$ > | $K^+$ > | $Li^+$ > | $Rb^+$ > | $Cs^+$ | 82 |
| | | | 32 | 66 | ND[c] | ND[c] | |
| **38** | 19C6 | $K^+$ > | $Na^+$ > | $Rb^+$> | $Li^+$ > | $Cs^+$ | 48 |
| | | | 4 | 14 | 21 | 67 | |

[a]Ratio of chloroform phase concentrations of best extracted metal ion and indicated metal ion. [b]At pH = 10.0. [c]Not detected.

**TABLE 6. Effect of Lipophilic Group Attachment Site on Selectivity and Efficiency of Competitive Alkali-Metal Cation Extraction from Aqueous Solutions into Chloroform by Lipophilic Dibenzo-16-crown-5-oxyacetic Acids.**

| cmpd | selectivity order and selectivity coefficients[a] | | | | | max metal (%) |
|---|---|---|---|---|---|---|
| **39** | $Na^+$ > | $K^+$ > | $Li^+$ > | $Rb^+$> | $Cs^+$ | 100 |
| | | 5 | 10 | 17 | 30 | |
| **33** | $Na^+$ > | $K^+$ > | $Li^+$ > | $Rb^+$> | $Cs^+$ | 68 |
| | | 5 | 17 | 36 | 83 | |
| **40** | $Na^+$ > | $K^+$ > | $Rb^+$> | $Cs^+$ > | $Li^+$ | 93 |
| | | 27 | 51 | 67 | 90 | |

[a]At pH = 10.0

**TABLE 7. Diameters of Crown Ether Cavities[a] and Alkali-Metal Cation in Angstroms**

| crown | cavity diam | alkali metal cation | diameter |
|---|---|---|---|
| 14-crown-4 | 1.2-1.5 | $Li^+$ | 1.20 |
| 16-crown-5 | 2.0-2.4 | $Na^+$ | 1.90 |
| 19-crown-6 | 3.0-3.5 | $K^+$ | 2.66 |
| 22-crown-7 | 4.7-5.0 | $Rb^+$ | 2.96 |
| | | $Cs^+$ | 3.38 |

[a]Estimated from CPK space-filling models.

and the attachment site of the lipophilic groups. Attachment of the lipophilic group to the polyether ring carbon atom that bears the oxyacetic acid group orients the sidearm over the crown ether cavity, and produces lipophilic dibenzocrown ether carboxylic acids **41**, **42** and **6** and carboxylate crowns **43-49** that show high extraction selectivity for lithium(I), sodium(I), and potassium (I).[21, 22] The selectivities observed for these compounds in liquid-liquid extraction are collected in Table 8. In each case, except for compound **6**, a preference for lithium(I) is observed. This preference for liquid-liquid extraction into chloroform is particularly high for lithium(I) over sodium(I), with ratios of up to 19-20 being observed. These highest selectivities are observed for **46** and **47**. Each of these compounds is a 14-crown-4 derivative with a lipophilic carboxylate group as the pendant arm. This crown size correlates with the ionic diameter of lithium(I), which is in accordance with this observed selectivity.

Compound **6** shows different selectivity because it is a *bis*-(benzocrown ether) linked by a poly(oxyethylene) chain. As a result this compound shows a "*bis*- crown" effect that results in different extraction characteristics for alkali metal ions.[23]

The highest extraction percentages are observed from aqueous solutions having a pH in the 10-11 range. Both acidic and neutral solutions result in little or no extraction in this series of competitive solvent extraction experiments.[23]

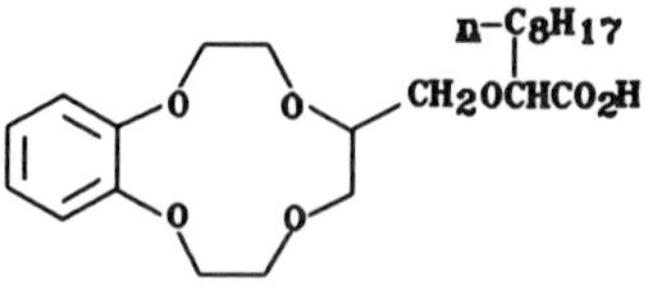

41

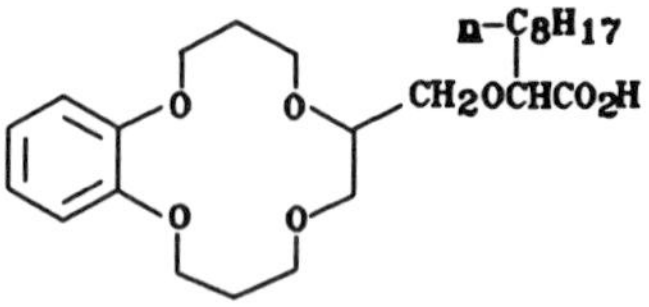

42

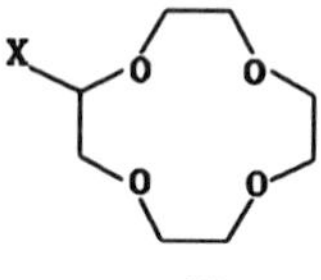

43

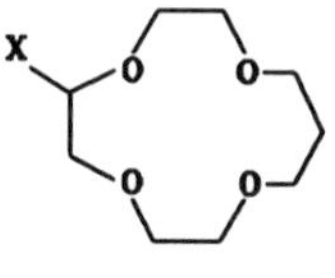

44

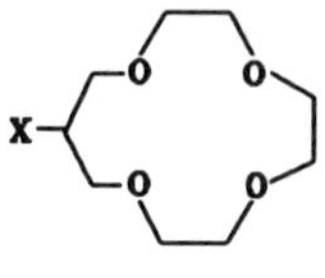

45

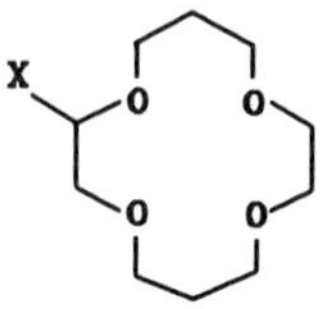

46

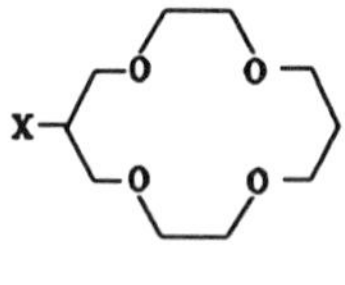

47

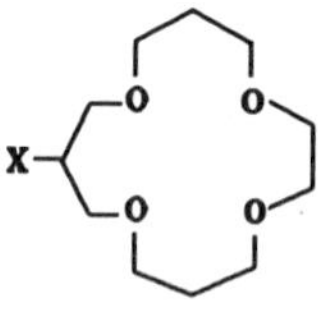

48

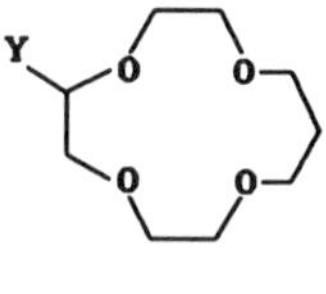

49

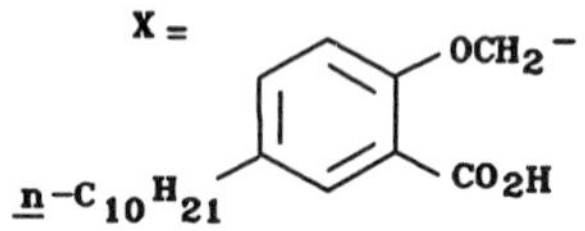

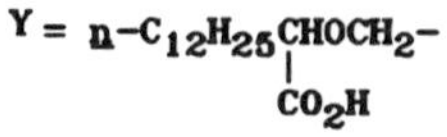

**TABLE 8. Competitive Extraction Selectivity Orders and $Li^+$ : $Na^+$ Ratios**

| cmpd | ring type | selectivity order | max $Li^+$ : $Na^+$ ratio |
|---|---|---|---|
| **41** | B12C4 | $Li^+>Na^+>K^+>Rb^+>Cs^+$ | 1.8 |
| **42** | B14C4 | $Li^+>Na^+>K^+>Rb^+$ | 4.7 |
| **6** | DB14C4 | $Na^+>Li^+>K^+>Rb^+>Cs^+$ | 0.6 |
| **43** | 12C4 | $Li^+>Na^+>K^+>Rb^+,Cs^+$ | 1.7 |
| **44** | 13C4 | $Li^+>Na^+>K^+>Rb^+>Cs^+$ | 2.3 |

| | | | |
|---|---|---|---|
| **45** | 13C4 | $Li^+ > Na^+ > K^+ > Rb^+, Cs^+$ | 2.5 |
| **46** | 14C4 | $Li^+ >> Na^+$ | 20 |
| **47** | 14C4 | $Li^+ >> Na^+$ | 19 |
| **48** | 15C4 | $Li^+ > Na^+ > K^+, Cs^+ > Rb^+$ | 3.5 |
| **49** | 13C4 | $Li^+ > Na^+ > K^+ > Rb^+ > Cs^+$ | 1.6 |
| [a]B = benzo, DB = dibenzo, [b]Reproducibility is ± 5% of the listed value | | | |

An analogous series of lipophilic acyclic proton-ionizable polyethers have been synthesized that are good extractants for alkali metal ions. These compounds **50-53** (n = 0-3 respectively) represent a

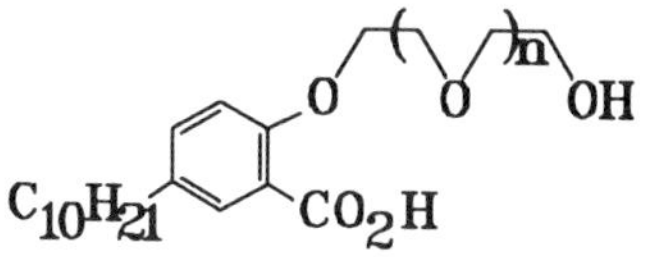

$C_{10}H_{21}$ O O OH n $CO_2H$

50–53

series of chelating agents for which the number of ether oxygens is systematically varied from one to four. For **50** and **51** the selectivity order is lithium(I) > sodium(I) > cesium(I) ≥ potassium(I), rubidium(I). For **52** the order becomes lithium(I) > sodium(I) > potassium(I) > rubidium(I) > cesium(I), with **53** being selective for potassium(I).

## 2.6. Phosphonate Crowns

Phosphonate groups have been appended to crown ethers to generate ionizable crowns that have an anion intramolecularly available for charge compensation to an alkali metal cation. Although this strategy mirrors that followed with carboxylate crowns, a difference between phosphonate and carboxylate groups is that the former have a much higher acidity, and are therefore deprotonated at lower pH than are their carboxylic acid analogs. As a result, it is expected that they will be function as extractants from solutions of higher acidity than do their carboxylic acid counterparts.

A series of crowns with alkylphosphonic acid groups appended through methylene spacers of different lengths **54-57** (n = 1-4 respectively) have been used as extractants for alkali metal ions into chloroform. The extraction order follows the sequence sodium (I) >

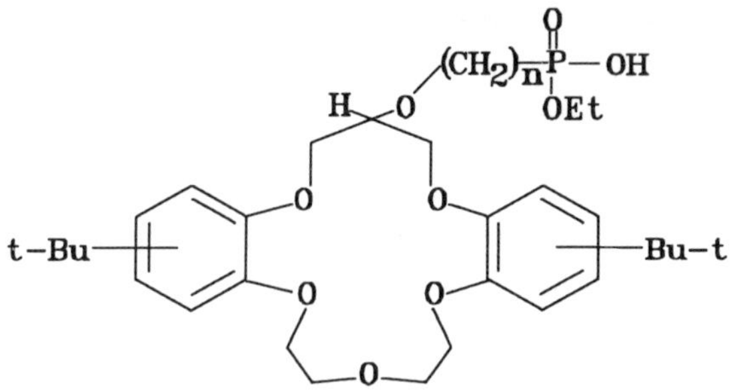

54–57

lithium (I) > potassium (I) > rubidium(I), cesium(I) for **54**, sodium(I) >> potassium(I) > lithium(I) > cesium(I) > rubidium(I) for **55**, and lithium(I) > sodium(I) > potassium(I) > rubidium(I), cesium(I) for **56** and **57**.[24] An acyclic phosphonic acid analog **58** shows a selectivity order of lithium(I) > sodium(I) > potassium(I), rubidium(I), cesium(I).[25-27]

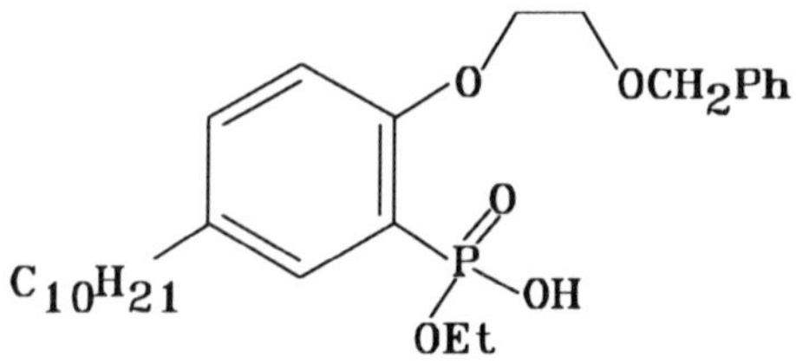

58

## 2.7. Lariat Crowns

Although these carboxylic acid and phosphonic acid substituted crowns can be classified as "proton-ionizable" crowns, they are also members of a class of macrocycles termed lariat crowns. Instead of their being proton-ionizable functionalities appended to their periphery, these lariat crowns can also have uncharged pendant groups.[28] This section is principally devoted to crown ethers that have an uncharged functional group appended to their periphery.

For lariat ethers a positive participation of the lariat side-arm in the extraction of metal ions is found with *C*- and *N*-lariat ethers.[29] The lariat crown **59** ($R^1$ = Me, $R^2$ = $MeOCH_2CH_2OCH_2$) has been studied both in terms of its complexation with alkali metal ions and their liquid-liquid extraction into an organic phase. This compound with its lariat side-arm shows improved extraction properties into dichloromethane as compared to **60** ($R^1 = R^2$ = H) and **61** ($R^1 = R^2$ = Me), but this feature is

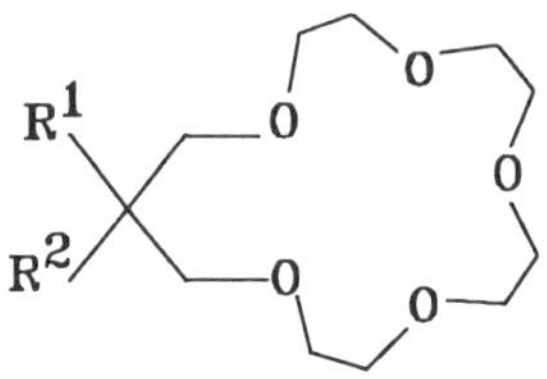

R1 O O O R2 O O

**59–61**

not reflected in its homogeneous phase complexation properties with alkali metal ions. If, however, the lariat effect occurs predominantly in the organic phase in the extraction process in the presence of a small quantity of participating water, this can lead to a different result than is observed in the homogeneous phase complexation in methanol.[30] The relationship between homogeneous binding constants and alkali metal ion transport has been investigated for a group of sixteen lipophilic 12-, 15-, and 18-membered ring ester and amide-side-armed lariat ethers. The cation transport rate in a bulk chloroform membrane correlates well with both the binding constant that is determined in methanol, and the picrate extraction constants determined in the membrane solvent mixture.[31]

## 2.8. Amino Crowns

Amino arms have also been appended to crowns, and the compounds used for the extraction of alkali metal ions. The compounds **62**, **63**, and **64** (R = $(CH_2)_2NH_2$) have been used as extractants for

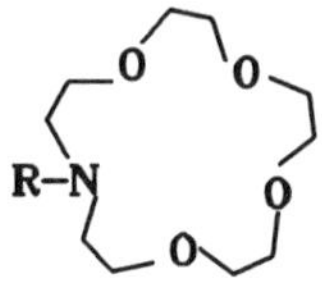

62 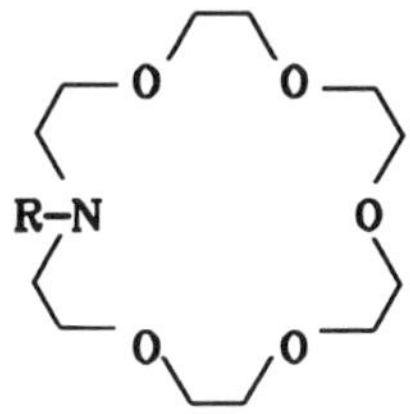 63 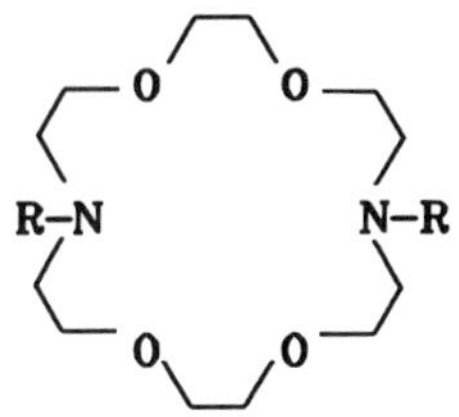 64

lithium(I), sodium(I), potassium(I), and cesium(I). These lariat crowns are much better extractants for these metal ions than are their parent crowns without the lariat group attached. In general the compounds show poor extraction selectivity, but for **62** which has the smallest ring, the respective cesium(I) : lithium(I), cesium(I) : sodium(I), and cesium(I) : potassium(I) selectivity ratios are 2.2, 2.5 and 2.3 respectively.[32] Other crown systems with a range of side-arms have been evaluated for their liquid-liquid extraction properties for sodium(I), and the effect of the side arm, complex formation, and the role of donor atom on cation-binding ability and selectivity for sodium(I) evaluated.[33]

Crowns with 4-picrylaminobenzo groups **65** have been used as extractants for alkali metal ions. They compounds show an extraction order of potassium(I) > rubidium(I) > cesium(I) > sodium(I) >>

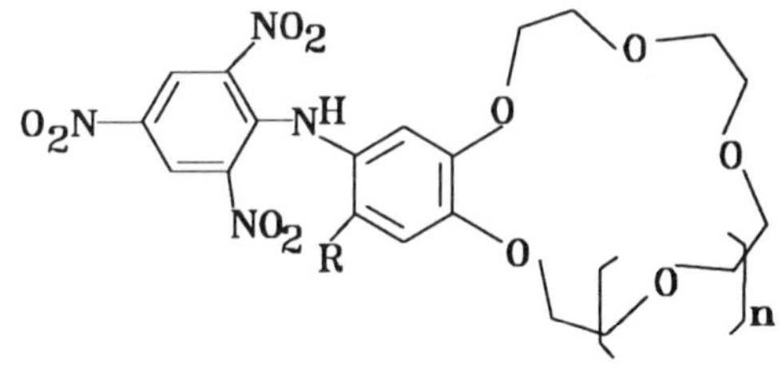

65

lithium(I). These extractants that have the picryl chromophore incorporated into the 18-crown-6 can be used for the colorimetric determination of potassium in the 4 to 40 ppm range in the presence of other alkali and alkaline earth metals.[34] The binding and extraction properties of alkali metals with macrocyclic tetralactams have been investigated, and compared with the alkaline earths. The large ring lactam **66** is a particularly good extractant for both sodium(I) and

**66**

potassium(I). This compound is a slightly better extractant for calcium(II) and barium(II), but a slightly poorer extractant for magnesium(II).[35] The cage-functionalized diaza-(1-crown-5) ethers **67** and **68** have been used as extractants for alkali metal ions. Complexation

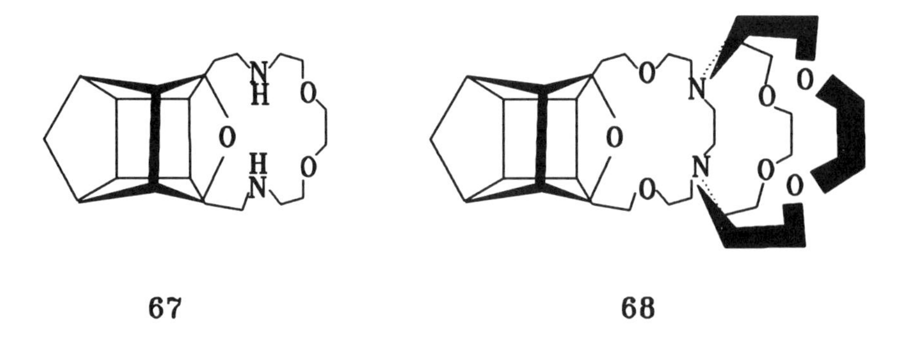

**67** **68**

is influenced by both cage-annulation and the *N*-lariat side arms. As a result, compound **67** favors all alkali metal ions, and the highly preorganized cryptand **68** has a strong preference for potassium(I) and rubidium(I).[36]

Ionizable hydrazone groups have been appended to crown ethers for use as extractants for alkali metal ions. These compounds **69** (n = 1, 2) are liquid-liquid extractants with chloroform, dichloromethane and 1,1-dichloroethane as the organic phase, although the selectivities are low.[37]

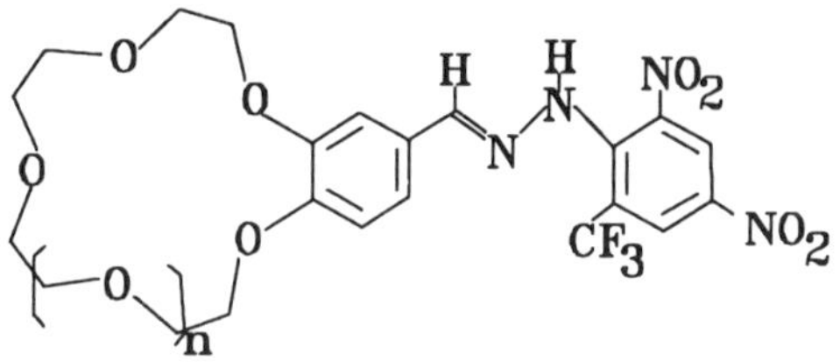

69

## 2.9. Aza Crowns and Photocrowns

Crowns with nitrogen functionalities in the ring have been used as extractants for alkali metal ions. In some cases these chromophores have been appended to the crown in order that the extraction of the metal ion can be observed, while in other cases photoresponsive functionalities have been incorporated into the crown in order that geometric changes can be induced by light. This latter situation is important when the binding properties of the alkali metal ion are sensitive to the geometry of the crown.

The two chromogenic crowns *N*-(2-hydroxy-5-nitrobenzyl)-aza-15-crown-5 and *N*-(2hydroxy-5-nitrobenzyl)-aza-18-crown-6 have been used as extractants, with the selectivity order following the sequence lithium(I) > sodium(I) > potassium(I), rubidium(I) for the crown-5, and potassium(I) > sodium(I) > rubidium(I) > lithium(I) > cesium(I) for the crown-6. The crown-5 analog can be used for determining lithium (I) in solutions in the 10-100 ppm range, and for concentrating lithium from dilute solutions.[38] Chromogenic groups have also been appended as lariat groups onto crowns. Examples are **70** (X = 1, R = $CF_3$) and **71** (X = 2, R = $NO_2$) that have nitroaromatic groups attached as a lariat to the 12-crown-4 macrocycle. These compounds extract alkali metal ions into a solution of triethylamine in chloroform.[39]

**70–71**

A series of articles have been published describing the function of photoresponsive crowns for the complexation and extraction of alkali metals. An early example involves the use of a *bis*-crown with an azo linkage. This compound efficiently extracts potassium(I) in its photoisomerized *cis*-form, but not in its *trans*-form. The *trans* : *cis* selectivity for sodium(I) : potassium(I) is 238.[40] A system has also been prepared that can be used for the photo control of alkali metal extraction and ion transport. This photo control transport mechanism can be viewed as a type of "butterfly-like motion" as shown in Figure 1. The

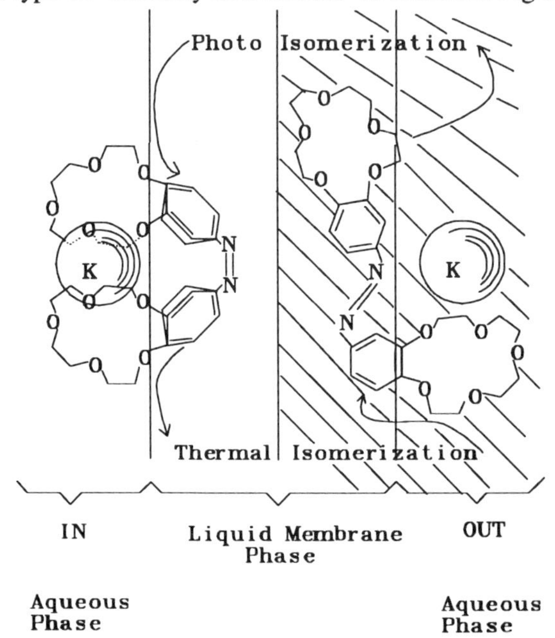


Figure 1 Photo Control Transport

potassium(I) is bound to the *cis* form in the IN aqueous phase, and released from the *trans* form in the OUT aqueous phase. The potassium(I) is rapidly extracted in the IN aqueous phase, and while moving in the liquid membrane to the OUT aqueous phase it is thermally isomerized to the *trans* form. The choice of counteranion is important with potassium(I) transport across a liquid membrane being suppressed by light when a hydrophobic counterion is used, and accelerated in the presence of a relatively hydrophilic counterion.[41] Another example involves the **72**. For the *cis* isomer, the selectivity order shows a preference for potassium(I), whereas for the *trans* isomer the smaller lithium(I) and sodium(I) are also extracted.[42,43] Tetralactams have also been used for the extraction of alkali and alkaline earth picrates from water into chloroform.[44]

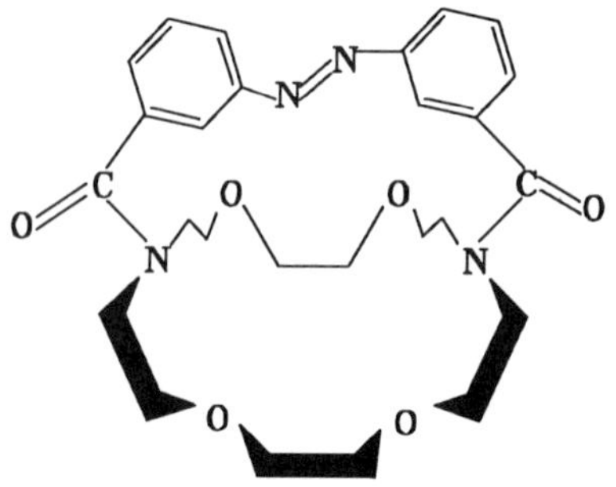

**72**

A modification to the butterfly-crown approach uses an aza-*bis*-(crown), again with ketone functionalities between the aryl group and the crown. This *cis*-azo compound linked to two *mono*-aza-15-crown-5 moieties (**73**) extracts potassium(I) better than rubidium(I). Size considerations with these *bis*- (crowns) becomes interesting, however,

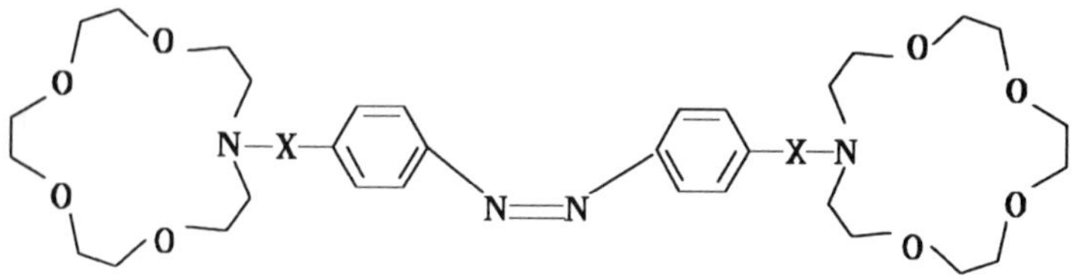

**73**

because the *cis-endo* stereochemistry **74** places the crowns further apart than does the *cis-exo* form **75**.[45] For a series of photoresponsive crown

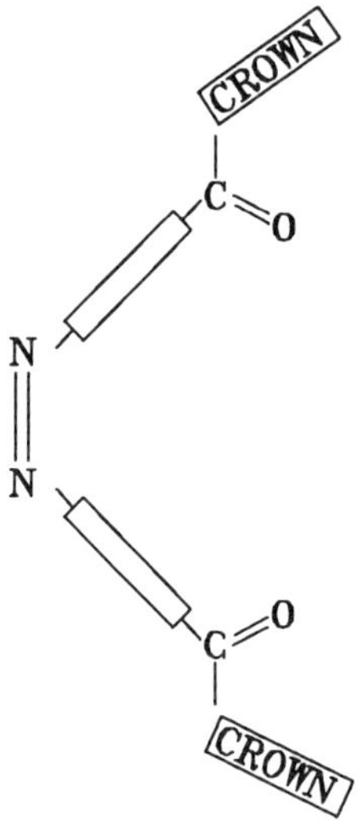

74 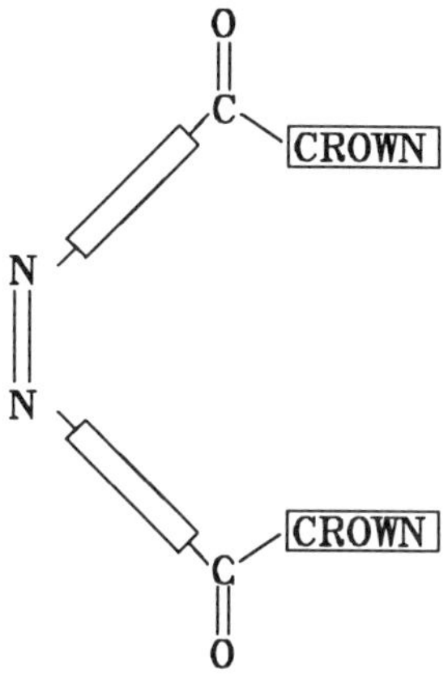 75

ethers, it has been found that in the presence of alkali metal ions the concentration of *cis*-isomer under the photostationary state is enhanced, and the rate of thermal *cis* to *trans* isomerization is decreased. These effects are explained on the basis of a “tying effect” of the complexed cations whereby additional energy is required to disrupt the crown-metal cation interaction. Correlation of the extraction data with these two factors implies that the alkali metal ions are extracted under photo irradiation as intramolecular 1 : 2 metal ion : crown complexes, and that the selectivity is associated with the fit between the ion size and the size of the spatial cavity generated by the *cis*-crown ethers. Thus the selectivity is affected both by the crown ring size and the steric crowding around the azo linkage.[46] A crown ether with a photoresponsive lariat phenolate group **76** has been used for the extraction of alkali metal ions. In the *cis*-form the phenolate group can interact with a metal ion bound into the crown ether moiety, and this is reflected by the enhanced extractability of sodium(I) and calcium(II). The enhancements are factors of 4 and 276 respectively.[47] A photoresponsive crown ether with

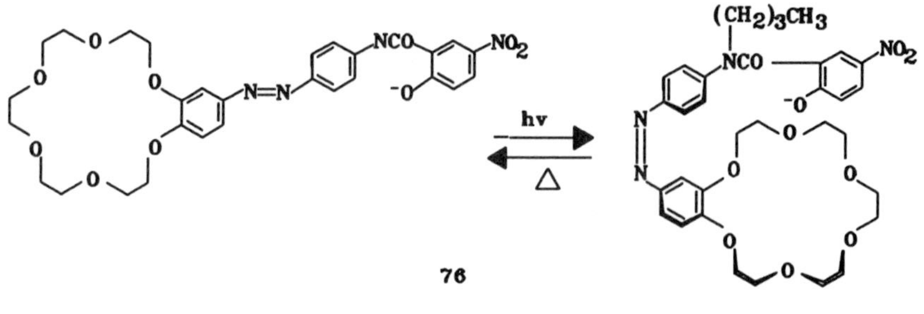

76

an anionic phosphoric acid group cap has also been used for the extraction of alkali metal ions. Again, sodium(I) suppresses the thermal isomerization of the *cis*-isomer to the *trans*-isomer.[48] The fluorescent crown ether *N*-(4-methylumbelliferone-8-methylene)-monaza-15-crown-5 and its -18-crown-6 analog extracts both alkali and alkaline earth metal ions into 1,2-dichloroethane.[49]

## 3. SPECIALTY CROWNS

Spiro crown ethers **77** (m = 2, n = 1), **78** (m = n = 2), **79** ( R = Me), **80** ( R = Me) and **81** (n = 3), especially the 16-crown-5 derivatives, show good extractions of alkali metal picrates. The data in Table 9 for

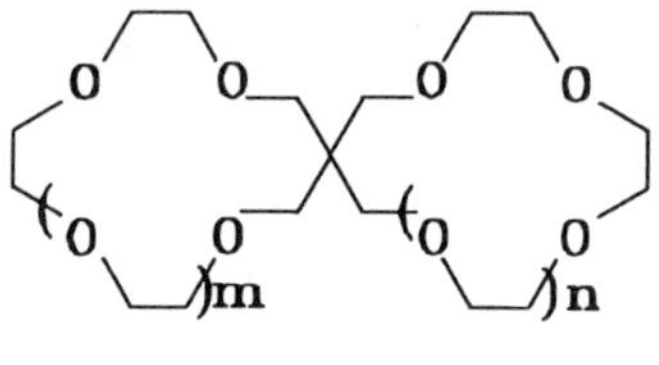

77, 78

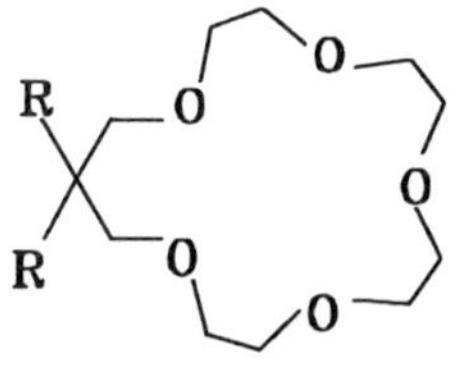

79

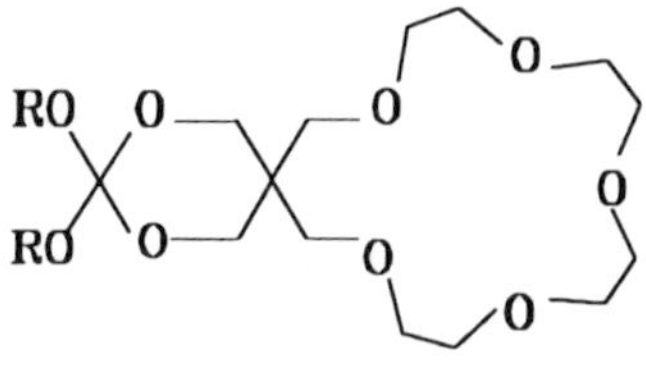

80

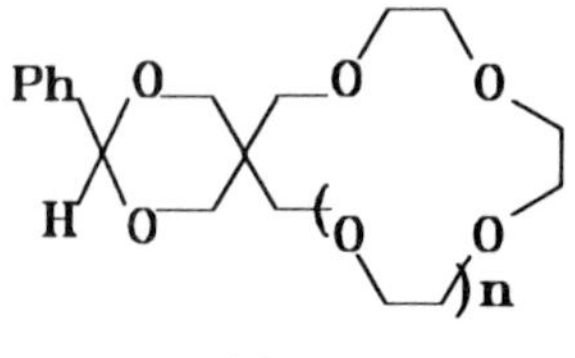

81

for these compounds show that for the crown-5 derivatives there is a preference for the lighter alkali metals. By contrast, the crown-6 derivatives show a preference for the heavier congeners.[50]

**TABLE 9. Extraction by Spiro Crown Ethers**

| Cmpd | Extraction (%) | | | |
|---|---|---|---|---|
| | Na(I) | K(I) | Rb(I) | Cs(I) |
| **77** | 14.5 | 3.4 | 2.9 | 1.9 |
| **78** | 24.0 | 10.2 | 9.0 | 7.6 |
| **79** | 39.7 | 8.2 | 8.2 | 4.8 |
| **80** | 32.8 | 9.6 | 6.8 | 4.2 |
| **81** | 1.1 | 33.2 | 27.3 | 11.4 |

Several macrocycles that are analogs of nactirs are effective for the extraction and transport of alkali metal ions. Two that are particularly effective (**82** and **83**) have furan rings within the macrocycle.

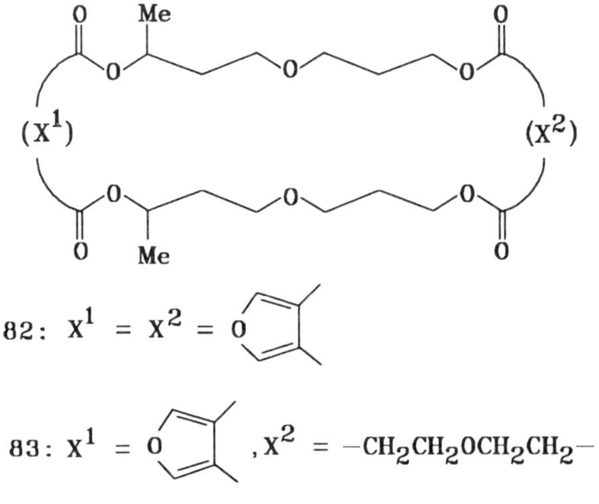

82: $X^1 = X^2 =$ (furan-2,3-diyl)

83: $X^1 =$ (furan-2,3-diyl), $X^2 = -CH_2CH_2OCH_2CH_2-$

For **82** the selectivity order is lithium(I) > rubidium(I) > potassium(I) >

sodium(I) > cesium(I) (lithium(I) : cesium(I) = 128), and for **83** it is lithium(I) > potassium(I) = rubidium(I) > sodium(I) > cesium(I) (lithium(I) : cesium(I) = 19).[51]

The effect of substituents on 16-crown-5 on the extraction of alkali metal ions shows a number of trends. The extraction depend on the position, number, type and stereochemistry of the substituents that are appended. Increasing substitution leads to decreased extractability, which is attributed to the limited access of counter anions and/or to the lack of conformational mobility that is required to make complex formation favorable. The effect of stereochemistry is shown with 14, 16-dimethyl-16-crown-5 where the *trans* isomer is a better extractant of alkali metal ions than is the *cis* isomer. The bridged 14, 16-vinylene-and 14, 16-ethylene-16-crown-5 compounds show high extractabilities which is interpreted on the basis of the structural freezing by the axial bridge, which prohibits the flip-flop motion and fixes two adjacent oxygens in a position favorable for complexation.[52]

Fluorinated crown ethers and cryptands have been used for the extraction of Group I and II metal ions. These fluorine-containing macrocycles form more stable metal complexes than do the analogous hydrocarbons. This stabilizing effect is only observed when the radii of the metal ion and the macrocyclic cavity are complementary, since only then can the metal; ion contact the C-F unit. For extraction of picrate salts into chloroform, **84** is matched for lithium(I), and **85** for sodium(I) and potassium(I).[53]

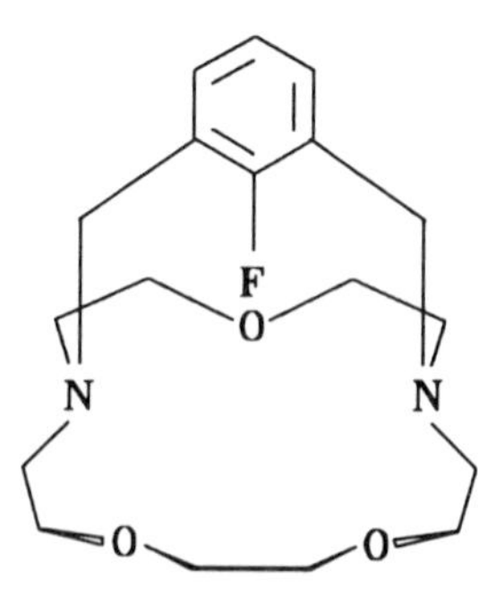


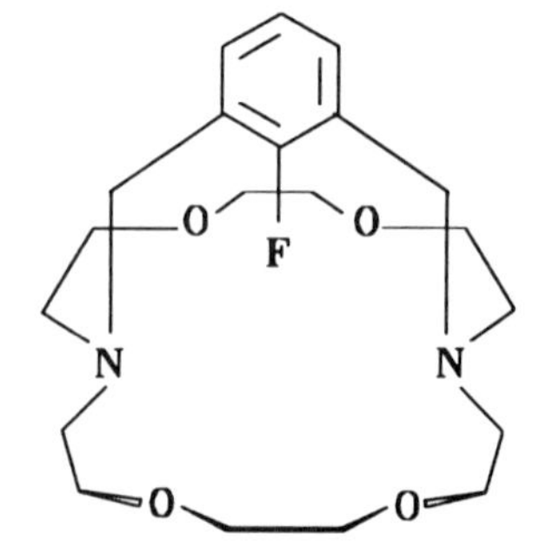


84 85

# 4. CAVITANDS AND CRYPTANDS

Cavitands can also be used to bind and extract Group I metal ions. An *N,O*-cryptand **86** is such an extractant. This cryptand extracts

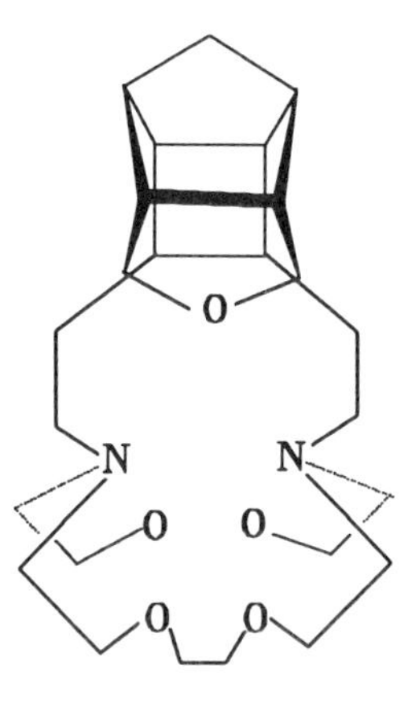


86

sodium(I) essentially quantitatively, with the sequence for alkali metal picrates being sodium(I) > potassium(I) > rubidium(I) > lithium(I) > cesium(I). This cryptand is a better extractant than are the *N,O*-crowns with which it has been compared.[54]

A series of hemispherands with self-organizing units have been used for the extraction of alkali metal ions into chloroform. These hosts contain four self-organizing methoxybenzene or ethoxybenzene units attached to one another, with their ends being attached to other units. These *O*- and *N,O*-donor hemispherands. The selectivities of these hosts for the alkali metal ions are given, and interpreted in terms of the principles of complementarity and of preorganization.[55] The cryptand **87**

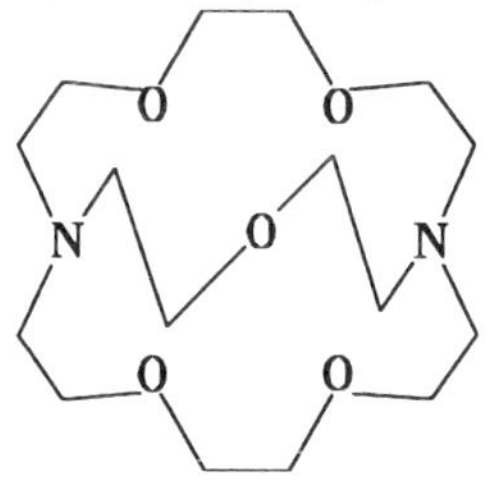


87

also extracts alkali metal ions into dichloromethane, with a selectivity order sodium(I) > potassium(I) >> lithium(I). This cryptand is a better extractant for both sodium and potassium than are the lariat crowns **88** (R = H, $CH_2CO_2Et$, $CH_2C(O)NEt_2$, $CH_2py$, $CH_2CN$, $CH_2Ph$).[56] A cryptand constructed around tertiary amines and containing fluorine substituents

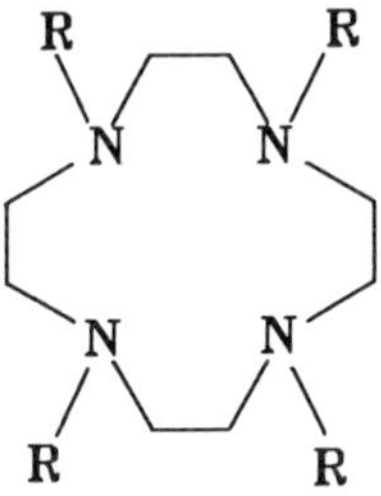

**88**

on the aryl groups (**89**) has been used as an extractant for alkali metal

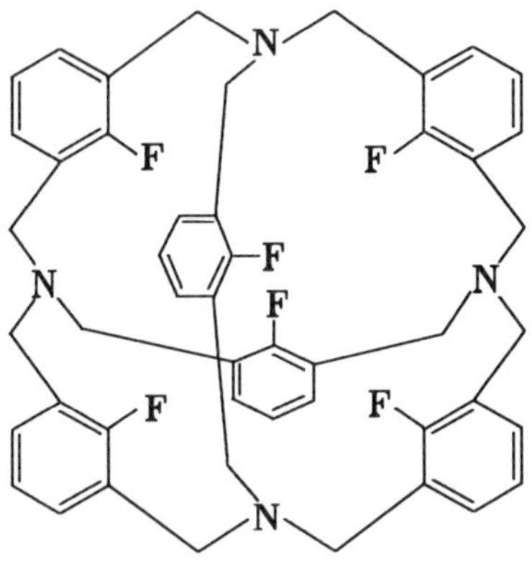

**89**

ions. Although the compound is an excellent extractant for silver(I), it is also a good extractant for potassium(I). The non-fluorinated analog is not an extractant for either of these metal ions. A crystal structure of the potassium complex of **89** reveals $K^+$- - F interactions, and this action of fluorine as a donor atom to the potassium ion is a likely explanation for **89** acting as an extractant.[57] Computational binding, free energy, and extraction selectivity calculations have been carried out on a group of anisole (**90** and **91**) and phenanthroline (**92**) spherands. The calculated

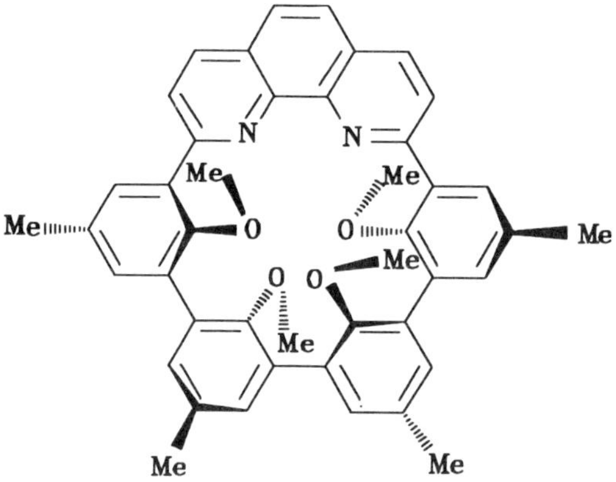

90

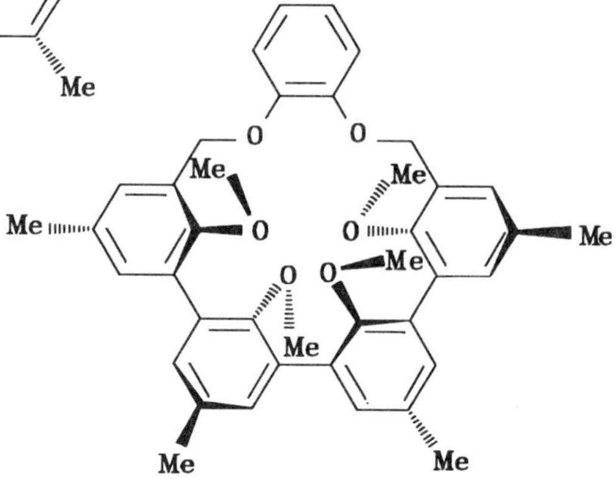

91

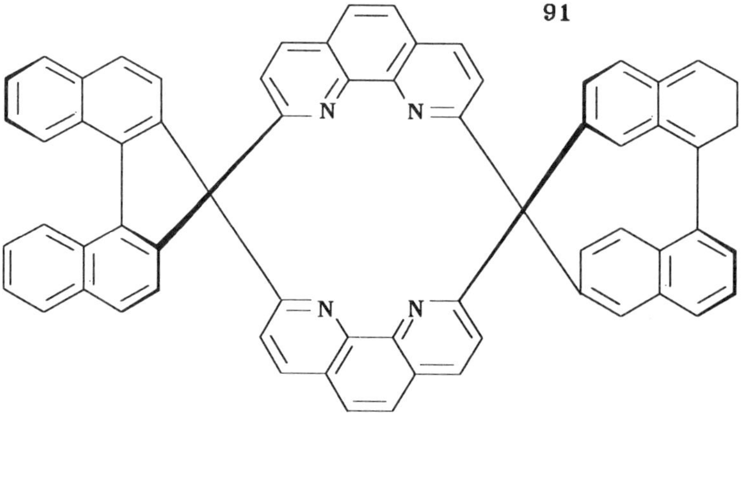

92

free energies all favor sodium(I) over lithium(I) and potassium(I). For **90** and **91** these calculations are in agreement with the experimental findings. For **92** the calculations again show a preference for sodium(I), whereas the experimental data show no such selectivity. No explanation is offered to explain the failure of the computational results for **92** to correlate with the experimental data.[58]

# 5. OTHER OXYGEN DONORS

Compounds **93**, **94** and **95** have been used as extractants for alkali and alkaline earth metal ions. The compounds **93** and **94** are liquid-liquid extractants for sodium(I) and potassium(I). Compound **93** is also an extractant for lithium(I), and **95** is an extractant for rubidium(I). The extraction selectivity of these compounds toward different cations parallels the potentiometric selectivity factors obtained for liquid membrane electrodes with these compunds as the membrane components.[59]

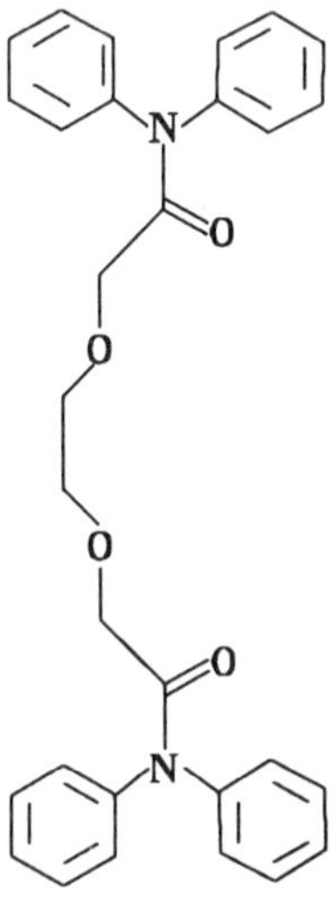

N
O
O
O
O
N

**93**

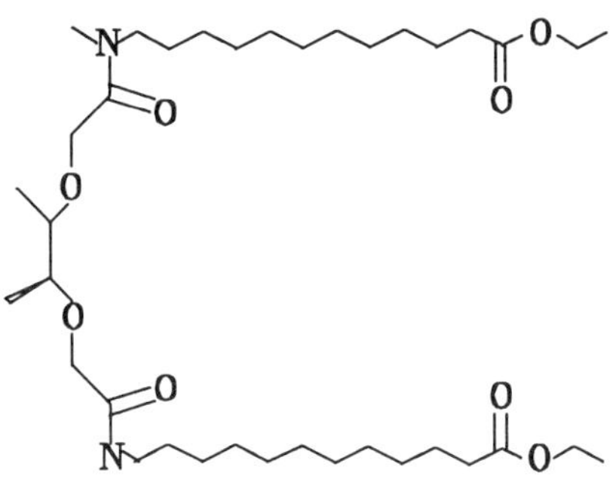

**94** 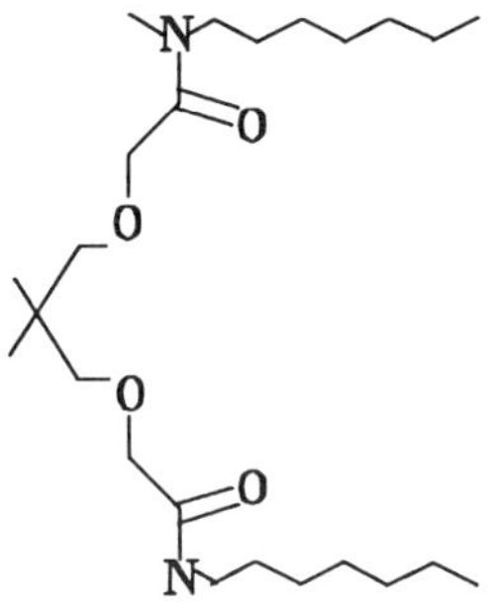 **95**

A cyclic alkene incorporated into a crown ether (**96**) can undergo *cis-trans* photoisomerization in the presence of benzophenone. For both liquid-liquid extraction and metal cation transport the *cis* isomer has a preference over the *trans* isomer for alkali metal ions.[60]

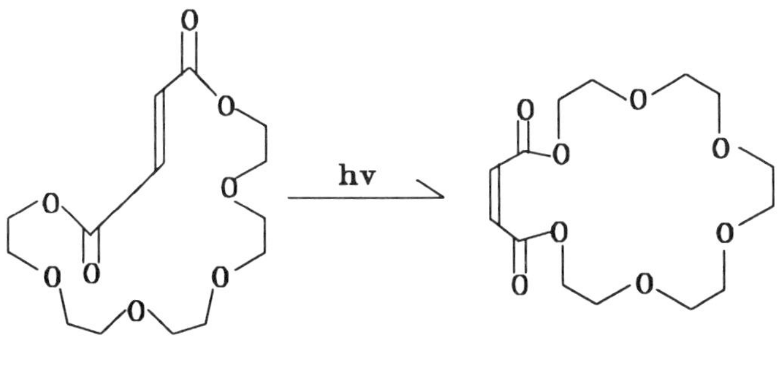

hv

**96**

Preorganized nickel complexes **97**- **99** can be used to transport alkali metal ions. The preorganization is accomplished by using the nickel complex as a template to keep the oxyethylene groups in the preferred configuration for binding.[61] Similarly, a palladium(II) complex

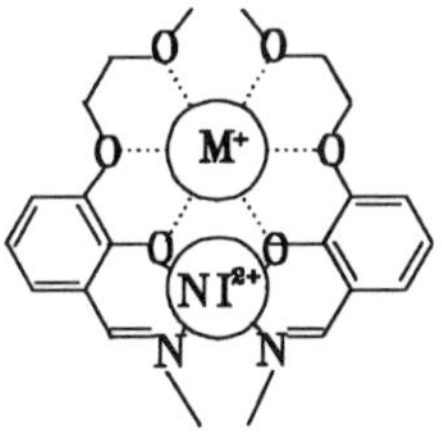

97 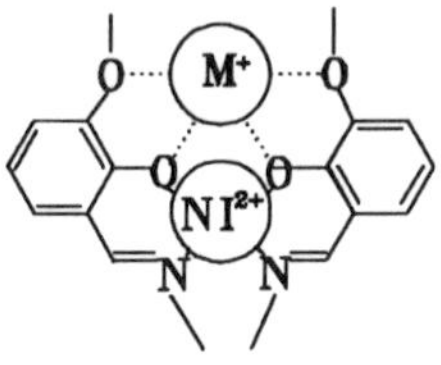 98 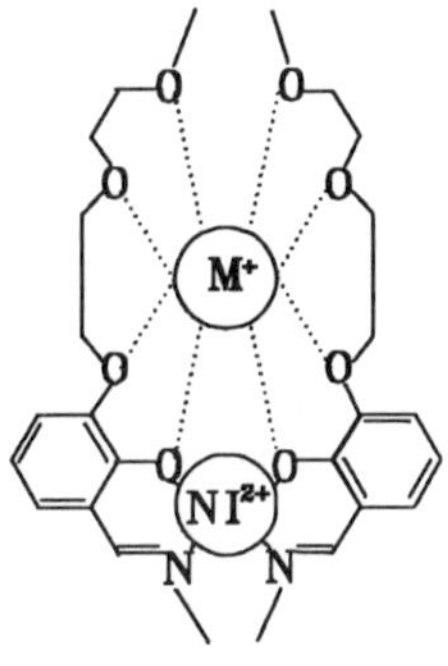 99

**100** (n = 1-3) has been used to link a pair of complexants together in a *trans* arrangement, and this complex is then used as an extractant for alkali metals. In each case the palladium(II) complex extracts alkali

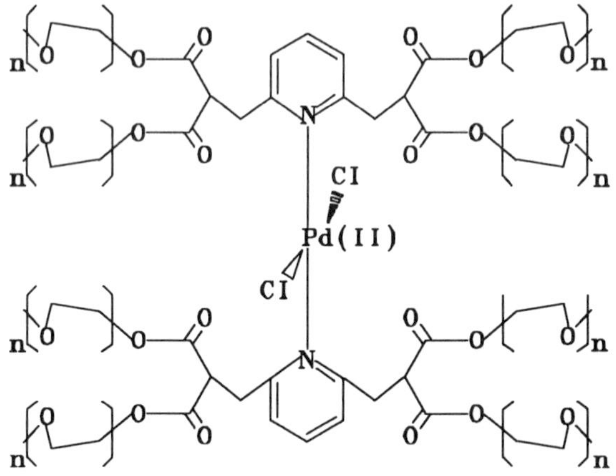

100

metals from aqueous solution, whereas the free complexants do not. The explanation offered to explain this difference is that the juxtaposition of functional groups on opposite ligands results in their being placed in a preferential arrangement for binding with alkali metals.[62]

A series of ethyl acetate and trioxyethylene ether derivatives of linear all-*ortho* methylene-linked oligomers of *p-tert*-butylphenol (**101**; m = 0-5; R = $CH_2CO_2Et$ or $(CH_2CH_2O)_3Me$) are extractants for alkali metals. The extractability for the ester series increases with increasing

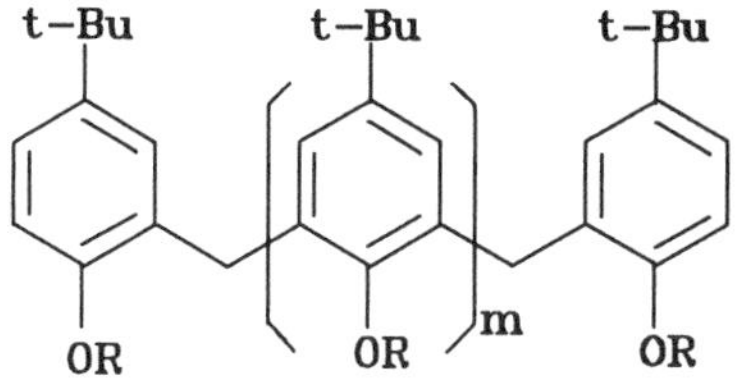

**101**

chain length. A different pattern is observed for the ether derivatives, however, with the affinity of the even-membered congeners being higher than the odd-membered ones for alkali metals.[63]

# 6. METAL OXIDES

Several types of metal oxide have been used for the extraction of alkali metal ions. One such is a hollandite-type manganese oxide prepared by reacting lithium permanganate with manganese(II) in sulfuric acid solution. The extracted metal ions can insert into the lattice by both redox-type and ion-exchange-type reactions. This lattice functions as an ion-sieve material with an effective pore radius of 1.41Å for the adsorption of alkali and alkaline earth metal ions.[64] Other manganese oxides that have been used are spinel-type lithium manganese oxide, birnessite-type sodium manganese oxide,[65] and todorokite-type magnesium manganese oxide. This todorokite-type material shows dibasic behavior toward lithium(I), sodium(I), and potassium(I), but monobasic behavior toward cesium(I). The relative affinity order is cesium(I) = lithium(I) < sodium(I) < potassium(I) at pH 5, and cesium(I) < potassium(I) < sodium(I) < lithium(I) at pH 10. The effective pore radius for the material is 2.7Å.[66]

# 7. PHYSICAL METHODS

For separation and analysis, the role and uses of ion-exchange chromatography in neutron activation analysis have been reviewed. Examples of post-irradiation group separations and isolation of individual radionuclides, and also pre-irradiation separations are presented. Special emphasis is given to difficult cases such as micro-macro separations of ions with similar properties, such as alkali metals and rare earths. The importance of the correct choice of chromatographic system, including such factors as the type of ion exchanger-solution combinations, resin cross-linking, and temperature, is covered in this review.[67]

The properties of the picrate anion have been reviewed, especially with respect to its structural motifs with metal cations. In many structures the picrate ion interacts as a mono and bidentate ligand. The preferred binding site in the solid state is the phenolate oxygen because of ionic interactions. The second choice is an oxygen of the *ortho*-nitro group. With respect to cation extraction, and membrane transport and selectivity, the effective delocalization of the negative charge over the aromatic polynitro system leads to the existence and stability of picrate complexes where the picrate anion is completely excluded from the cation coordination sphere. On the other hand, this versatility may be disadvantageous because the picrate anion may provide additional binding sites for unfavored cations to saturate their coordination sphere. The highly polarizable aromatic polynitro system in the picrate anion can be advantageous in increasing the extractability and transport rate of alkali and alkaline earth metal cations in hydrophobic organic solvents and membranes, and may not be representative of the behavior of other salts of these cations.[68] As a result, extraction values obtained with picrate salts may not be representative of those obtained with hydrophilic anions such as chloride and nitrate, which are common anions in solutions from which metal ions are commonly extracted.

The thermodynamics involved in the liquid-liquid extraction of alkali metal cations have been considered. Gibbs functions, enthalpies, entropies, and volumes of reaction for the reaction of the bacterial ionophore monensin with alkali metal cations in a range of water-organic biphasic solvent systems have been obtained. The selectivity sequence is not solvent dependent, even though the enthalpic and entropic contributions for the different cations vary with the solvent system involved. Instead, the selectivity results from small differences between two strongly cation-dependant parameters: the chemical potential of the cation in water, and the chemical potential of the lasalocid alkali metal salt in the organic phase. Changes with cation of the Gibbs functions of formation from the monesin acid of the complex neutral salt, which determines this selectivity, are mainly governed by enthalpic effects.[69]

# 8. CESIUM

Cesium has become of particular interest recently because of its presence as the radioactive cesium-137 in wastes generated by the nuclear industry. The challenge therefore is to find extractants that are very selective for cesium(I) under the particularly demanding conditions common in such wastes. These conditions may be strongly basic or acidic, they often have high concentrations of other salts, and are highly radioactive. A useful extractant must be able to tolerate these extreme conditions.

In targeting selective extractants for cesium(I), crown ethers are a viable choice because their cavity size can be tailored to match that of the cation. An early study of the extraction of alkali metal picrates by crowns revealed that for 18-crown-6 ethers the selectivity order is potassium(I) ~ cesium(I) > rubidium(I) > sodium(I).[70] Subsequently crown ethers have been synthesized that have higher cesium selectivity. One of these, a *bis*-[4(5)-*tert*-butylbenzo]-21-crown-7 (**102**), has a high

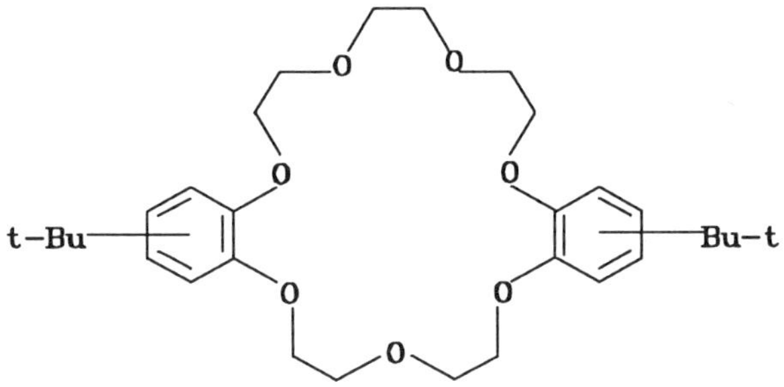

102

liquid-liquid extraction ratio for cesium(I) into toluene, and the selectivity is sufficiently high that it can be used to recover cesium(I) from large amounts of sodium(I) and potassium(I).[71] The analogous dibenzo-21-crown-7 also has a high capability of removing cesium(I)-137 from both acidic and alkaline solutions. However, high radiation does can cause a decrease in both the extraction efficiency and the selectivity. A high resistance to radiolysis can nevertheless be achieved by dissolving the crown ether in fluorinated alcohols.[72] The effect of different substituents on the 21-crown-7 crown ethers **103-106** in relation

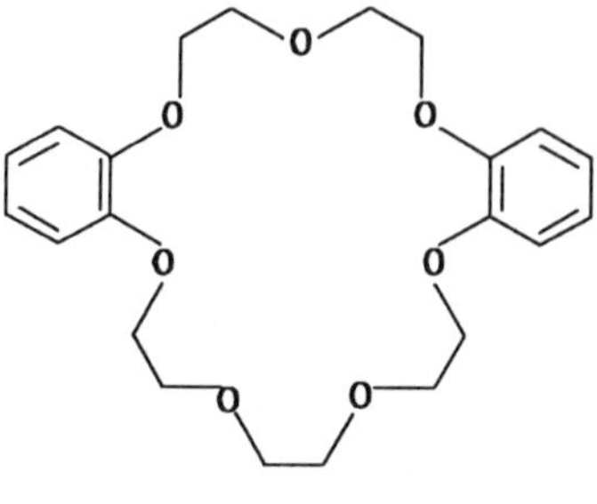

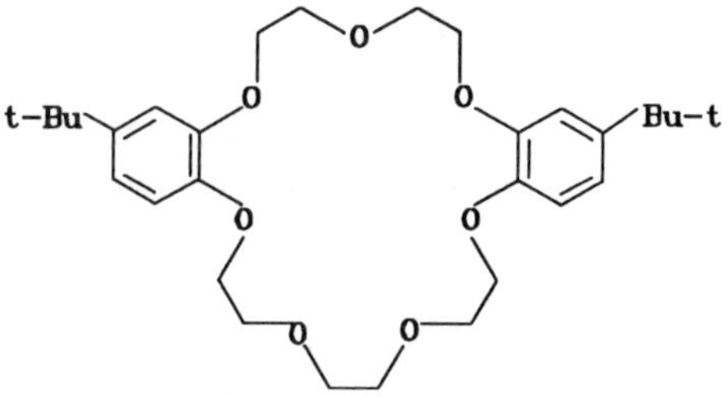

103

104

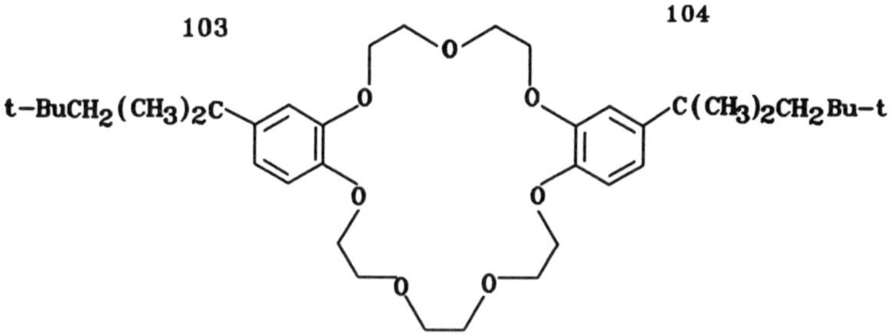

105

to cesium(I) extraction has revealed a number of different effects. The solvent extraction of cesium(I) nitrate occurs with the formation of the species $CsLNO_3$ (L = crown ether), and partial dissociation to $CsL^+$ and nitrate ion in the organic phase. Alkyl substitution on the benzo groups (**104**, **105**) does not significantly increase the extraction ability. However, the dicyclohexano-21-crown-7 (**106**) is a much improved

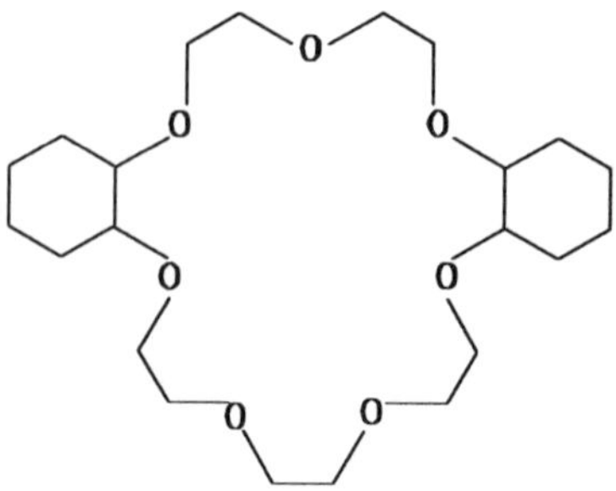

106

extractant. A computer analysis of the extraction data reveals the additional formation of a 1 : 2 metal : crown species in the case of **106**.[73] The larger crown-8 derivatives have also been used as extractants for cesium(I). One such compound is tetrabenzo-24-crown-8, which as its 4,4-and -4,5-*bis*-(*tert*-octylbenzo) derivatives (**107** and **108**) are good extractants. The liquid-liquid distribution data suggest that in the 1,2-dichloroethane (DCE) phase the anions are in the outer coordination sphere of cesium(I), apparently due to coordination by DCE.[74] The addition of benzene-1,3,5-tricarboxamide enhances cesium(I) nitrate by tetrabenzo-24-crown-8 because of its hydrogen bonding to the nitrate.[75]

Cesium K- and 1,3-edge EXAFS (Extended X-ray Absorption Fine Structure) have been evaluated as a method of probing the interaction of cesium(I) with dibenzocrown ethers in a tributyl phosphate

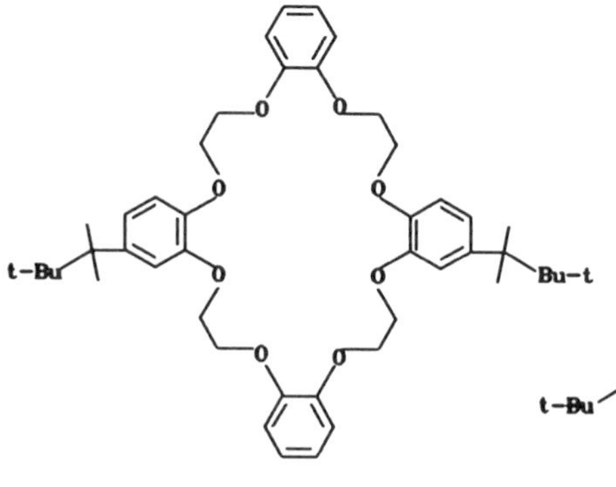

**107**

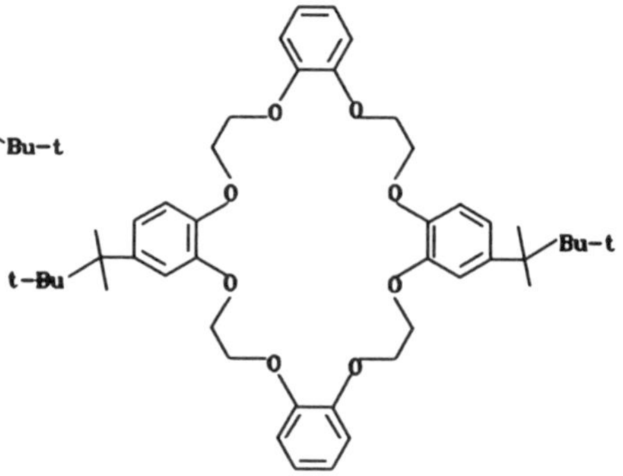

**108**

liquid phase under conditions used for the extraction of radiocesium from acidic nuclear wastes. The K-edge data are found to be less satisfactory than the 1,3-edge data. The cesium(I) coordination to the crowns is unaffected by the presence of water, or the nature of the counterion. The average Cs-O bond length decreases from 3.23(3) Å in the cesium(I) complex of *bis*-(*tert*-butylbenzo)-18-crown-6, to 3.01(4) Å in the corresponding 21-crown-7 and 24–crown-8 compounds, a result that is consistent with extraction studies indicating 1:1 complex formation in the 21-crown-7 and 24-crown-8 compounds, and a mixture of 1:1 and 1:2 complexes in the 18-crown-6.[76] Since both crowns and calixarenes are good extractants for cesium, a compound that combines both of these structural motifs has the potential to be an excellent extractant for cesium(I). This potential has been realized with calixcrown **109**, which is a highly selective extractant for cesium(I).[77]

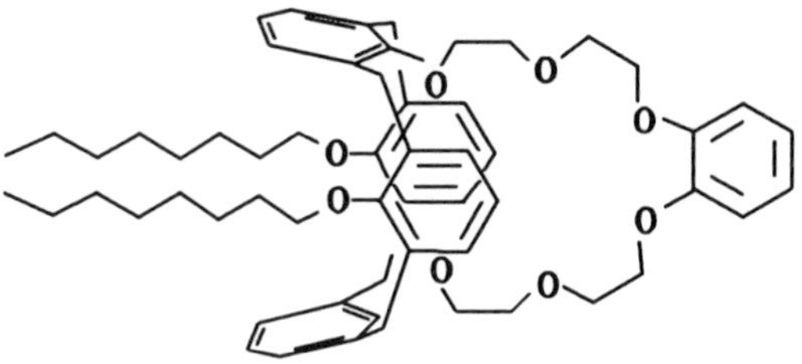

**109**

Another type of cage compound that can be used as an extractant is a carborane.. As a result, aromatic substituted metallacarboranes have been used as extractants for cesium-137 and strontium-90 from nuclear wastes. The two compounds that have been used are the sandwich compounds $[3,3\text{-Co}(1\text{-Ph-}1,2\text{-}C_2B_9H_{10})_2]^-$ (**110**) and $[3,3\text{-Co}(1,7\text{-Ph}_2\text{-}1,7\text{-}C_2B_9H_9)_2]^{2-}$ (**111**). Good extraction of cesium-137 is observed at pH 3, but is decreased at pH 1 for **110**, while **111** is unaffected in its extraction properties by the pH lowering.[78]

# REFERENCES


1. Y. Marcus, L. E. Asher, *J. Phys. Chem.*, **1978**, *82*, 1246.
2. F. Wada, Y. Wada, T. Goto, K. Kikukawa, T. Matsuda, *Chem. Lett.*, **1980**, 1189.
3. K. Kikukawa, H. Gong-Xin, A. Abe, T. Goto, R. Arata, T. Ikeda, F. Wada, T. Matsuda, *JCS, Perkin Trans. II*, **1987**, 135.
4. A. S. Khan, W. G. Baldwin, A. Chow, *Can. J. Chem.*, **1981**, *59*, 1490.
5. U. Olsher, M. G. Hankins, Y. D. Kim, R. A. Bartsch, *J. Am. Chem. Soc.*, **1993**, *115*, 3370.
6. Y. Inoue, Y. Liu, F. Amano, M. Ouchi, A. Tai, T. Hakushi, *JCS, Dalton Trans.*, **1988**, 2735.
7. Y. Inoue, F. Amano, N. Okada, H. Inada, M. Ouchi, A. Tai, T. Hakushi, Y. Liu, L.-H. Tong, *JCS, Perkin Trans 2*, **1990**, 1239.
8. Y. Liu, L.-H. Tong, Y. Inoue, T Hakushi, *JCS, Perkin Trans. 2*, **1990**, 1247.
9. I. M. Kolthoff, M. K. Chantooni, Jr., *J. Chem. Eng. Data,* **1997**, *42*, 49.
10. U. Olsher, J. Jagur-Grodzinski, *JCS, Dalton Trans.*, **1981**, 501.
11. R. A. Bartsch, M. D. Eley, A. P. Marchard, R. Shukla, K. A. Kumar, G. M.

Reddy, *Tetrahedron*, **1996**, *52*, 8979.
12. R. D. Rogers, A. H. Bond, C. B. Bauer, *Pure Appl. Chem.*, **1993**, *65*, 567.
13. Y. Marcus, L. E. Asher, *J. Phys. Chem.*, **1978**, *82*, 1246.
14. J. D. Lamb, J. J. Christensen, S. R. Izatt, K. Bedke, M. S. Astin, R. M. Izatt, *J. Am. Chem. Soc.*, **1980**, *102*, 339.
15. M. Burgard, L. Jurdy, H. S. Park, R. Heimburger, *Nouv. J. Chem.*, **1983**, *7*, 575.
16. T. M. Fyles, *Can. J. Chem.*, **1987**, *65*, 884.
17. J. Strezelbicki, R. A. Bartsch, *Anal. Chem.*, **1981**, *53*, 2251.
18. J. Strezelbicki, R. A. Bartsch, *Anal. Chem.*, **1981**, *53*, 1984.
19. W. A. Charewicz, R. A. Bartsch, *Anal. Chem.*, **1982**, *54*, 2300.
20. W. A. Charewicz, W. Walkowiak, R. A. Bartsch, *Anal. Chem.*, **1987**, *59*, 494.
21. W. Walkowiak, W. A. Charewicz, S. I. Kang, I.-W. Yang, M. J. Pugia, R. A. Bartsch, *Anal. Chem.*, **1990**, *62*, 2018.
22. W. Walkowiak, S. I. Kang, L. E. Stewart, G. Ndip, R. A. Bartsch, *Anal. Chem.*, **1990**, *62*, 2022.
23. R. A. Bartsch, B. P. Czech, S. I. Kang, L. E. Stewart, W. Walkowiak, W. A. Charewicz, G. S. Heo, B. Son, *J. Am. Chem. Soc.*, **1985**, *107*, 4997.
24. M. J. Rugia, G. L. Ndip, H. K. Lee, I. W. Yang, R. A. Bartsch, *Anal. Chem.*, **1986**, *58*, 2723.
25. W. Walkowiak, G. M. Ndip, D. H. Desai, K. H. Lee, R. A. Bartsch, *Anal. Chem.*, **1992**, *64*, 1685.
26. B. P. Czech, H. Huh, R. A. Bartsch, *J. Org. Chem.*, **1992**, *57*, 725.
27. R. A. Bartsch, I. W. Yang, E. G. Jeon, W. Walkowiak, W. A. Charewicz, *J. Coord. Chem.*, **1992**, *27*, 75.
28. R. B. Davidson, R. M. Izatt, J. J. Christensen, R. A. Schultz, D. M. Dishong, G. W. Gokel, *J. Org. Chem.*, **1984**, *49*, 5080.
29. Y. Inoue, C. Fujwara, K. Wada, A. Tai, T. Hakushi, *JCS. Chem. Comm.*, **1987**, 393.
30. Y. Inoue, M. Ouchi, K. Hosoyama, T. Hakushi, Y. Liu, Y. Takeda, *JCS, Dalton Trans.*, **1991**, 1291.
31. J. C. Hernandez, J. E. Trafton, G. W. Gokel, *Tet. Letts.*, **1991**, *32*, 6269.
32. M. Zinic, S. Alihodzic, V. Skaric, *JCS, Perkin 1*, **1993**, 21.
33. M. Ouchi, K. Mishima, R. Dohno, T. Hakushi, *ACS Symp. Ser.*, **1997**, *659*, 293.
34. H. Nakamura, M. Takagi, K. Ueno, *Anal. Chem.*, **1980**, *52*, 1668.
35. T. Pigot, M.-C. Duriez, C. Picard, L. Cazaux, P. Tisnes, *Tetrahedron*, **1992**, *42*, 4359.
36. A. P. Marchand, H.-S. Chong, *Tetrahedron*, **1999**, *55*, 9697.
37. H. Sakamoto, H. Goto, K. Doi, M. Otomo, *Chem. Lett.*, **1992**, 1535
38. H. Nakamura, H. Sakka, M. Takagi, K. Ueno, *Chem. Lett.*, **1981**, 1305.
39. B. P. Bubnis, C. E. Pacey, *Tet. Lett.*, **1984**, *25*, 1107.
40. S. Shinkai, T. Ogawa, Y. Kusano, O. Manabe, *Chem. Lett.*, **1980**, 283.
41. S. Shinkai, T. Nakaji, T. Ogaura, K. Shigematsu, O. Manabe, *J. Am. Chem. Soc.*, **1981**, *103*, 111.
42. S. Shinkai, T. Ogawa, T. Nakaji, Y. Kusano, O. Nanabe, *Tetrahedron. Lett.*, **1979**, *47*, 4569.
43. M. Shiga, M. Takagi, K. Ueno, *Chem. Lett.*, **1980**, 1021.
44. T. Pigot, M.-C. Duriez, C. Picard, L. Cazaux, P. Tisnes, *Tetrahedron*, **1992**, *48*, 4359.

45. S. Shinkai, K. Shigematsu, Y. Kusano, O. Manabe, *J. Chem. Soc, Perkin Trans. I*, **1981**, 3279.
46. S. Shinkai, T. Ogaura, Y. Kusano, O. Manabe, K. Kikukaura, T. Goto, T. Matsuda, *J. Am. Chem. Soc.*, **1982**, *104*, 1960.
47. S. Shinkai, T. Minanin, Y. Kusano, O. Manabe, *J. Am Chem. Soc.*, **1982**, *104*, 1967.
48. S. Akabori, Y. Miura, N. Yotsumoto, K. Uchida, M. Kitano, Y. Habata, *JCS, Perkin Trans. I*, **1995**, 2589.
49. H. Nishida, Y. Katayama, H. Katsuki, H. Nakamura, M. Takagi, K. Ueno, *Chem. Lett.*, **1982**, 1853.
50. M. Ouchi, Y. Inoue, H. Sakamoto, A. Yamahira, M. Yoshinaga, T. Hakushi, *J. Org. Chem.*, **1983**, *48*, 3168.
51. A. Samat, M. El. Malouli Bibout, J. Elguero, JCS, *Perkin Trans. I*, **1985**, 1717.
52. Y. Inoue, K. Wada, Y. Liu, M. Ouchi, A. Tai, T. Hakushi, *J. Org. Chem.*, **1989**, *54*, 5268.
53. H. Plenio, R. Diodone, *J. Am. Chem. Soc.*, **1996**, *118*, 356
54. Marchand, S. Alihodzic, A. S. McKim, K. A. Kumar, K. Mlinaric-Majerski, T. Sumanovae, *Tetrahedron. Lett.*, **1998**, *39*, 1861.
55. S. P. Artz, D. J. Cram, *J. Am. Chem. Soc.*, **1984**, *106*, 2160.
56. H. Tsukube, Y. Mizutani, S. Shinoda, T. Okazaki, M. Tadokoro, K. Hori, *Inorg. Chem.*, **1999**, *38*, 3506.
57. H. Takemura, N. Kon, M. Yasutake, H. Kariyazono, T. Shinmyozu, T. Inazu, *Angew. Chem., Int. Ed. Engl.*, **1999**, *38*, 959.
58. J. Vacek, P. A. Kollman, *J. Phys. Chem. A.*, **1999**, *103*, 10015.
59. N. L. Kirsch, R. J. J. Funck, W. Simon, *Helv. Chim. Acta.*, **1978**, *61*, 2019.
60. H. Sasaki, A. Veno, T. Osa, *Chem. Lett.*, **1986**, 1785.
61. A. Schepartz, J. P. McDevitt, *J. Am. Chem. Soc.*, **1989**, *111*, 5976.
62. T. Mutou, K. Amimoto, H. Kanatomi, H. Koyama, T. Kawato, *Chem. Lett.*, **1999**, 1231
63. T.-I. Yamagishi, K. Tani, K. Shirano, S.-I. Ishida, Y. Nakamoto, *J. Polym. Sci: Part A: Polym. Chem.*, **1996**, *34*, 687.
64. Q. Feng, H. Kanoh, Y. Miyai, K. Ooi, *Chem. Mater.*, **1995**, *7*, 148.
65. Q. Feng, H. Kanoh, Y. Miyai, K. Ooi, *Chem. Mater.*, **1995**, *7*, 1226.
66. Q. Feng, H. Kanoh, Y. Miyai, K. Ooi, *Chem. Mater.*, **1995**, 7, 1722.
67. R. Dybezynski, *J. Chromatogr.*, **1992**, *600*, 17.
68. U. Olsher, H. Feinberg, F. Frolow, G. Shoham, *Pure & Appl. Chem.*, **1996**, *68*, 1195.
69. M. Hebrant, Y. Pointud, J. Juillard, *J. Phys. Chem.*, **1991**, *95*, 3653.
70. K. H. Wong, H. L. Ng, *J. Coord. Chem.*, **1981**, *11*, 49.
71. W. J. McDowell, G. N. Case, J. A. McDonough, R. A. Bartsch, *Anal. Chem.*, **1992**, *64*, 3013.
72. V. M. Abashkin, E. A. Filippov, A. K. Nardova, I. V. Mamakin, G. F. Egorov, *Mater. Res. Soc, Symp. Proc. 556 (Scientific Basis for Nuclear Waste Management XXII)*, **1999**, 1291.
73. Y. Deng, R. A. Sachleben, B. A. Moyer, *JCS, Farad. Trans.*, **1995**, *91*, 4215.
74. T. G. Levitskaia, J. C. Bryan, R. A. Sachleben, J. D. Lamb, B.A. Moyer, *J. Am. Chem. Soc.*, **2000**, *122*, 554.
75. K. Kavallieratos, R. A. Sachleben, G. J. Van Berkel, B. A. Moyer, *JCS, Chem. Comm.*, **2000**, 187.
76. M. R. Antonio, M. L. Dietz, M. P. Jensen, L. Soderholm, E. P. Horwitz, *Inorg.*

*Chim. Acta*, **1997**, *255*, 13.

77. V. Bonnesen, J. Haverlock, L. Engle, A. Sachleben, B. A. Moyer *ACS Sympos. Ser.* **2000**, *757*, 26.
78. C. Vinas, J. Bertran, S. Gomez, F. Teixidor, J-F. Dozol, H. Rouquette, R. Kivekas, R. Sillanpaa, *JCS, Dalton Trans.*, **1998**, 2849.

# 12

# PHYTOREMEDIATION AND BIOREMEDIATION OF SOILS AND WATERS

## 1. INTRODUCTION

Toxic metals are problematic in soils because not only are they adsorbed into the zeolite soil structure, but they are also absorbed into the humus and biomass present in soils. These humic and fulvic substances in soils contain compounds that act as chelating agents to these metals, thereby contributing to the difficulty of their removal. In addition, these substances often contain redox active agents that can convert the adsorbed metals into ones that have different oxidation states, or reduce metal ions down to the metallic state. Although the removal of metals from soils poses several challenges, the presence of good chelating agents in humic and fulvic substances is one of the obstacles that must be overcome if phytoremediation or bioremediation is to be the method of choice for metal removal. Phytoremediation is the use of green plants to remove pollutants from the environment. Two recent reviews have been written on this subject.[1,2] Among the types of phytoremediation currently in use are phytoextraction and rhizofiltration. Phytoextraction is defined as the use of metal-accumulating plants that concentrate them into the harvestable parts. Rhizofiltration is the use of plant roots to absorb metals from aqueous waste streams. Phytoextraction can be carried out either with or without added chelate complexant to assist in removing the metals. In certain cases the addition of chelating agents enhances the accumulation of metals by plants, especially if the chelate has a strong affinity for the targeted metal. Nevertheless, a consideration when using this method is the requirement that the chosen chelate must be biodegradable or readily removed from the contaminated site. Alternatively, phytoremediation can rely only on the physiological processes that allow plants themselves to accumulate metals. A disadvantage of this approach is that growth rates are slow, and the selectivity for particular metals is likely to be low. In the future, however, genetic engineering could be useful in producing plants that have both higher growth rates and metal selectivities. Bioremediation involves the use of biological remedies for pollution reduction.[3] For metals this detoxification process must involve processes such as the oxidation or reduction of the metal center to make it either more water soluble, so that it precipitates and can be removed in solid form, or converted to a more volatile form that can be removed in the gas phase. In choosing a bioremediation strategy for metals, the biological system must be able to tolerate the concentration of metal that is present at the site.

Among the toxic metals that are common soil contaminants and need to be removed are cadmium, lead, copper, chromium, and uranium. The challenge therefore is to find a phytoremediation or bioremediation strategy that is competitive with the metal binding function of humus and biomass. When considering phytoremediation or bioremediation strategies for these metals several factors need to be addressed. Among these factors are the oxidation state of the metal, the presence of preferred chelating functionalities for these metals, a high selectivity for complexing these particular metals against others such as sodium and calcium, metals that are themselves common components of soils, and a tolerance for high concentrations these metals that can be toxic to plants as well as to mammals. For uranium and other transuranic elements, the tolerance to radioactivity is another factor that needs to be considered.

## 2. PHYTOREMEDIATION

Certain plant materials, like fish and shellfish, can accumulate toxic metals. If these plants are part of the human food chain, it is important that they be grown under conditions where these metals are absent, or that they are subsequently removed. Indeed, the primary risk pathway associated with cadmium contaminated soils has been identified as the transfer of the metal to humans *via* accumulation in the plant food cycle.[4] Lettuce accumulates particularly high levels of cadmium. Nevertheless, although this feature can be problematic, it can also be useful in the phytoremediation of metal contaminated soils. Phytoremediation uses metal-tolerant plants such as lettuce that can hyperaccumulate metals in their shoots or foliage.

Phytoextraction involves the use of plants to remove toxic metals from soil. Early phytoextraction studies have focused on metallophytes from predominantly non-myocorrhizal plant families such as *Brassicaceae* and *Caryophyllaceae.* Nevertheless, the arbuscular mycorrhizal (AM) fungi (Glomales, Zygomycetes) also deserve special attention because about 95% of all plant species belong to characteristically mycorrhizal families, and potentially benefit from AM fungus-mediated mineral nutrition. A study has been made on the effects of phytoextraction practices with three plant species (*Silene vulgaris, Thlaspi caerulescens,* and *Zea mays*) and a fractional variation of soil amendments (either an ammonium or nitrate source of nitrogen, and the presence or absence of an elemental sulfur supplement) on AM fungi. The focus of the study is to test whether the treatments affect the density of glomalean spores and AM colonization in maize. The results show that the choice of plant species and soil amendments can have a great impact on the quantity and species composition of glomalean propagules, as well as on mycorrhiza functioning during long-term metal-remediation treatments.[5]

As an example of phytoremediation, *Thlaspi caerulescens* accumulates both cadmium and zinc, with toxicity being observed for shoot concentrations of 1270 mg $kg^{-1}$.[6] A similar approach uses chemically modified biomass. Two marine algae, *Sargassum fluitans* and *Ascophyllum nodosum*, cross-linked with aldehyde or embedded in polyethyleneimine, uptake heavy metals. The absorption order for the former is lead > cadmium > copper > nickel > zinc, and for the latter it is lead > copper > cadmium > nickel > zinc.[7] Both particle size and the type of chemical modification have a significant influence on the biosorption of these metals. Cadmium, lead, and zinc are complexed by a monorhamolipid biosurfactant produced by *Pseudomonas aeruginosa* ATCC 9027. This material binds cadmium and lead stronger then does fulvic acid or activated sludge solids.[8] In a study of the binding of mercury salts to natural substances, chitosan is the most effective of the group studied which includes bark, activated sludge (Milorganite), chitosan, redwood leaves, pine needles, senna leaves, sulfonated lignin, peet moss, and orange peel. The high binding capactiy of chitosan is considered to result from the greater basicity of its aliphatic primary amine group.[9] In a complexation study of humic acid with cadmium and lead it has been shown that the complexes are labile, that lead complexes are more stable than cadmium complexes, and that the humic acid has a more heterogeneous behavior with respect to lead than to cadmium.[10]

## 2.1. Chromium(VI)

Phytoremediation is becoming an important method for the removal of metals from soils and waters, and it has been used for the removal of chromium. When seeking phytoremediation for this metal the oxidation state is important. Such a situation is observed with alfalfa, which binds the metal in its chromium(III) form, but not when it is present as chromium(VI). In an approach to developing strategies for chromium phytoremediation, thirty-six plant species have been evaluated. This evalution has been done for both chromium(III) and chromium(VI) uptake and accumulation. There is a significant difference in the degree of tolerance, uptake, and accumulation of these chromium forms among the plant species. *Helianthus annus* (sunflower) is the least tolerant to chromium, and *Cynodon dactylon* (bermudagrass) and *Panicum virgatum* (switchgrass) are the most tolerant. *Brassica juncea* (indian mustard) and sunflower are the best accumulators of chromium. The majority of the plant species that were treated with chromium(VI) hyperaccumulated chromium and died. In some plants, such as those from the *Brassica* family, the addition of EDTA enhances chromium(III) uptake.[11] The greater uptake of chromium(III) in the presence of EDTA occurs because the chromium(III)-EDTA complex is not retarded by ion exchange in plant tissues.[12] Similarly, since chromium(VI) is anionic, it too is not held by ion exchange. However, both the accumulation of chromium(VI) in plants, and the chelate-induced uptake of chromium(III), can injure and kill the plant. The toxicity of chromium(VI) is due to its being a strong oxidant. As a result, a phytostablization technique for chromium(VI) has been suggested where it is reduced to chromium(III) in the plant rhizophere. This reduction has been proposed to occur in the roots of several plants.[13, 14] By using X-ray spectroscopy it has been shown that *Eichhornia crassipes* (water hyacinth) supplied with chromium(VI) accumulates chromium(III) in the root and shoot tissues. The chromium(VI) reduction appears to occur in the fine lateral roots. Following reduction, a portion of the chromium is transported to the leaves, where it may be bound as a chromium(III) oxalate complex.[15]

The efficiency of wheat, rape, and buckwheat for the removal of chromium from waste waters has been compared, with buckwheat being the most efficient. The plant roots are powerful agents for chromium(III) removal, inducing its subsequent precipitation on their surface. Although the chromium(VI) form of chromium is more toxic than is the chromium(III) form, some plants retain their capacity for chromium removal despite showing strong toxicity symptoms.[16] Other plants, however, do not show the symptoms of chromium(VI) damage. Such is the case for *Scirpus lacustris*, *Phragmites karka*, and *Bacopa monnieri*, which accumulate substantial amounts of chromium during the period of one week.[17]

## 2.2. Cadmium, Lead, Copper, and Zinc

The success of phytoremediation depends on choosing plant species that maximize metal removal to the plant shoots so that harvesting can effectively reduce the metal residues that were present in the soil. The overall effectiveness of the method depends on the growth rate of the plant species, its efficiency in accumulating the metal, and the quantity of biomass that is produced during the growth period. Early phytoremediation strategies for metals have used hyperaccumulator plant species such as *Thlaspi caerulescens*, but these plants suffer from slow growth and the production of only small amounts of biomass. Indian mustard (*Brassica juncea*), which accumulates metals only moderately, can however be more useful for phytoremediation. For example, in a test for zinc removal, *B. juncea* and *T. caerulescens* have been grown for six weeks in a soil that is contaminated with zinc. After this time period it is found that *B. juncea* has removed four-fold more zinc than has *T. caerulescens*. This result is found because the former plant species produces ten times more biomass than does the latter.[18]

*Arabidopsis thaliana*, a plant species with a relationship to the *Brassicaceae* family, can also be used for lead(II) accumulation. Different mutants are also effective, but they also accumulate many other metals in their shoots. An interesting observation is that lead accumulation is not necessarily linked with lead tolerance, suggesting that different genetic factors control these two processes.[19] The metal tolerant plants *Thlaspi caerulescens* and *Silene vulgaris* have been grown with Paris Island Cos Romaine lettuce on plots containing cadmium and zinc. For *S. vulgaris* and lettuce, cadmium and zinc uptake are as expected, with increasing concentration in plants that are grown in the more contaminated regions of the soil and at lower soil pH levels. The *T. caerulescens* follows the same pattern for cadmium, but not for zinc, with zinc concentrations being 10-fold greater than in *S. vulgaris*. These levels show no relation to available soil zinc.[20] *Eichhornia crassipes* (water hyacinth) effectively translocates cadmium from the roots to the tops.[21]

An important pathway by which plants detoxify heavy metals is through sequestration with heavy-metal binding peptides called phytochelatins, or with their precursor glutathione. In order to develop transgenic plants with an increased capacity for the accumulation or tolerance of heavy metals, the *E. coli* gsh II gene encoding glutathione synthetase is overexpressed in the cytosol of *B. juncea* (Indian mustard). The transgenic GS plants accumulate significantly more cadmium than the wild type. In addition the GS plants show enhanced tolerance to cadmium at both the seedling and mature-plant stages. The cadmium treated GS plants have higher concentrations of glutathione, phytochelatins, thiol, sulfur, and cadmium. Thus in the presence of cadmium, the GS enzyme is rate limiting for the biosynthesis of glutathione and phytochelatins, and that overexpression of GS offers a promising strategy for the production of plants with superior heavy-metal phytoremediation strategy.[22] Metals are unavailable to plants when fixed within relatively inert mineral matrices in soil. In a study of the effect of added EDTA in phytoremediation with *B. juncea*, increased plant uptake of cadmium, zinc, lead, nickel, and copper is observed, but this non-labile pool of cadmium in the soil remains immobilized.[23] Other chelates of the EDTA-type have been used. Their relative effectiveness in enhancing lead phytoextraction follows the sequence EDTA > HEDTA > DTPA > EGTA > EDDHA. These chelates facilitate lead transport into the xylem, and increase lead translocation from roots to shoots.[24] The effect of added elemental sulfur or nitrilotriacetate on the phytoextraction of cadmium, zinc and copper from calcareous soils leads to a two-three fold improvement in their extraction.[25]

Several grass varieties that have high metal tolerances also show moderate to high metal accumulations. Since these grasses produce large quantities of biomass, they have also been screened as potential phytoremediation candidates. Such a screening of twenty-two grasses shows that *Avena sativa* (oat) and *Hordeum vulgare* (barley) tolerate high concentrations of cadmium(II), copper(II), and zinc(II), and accumulate elevated concentrations of these metals in their shoots. Although zinc accumulation by *B. juncea* is enhanced by the addition of the complexant EDTA, it has no effect on zinc removal by these grasses.[18]

Phytoremediation is also an effective method for the removal of copper. One method of accomplishing its removal is to use *Medicago sativa* (alfalfa). Alfalfa has been considered for this particular application because metal accumulation by this plant has been observed at concentrations that are much higher than are the tolerance levels for other plants. In addition, alfalfa has a high protein concentration. Since proteins contain metal binding functionalities in their structures, their presence in plants and grasses may be advantageous for phytoremediation. From a study involving seven different varieties of alfalfa, it is observed that they all have a high affinity for copper(II). Optimum copper(II) binding is found at a pH of 5.0, with the adsorption of copper from solution being rapid. The shoots have higher binding affinities than do the roots, and as much as 98% of the copper can be subsequently recovered by treatment with 0.1 M hydrochloric acid.[26] For application under flow conditions the alfalfa cells have been immobilized on silica. This material can be packed into a column and used for the extraction of copper(II) from solution.[27] Nickel(II) is also rapidly bound to this silica-immobilized alfalfa, allowing therefore for a high recovery of this metal from solution. Again, the nickel(II) can be removed from the biomass by elution with 0.1 M hydrochloric acid.[28] The presence of calcium(II) and magnesium(II) does not significantly affect the binding of copper(II) to alfalfa, but high concentrations of these metal ions do reduce the binding of nickel(II).[29] Although the focus of theses studies have been copper and nickel, alfalfa is also useful for the removal of lead(II), cadmium(II), zinc(II), and chromium(III). While it is apparent that the methods used in the phytoremediation of metals need to be optimized, it is clear that the strategy of using plant species for the remediation of metals in the environment has considerable potential.[30]

The adsorption of metals by biomass, or extracts derived from it, is closely related to phytoremediation. This is because whether the plant material is in a living or dead form, it still has the same chemical functional groups that can bind to metal ions. Canadian *Sphagnum* peat moss, and the humic acid and humin that is extracted from it, are all very good adsorbents for copper(II). The carboxylate groups in these biomaterials are involved in complexation with the copper(II), since it is found that converting these groups into the methyl ester functionality reduces their adsorptive capacity for copper(II).[31] Inactivated cells of *Mucor rouxii* also adsorb copper(II) from solution. Esterification of the carboxylate groups again reduces the adsorption capacity for the metal, but it is recovered again following saponification of the ester back to the carboxylic acid state.[32] Similarly, the inactivated biomass of *Larrea tridentata* (creosote bush) absorbs copper(II) from aqueous solution. This material also has a high capacity for lead(II) and chromium(III). In general, the leaves of the bush bind metals better than do the stems and roots, and the metal can again be released by treatment with 0.1 M hydrochloric acid.[33] The biomass of *Medicago sativa* (alfalfa) has also been used for the phytofiltration of cadmium(II), lead(II), zinc(II), and chromium(III). The method is analogous to that used for copper(II) extraction, where immobilization is achieved on silica.[28] This biomass has the adsorptive capacity for these metals in the sequence: lead(II) > chromium(III) ~ cadmium(II) > zinc(II).[34] Competition experiments with solutions containing multiple metals show that alfalfa exhibits selective binding. For an aqueous solution having each metal ion at a concentration of 1 mM, the uptake follows the sequence: copper(II) > chromium(II) > lead(II) > zinc(II) > nickel(II) > cadmium(II). A similar sequence is found with the silica-immobilized material.[35]

Spectroscopic methods have been used to probe the structure of these adsorbed metals in biomass. Such a direct approach involves carrying out an X-ray adsorption spectroscopic analysis (XANES and EXAFS) of the metal-bound alfalfa shoot biomass. Such an analysis shows that nickel(II), iron(II), iron(III) and chromium(III) are bound *via* oxygen donor ligands, probably carboxylates.[36, 37] Similarly, X-ray microfluorescence has been used to confirm that gold binds to alfalfa biomass in its gold(III) oxidation state.[38]

## 2.3. Mercury

Transgenic plants have been used in the phytoremediation of mercury(II). Mercuric ion reductase, MerA, converts the mercury(II) to elemental mercury. The bacterial merA sequence is rich in CpG dinucleotides and has a highly skewed codon usage, both of which are particularly unfavorable to efficient expression in plants. A mutagenized merA sequence, merApe9, modifying the flanking region, has been constructed and placed under control of plant regulatory elements. *Arabidopsis thaliana* seeds expressing merApe9 grow in mercury(II)-containing medium, and these seedlings evolve considerable amounts of elemental mercury relative to control plants.[39]

## 2.4. Radionuclides

Soil contaminated with radionuclides offers an excellent opportunity for phytoremediation. This opportunity arises because the contamination is often at low level, but spread over a wide surface area. As a result, soil excavation and washing is economically unfeasible, and an *in situ* remediation method must be used. Among the radionuclides that are problematic are cesium-137 and strontium-90. A large number of plant species accumulate these radionuclides over a time period of 5 to 20 years. Example of such plants are *Festuca arundinacea*, *Lolium perenne*, *Festuca rubra*, *Trifolium repens* and *Cerastium fontanum*, all of which accumulate cesium-137. Trees can also accumulate substantial quantities of radionuclides. Among these are *Acer rubrum*, *Liquidambar stryaciflua*, and *Liriodendron tulipifera*, which accumulate curium-244, cesium-137, plutonium-238, radium-226, and strontium-90.[40]

# 3. BIOREMEDIATION

Bioremediation is a technology that differs from the simple use of biomass for the adsorption of metals.[41-49] Instead, it is a technology that focuses on developing problem-specific organisms. As a result, bioremediation, which uses these living microorganisms for the degradation of pollutants, is becoming increasingly important in the search for a clean environment. Genetically engineered microbes can be designed that target specific pollutants and environmental conditions. Although bioremediation is primarily used for organic pollutants, there are also possibilities for the development of biotechnologies that are focused on the remediation of metals.

Bioremediation methods fall into several categories. In some cases the methods are focused on a single metal or small group of metals, whereas in other cases the method has wide applicability to a broad spectrum of metals. Bioremediation options for metals cover a wide range of types of biochemical mechanisms. Among these are ones that lead to mineralization or altered redox state. Redox pathways for bioremediation are targeted where the changed oxidation state is less toxic, more water soluble for subsequent removal by leaching, or less water soluble for removal by precipitation methods. Materials that are contaminated by metal sulfides can also be cleansed by redox reactions. This bioleaching cleansing process can be carried out using bacteria such as *Thiobacillus ferrooxidans*, which oxidizes insoluble metal sulfides into their water soluble sulfates. Metal ions can also become accumulated by having them bind to cell walls. This non-specific binding by cells is analogous to an ion exchange resin with the carboxylate functional groups on the cell wall providing the anionic functionality onto which the metal cations bind. Bioremediation is also an option for heavy metals in cases where specific binding of these metals can occur to sulhydryl residues that are present at the cell wall.[50]

A variety of hyperthermophilic microorganisms are capable of reducing metals ions by using molecular hydrogen as the electron donor. One such organism, *Pyrobaculum islandicum*, reduces several metal ions when hydrogen is present, and the system is kept at 100 °C. The metals that are reduced by this system are uranium(VI), technetium(VII), chromium(VI), cobalt(III), and manganese(IV). The respective reduction products are the insoluble uranium(IV) mineral uraninite, the insoluble technetium(IV) or (V) form, the less toxic and soluble chromium(III), cobalt(II), and manganese(II) carbonate. This microorganism, therefore, not only is widely applicable, but it has the potential to be used for the bioremediation of these metals when they are present in a high temperature environment.[51]

Fungi can also be used as bioremediation agents for metal ions. A recent review covers a range of topics that include the accumulation of metals by fungi, the translocation of metals by filamentous fungi, and their applications in practice. Fungi possess several mechanisms for developing resistance and tolerance to heavy metals, and they efficiently solubilize metal ions in soil. The translocation mechanisms of fungi can lead to their being a concentrator of metals in particular regions of the mycelium. Where translocation is very rapid, as with zinc, or where it occurs only through particular pathways within the mycelium, as with cadmium, the capacity for the concentration of pollutants is increased.[52]

As bioremediation becomes of increasing importance the engineering aspects must also be considered. For processes such as the bacterial leaching of metals from ores, the size of the reactor and the mixing efficiency must all be addressed in designing a cost-effective system.[53] The roadblocks to the implementation of biotreatment strategies have been recently reviewed and discussed.[54]

## 3.1. Mercury

Several methods are available for the bioremediation of mercury. For methylmercury(II), an organomercury lyase is available that catalyzes the protonation of the methyl-mercury bond to give a mixture of methane and an inorganic mercury(II) salt. Alternatively, mercuric ion reductase (MerA) can be used to catalyze the reduction of mercury(II) to elemental mercury according to equation 1.[55] Subsequently a mutagenized *merA* sequence

$$RSHg^{+} + NADPH \rightarrow Hg + RSH + NADP^{+} \quad (1)$$

(*merApe9*) has been developed. Transgenic *merApe9* seedlings evolve considerable amounts of elemental mercury relative to control plants. The rate of mercury evolution and the level of resistance are proportional to the steady-state mRNA level, confirming that resistance is due to expression of the merApe9 enzyme.[56] Mercury(II), however, forms complexes with components of Luria-Bertani (LB) broth that are used to grow the mercury-resistant *Pseudomonas aeruginmosa.* These particular mercury(II) complexes cannot be reduced with either meruric reductase or *P. aeruginosa.*[57]

## 3.2. Cadmium, Copper and Lead

Although cadmium can be chemically leached from wastes with sulfuric acid, if this reagent is environmentally undesirable, the solubilization of cadmium can be achieved using a microbial leaching process with iron(II) sulfate and an iron-oxidizing bacteria. This particular leaching method is also successful for the metals copper, manganese and zinc.[58, 59] In an investigation of the ability of a series of bacteria to remove cadmium(II) and copper(II) from solution, it has been found that *Escherichia coli* is the most effective one for cadmium(II), and that *Bacillus subtilis* removes the most copper(II). In general, the bacterial removal of these metals follows the sequence: copper(II) > cadmium(II).[60] Microbially-produced surfactants (biosurfactants) also complex cadmium(II). For example, an anionic monorhamnolipid biosurfactant produced by *Pseudomonas aeruginosa* ATCC 9027 rapidly binds cadmium(II) with a stability constant that is higher than that which is found with humic acids.[61] This particular strain of *P. aeruginosa* also complexes lead(II) and zinc(II).[62] Other strains bind cadmium(II).[63] A calcium alginate-immobilized biomass of *P. aeruginosa* can be used to remove cadmium(II), copper(II), and lead(II) from aqueous solution. By using a pH gradient elution, the bound metals can be subsequently removed from the biomass.[64, 65]

Copper(II) can also be bound to fungi. An example of such binding is found with *Mucor rouxii*. The copper can be subsequently recovered from the fungi by washing with 0.1M hydrochloric acid. It is likely that the copper(II) binds by complexation with carboxylate groups on the fungi.[66] Two recent applications of cadmiun bioremediation involve precipitating it as its sulfide, or the use of recombinant DNA technology to improve the metal binding capacity of a specific bacteria. Such a bacterial strain, *Klebsiella platicola*, grows in solutions containing concentrations of up to 15 mM cadmium(II) chloride. After thiosulfate is added to the growth medium, the cadmium is precipitated as its sulfide.[67] Although bacterial binding of toxic metals has been accomplished using non-engineered bacteria, recombinant DNA technology offers the possibility of increasing this binding capacity. Thus recombinant *Staphylococcus xylosus* and *Staphylococcus carnosus* strains have been generated with surface-exposed chimeric portions containing polyhistidyl peptides designed for binding to divalent metal ions. These recombinant cells have increased binding capacity for cadmium(II) and nickel(II), which is especially pronounced for the former metal ion. In order to create *staphylococci* stains with specific metal ion recognition, an ion-specific peptide or protein needs to be engineered, for example, by combinatorial library approaches.[68]

## 3.3. Actinides

The recovery of uranium and other actinides by biotreatment strategies is receiving increased attention. One enzymatic reaction that has been used for recovering uranium from water involves adding glycerol 2-phosphate, followed by treatment of the aqueous solution with a phosphatase-containing *Citrobacter sp.* In the process, the phosphatase releases phosphate from the glycerol 2-phosphate, which then forms a precipitate of uranium phosphate on the cell surface.[69,70] Several microorganisms have also been found that enzymatically reduce uranium(VI) to uranium(IV), which is obtained in the form of an extracellular precipitate of uraninite.[71-73] Uranium can also be removed from solution by reduction with *Desulfovibrio desulfuricans.* Uranium concentrations of up to 24 mM can be tolerated by this system. The process is, however, inhibited by high concentrations of copper(II). This anaerobic process leads to precipitation of the uranium, even when the initial uranium(VI) is complexed as its carbonate.[74] A technique that is described as microbially-enhanced chemisorption of heavy metals (MECHM) has also been used for the bioremediation of the actinides neptunium and plutonium. The immobilized cells of a *Citrobacter sp* that are used do not themselves remove these actinides until they are treated with lanthanium in the presence of an organophosphate substrate. The resulting cell-bound lanthanum phosphate material can give up to 90% removal of neptunium or plutonium.[75]

## 3.4. Chromium

The ability of a wide range of microorganisms to reduce chromium(VI) to chromium(III) has led to their use in chromium(VI) bioremediation.[76] One such enzyme, which acts by enhancing the rate of NADH reduction of chromium(VI), has been obtained from *Pseudomonas ambigua.*[77] An *E. cloacae* strain (HO1) has been isolated that is resistant to chromium(VI) under both aerobic and anaerobic conditions, although only the anaerobic culture causes its reduction. This reduction is faster in glycerol- or acetate-grown cells than it is in glucose-grown cells.[78] Washed cell suspensions of *Desulfovibrio vulgaris* rapidly reduce chromium(VI) to chromium(III) with molecular hydrogen as the electron donor. The $c_3$ cytochrome from this organism functions as a chromium(VI) reductase. An advantage of molecular hydrogen as the electron donor is that it does not leave any organic residue behind.[79] A consortium of bacteria named SRB III has also been used for the bioremediation of chromium(VI). These bacteria reduce the chromium(VI) to chromium(III), which then precipitates onto the bacterial surface. It is suggested that hydrogen sulfide is the electron donor in the chemical reaction.[80]

Allied Signal Corporation has developed a process for the bioremediation of groundwater and soils contaminated by metals that is successful for chromium(VI).[81] The process takes advantage of the ability of anaerobic sulfate-reducing bacteria to reduce metals to their lower oxidation states, which for chromium(VI) leads to its conversion to the less toxic chromium(III). The process circumvents the acute toxicity of chromium(VI) to microbes by using bacteria that produce hydrogen sulfide gradients, with the chromium(VI) being reduced to chromium(III) before it reaches the microbial cells. This technology has already been successfully field tested for chromium(VI). Chromium(VI)-resistant microbes of either the fungus or the bacteria type can also be used to effect the reduction of chromium(VI) to chromium(III). In comparative experiments, it has been found that the fungus tested is more tolerant to chromium(VI) than is the bacteria, and that all microbes are more effective reductants under anaerobic conditions.[82] In order to be able to apply these technologies, a two-stage bioreactor has been designed where *E. Coli* cells are grown in a first stage, and then pumped to a second stage plug-flow reactor where the anaerobic reduction of chromium(VI) occurs.[83]

An alternate approach for chromium remediation couples the microbial reduction of chromium(VI) with the anaerobic degradation of benzoate.[84] Chromium(VI) has also been removed from aqueous solutions by *Zoogloea ramigera,*[85] water hyacinth,[86, 87] and an *Enterobacter cloacae* strain that reduces chromium(VI) under anaerobic conditions.[88] With other systems, both chromium(VI) and organics have been simultaneously removed by an anaerobic consortium of bacteria.[89] The microbial reduction of chromium(VI), and its potential application for the removal of chromium from polluted sites, has been reviewed.[90] Polluted sites often contain mixtures of both chromium(VI) and toxic organic compounds, because industrial processes such as leather tanning, metals finishing, and petroleum refining discharge mixtures of chromium(VI) and aromatics. An important goal, therefore, is the simultaneous detoxification of both chromium(VI) and these phenolic compounds using aerobic bacteria or abiotic redox reactions.[91,92]

# REFERENCES


1. D. E. Salt, R. D. Smith, I. Raskin, *Ann. Rev. Plant Physiol. Plant Mol. Biol.,* **1998**, *49,* 643.
2. E. L. Kruger, T. A. Anderson, J. R. Coats, *ACS Sympos. Ser.*, Vol. 664, *Phytoremediation of Soil and Water Contaminants,* American Chemical Society, 1997.
3. M. J. R. Shannon, R. Unterman, *Ann. Rev. Microbiol.,* **1993**, *47,* 715.
4. S. L. Brown, R. L Chaney, C. A. Loyd, J. S. Angle, J. A. Ryan, *Environ Sci. Technol.,* **1996**, *30,* 3508.
5. T. E. Pawlowska, R. L. Chaney, M. Chin, I Charvat, *Appl. Environ. Microbiol.,* **2000**, *66,* 2526.
6. S. L. Brown, R. L. Chaney, J. S. Angle, A. J. M. Baker, *Soil Sci. Soc. Am. J.,* **1995**, *59,* 125.
7. A. Leusch, Z. R. Holan, B. Volesky, *J. Chem. Tech. Biotechnol.,* **1995**, *62,* 279.
8. D. C. Herman, J. F. Artiola, R. M. Miller, *Environ. Sci. Technol.,* **1995**, *29,* 2280.
9. M. S. Masri, R. W. Reuter, M. Friedman, *J. Appl. Polym. Sci.,* **1974**, *18,* 675.
10. J. P. Pinheiro, A. M. Mota, M. L. Simoes Goncalves, *Anal. Chim. Acta,* **1994**, *284,* 525.
11. H. Shahandeh, L. R. Hossner, *Int. J. Phytoremed.,* **2000**, 2, 31.
12. R. A. Sheffington, P. R. Shewry, P. Peterson, *Planta,* **1976**, *132,* 209.

13. R. J. Bartlett, *Adv. Environ. Sci. Technol.*, **1988**, *20*, 267.
14. D. E. Salt, M. Blaylock, N. P. B. A. Kumar, V. Dushenkov, B. D. Ensley, I. Chet, I. Raskin, *Biotechnol.*, **1995**, *13*, 468.
15. C. M. Lytle, F. W. Lytle, N. Yang, I-H. Qian, D. Hansen, A. Zayed, N. Terry, *Environ. Sci. Technol.*, **1998**, *32*, 3087.
16. I. D. Kleiman, D. H. Cogliatti, *Enivron. Technol.*, **1998**, *19*, 1127.
17. P. Chandra, S. Sinha, V. N. Rai, *ACS Sympos. Ser.*, **1997**, *664*, 274.
18. S. D. Ebbs, L. V. Kochian, *Environ. Sci. Technol.*, **1998**, *32*, 802.
19. J. Chen., J. W. Huang, T. Caspar, S. D. Cunningham, *ACS Sympos. Ser.*, **1997**, *664*, 264.
20. S. L. Brown, R. F. Chaney, S. J. Angle, A. J. M. Baker, *Environ. Sci. Technol.*, **1995**, *29*, 1581.
21. B. C. Wolverton, R. C. McDonald, *EHP, Environ. Health Perspect.*, **1978**, *27*, 161.
22. Y. L. Zhu, E. A. H. Pilon-Smits, L. Jovanin, N. Terry, *Plant Physiol.*, **1999**, *119*, 73.
23. K. G. Stanhope, S. D. Young, J. J. Hutchinson, R. Kamath, *Environ. Sci. Technol.*, **2000**, *34*, 4123.
24. J. W. Huang, J. Chen, S. D. Cunningham, *ACS Sympos. Ser.*, **1997**, *664*, 283.
25. A. Kayser, K. Wenger, A. Keller, W. Attinger, H. R. Felix, S. K. Gupta, R. Schulin, *Environ. Sci. Technol.*, **2000**, *34*, 1778.
26. J. L. Gardea-Torresdey, K. J. Tiemann, J. H. Gonzalez, J. A. Henning, M. S. Townsend, *Solv. Extr. Ion Exch.*, **1996**, *14*, 119.
27. J. L. Gardea-Torresdey, K. J. Tiemann, J. H. Gonzalez, J. A. Henning, M. S. Townsend, *J. Hazard. Mater.*, **1996**, *48*, 181.
28. J. L. Gardea-Torresdey, K. J. Tiemann, J. H. Gonzalez, I. Cano-Aguilera, J. A. Henning, M. S. Townsend, *J. Hazard. Mater.*, **1996**, *49*, 205.
29. J. L. Gardea-Torresdey, K. J. Tiemann, J. H. Gonzalez, O Rodriguez, *J. Hazard. Mater.*, **1997**, *56*, 169.
30. S. D. Cunningham, J. R. Shann, D. E. Crowely, T. A. Anderson, *ACS Sympos. Ser.* **1997**, *664*, 2.
31. J. L. Gardea-Torresdey, L. Tang, J. M. Salvador, *J. Hazard Mater.*, **1996**, *48*, 191.
32. J. L. Gardea-Torresdey, I. Cano-Aguilera, R. Webb, K. J. Tiemann, F. Gutiérrez-Corona, *J. Hazard. Mater.*, **1996,** *48*, 171.
33. J. L. Gardea-Torresdey, A. Hernandez, K. J. Tiemann, J. Bibb, O. Rodriguez, *J. Hazard. Substance Res.* Vol 1, **1997**, Kansas State Univ. p.3-1.
34. J. L. Gardea-Torresdey, J. H. Gonzalez, K. J. Tiemann, O. Rodriguez, G. Games, *J. Hazard Mater.*, **1998**, *57*, 29.
35. J. L. Gardea-Torresdey, K. J. Tiemann, G. Gamez, K. Dokken, *J. Hazard. Mater.*, **1999**, *B69*, 41.
36. K. J. Tiemann, J. L. Gardea-Torresdey, G. Gamez, K. Dokken, S. Sias, M. W. Renner, L. R. Furenlid, *Environ Sci. Technol.*, **1999**, *33*, 150.
37. K. J. Tiemann, J. L. Gardea-Torresdey, G. Gamez, K. Dokken, I. Cano-Aguilera, M. W. Renner, L. R. Furenlid, *Environ. Sci. Technol.*, **2000**, *34*, 693.
38. J. L. Gardea-Torresdey, K. J. Tiemann, G. Gamez, K. Dokken, N. E. Pingitore, *Adv. Environ. Res.*, **1999**, *3*, 83.

39. K. G. Stanhope, S. D. Young, J. J. Hutchinson, R. Kamath, *Environ. Sci. Technol.,* **2000**, *34*, 4123.
40. J. A. Entry, L. S. Watrud, R. S. Manasse, N. C. Vance, *ACS Sympos. Ser.,* **1997**, *664*, 299.
41. J. M. Tobin, J. C Roux, *Water Res.,* **1998**, *32*, 1407.
42. J. M. Chen, O. J. Hao, *J. Chem. Tech. Biotech.,* **1997**, *69*, 70.
43. N. Verma, R. Rehal, *J. Ind. Pollut. Control,* **1996**, *12*, 55.
44. M. M. Alves, C. G. G. Beca, R. G. De Carvalho, J. M. Castanheira, M. C. S. Pereira, *Water Res.,* **1993**, *27*, 1333.
45. D. C. Sharma, C. F. Forster, *Water Res.,* **1993**, *27*, 1201.
46. S. Niyogi, T. E. Abraham, S. V. Ramakrishna, *J. Sci. Ind. Res.,* **1998**, *57*, 809.
47. D. Kratochvil, P. Pimentel, B. Volesky, *Environ. Sci. Tech.,* **1998**, *32*, 2693.
48. M. Zhao, J. R. Duncan, *Biotech. Appl. Biochem.,* **1997**, *26,* 179.
49. S. Samantaroy, A. K. Mohanty, M. Misra, *J. Appl. Polym. Sci.,* **1997**, *66*, 1485.
50. M. J. R. Shannon, R. Unterman, *Ann. Rev. Microbiol.,* **1993**, *47*, 715.
51. K. Kashefi, D. R. Lovely, *Appl. Environ. Microbiol.,* **2000**, *66*, 1050.
52. S. N. Gray, *Biochem Soc. Trans.,* **1998**, *26*, 666.
53. G. Andrews, *Biotechnol. Prog.,* **1990**, *6*, 225.
54. J. W. Talley, P. M. Sleeper, *Ann. N. Y. Acad. Sci.,* **1997**, *829*, 16.
55. B. Fox, C. T. Walsh, *J. Biol Chem.*, **1982**, *257*, 2498.
56. C. L. Rugh, H. D. Wilde, N. M. Stack, D. M. Thompson, A. O. Summers, R. B. Meagher, *Proc. Natl. Acad. Sci.*, **1996**, *93*, 3182.
57. J. S. Chang, J. Hong, O. A. Ogunseitan, *Biotechnol. Prog.,* **1993**, *9*, 526.
58. J. K .Blais, R. D. Tyagi, J. C. Auclair, C. P. Huang, *Water Sci. Technol. Water Qual. Int.,* `92, Pt.1 **1992**, *26*, 197.
59. H. Seidel, J. Ondruschka, P. Morgenstern, U. Stottmeister, *Water Sci. Technol.*, **1998**, *37*, 387.
60. M. D. Mullen, D. C. Wolf, F. G. Ferris, T. J. Beveridge, C. A. Flemming, G. W. Bailey, *Appl. Environ Microbiol.,* **1989**, *55*, 3143.
61. H. Tan, J. T. Champion, J. F. Artiola, M. L. Brusseau, R. M. Miller, *Environ. Sci. Technol.,* **1994**, *28*, 2402.
62. D. C. Hermann, J. F. Artiola, R. M. Miller, *Environ Sci. Technol.,* **1995**, *29*, 2280.
63. C. L. Wang, P. C.' Michels, S. C. Dawson, S. Kitisakkul, J. A. Baross, J. D. Keasling, D. S. Clark, *Appl. Environ. Microbiol.,* **1997**, *63*, 4075.
64. J.-S. Chang, J.-C. Huang, *Biotechnol.Prog.,* **1998**, *14*, 735.
65. J.-S. Chang, J.-C. Huang, C.-C. Chang, T.-J. Tarn, *Water Sci. Technol.,* **1998**, *38*, 171.
66. J. L. Gardea-Torresdey, I. Cano-Aguilera, R. Webb, F. Gutierrez-Corona, *Environ. Toxicol. Chem.,* **1997**, *16*, 435.
67. P. K. Sharma, D. L. Bulkwill, A. Frenkel, M. A. Vairavamurthy, *Appl. Environ. Microbiol.,* **2000**, *66*, 3083.
68. P. Samuelson, H. Wernérus, M. Svedberg, S. Stahl, *Appl. Environ. Microbiol.,* **2000**, *66*, 1243.
69. L. E. Macaskie, *Crit. Rev. Biotechnol.,* **1991**, *11*, 41.
70. L. E. Macaskie, A. C. R. Dean, *Adv. Biotechnol. Processes*, **1989**, *12*, 159.

71. D. R. Lovley, E. J. P. Phillips, Y. A. Gorby, E. R. Landa, *Nature,* **1991**, *350,* 413.
72. Y. A. Gorby, D. R. Lovely, *Environ. Sci. Technol.,* **1992,** *26*, 205.
73. D. R. Lovley, E. J. P. Phillips, *Appl. Environ. Microbiol.,* **1992**, *58*, 850.
74. D. R. Lovley, E. J. P. Phillips, *Environ Sci Technol.,* **1992**, *26*, 2228.
75. L. E. Macaskie, G. Basnakova, *Environ. Sci. Technol.,* **1998**, *32*, 184.
76. D. R. Lovley, *Ann. Rev. Microbiol.*, **1993**, *47*, 263.
77. T. Suzuki, N. Miyata, H. Horitsu, K. Kawai, K. Takamizawa, Y. Tai, M. Okazaki, *J. Bacteriol.,* **1992**, *174*, 5340.
78. C. P. Wang, T. Mori, K. Komori, *Appl. Environ. Microbiol.*, **1989**, *55*, 1665.
79. D. R. Lovley, E. J. P. Phillips, *Appl. Environ. Microbiol.,* **1994**, *60*, 726.
80. L. Fude, B. Harris, M. E. Urrutia, T. J. Beveridge, *Appl. Environ. Microbiol.*, **1994**, *60,* 1525.
81. A. Trivedi, *Bioremediation of Hazardous Metals*, AiResearch, Los Angeles Division.
82. *The Hazard. Waste Consultant*, **1995**, (*Jan/Feb*), L1.
83. H. Shen, Y-T. Wang, *J. Environ. Eng.*, **1995**, *121*, 798.
84. H. Shen, P. H. Pritchard , G. W. Sewell, *Environ. Sci. Technol.*, **1996**, *30*, 1667.
85. C. Solisio, A. Lodi, A. Converti, M. Del Borgi, *Chem. Biochem. Eng. Quarterly,* **1998**, *12*, 45.
86. O. Saltabas, G. Akcin, *Toxicol. Environ Chem.*, **1994**, *43*, 163.
87. O. Saltabas, G. Akcin, *Toxicol. Environ Chem.*, **1994**, *41*, 131.
88. K. Komori, A. Rivas, K. Toda, H. Ohtake, *Biotech. Bioeng.*, **1990**, *35*, 951.
89. Y-T. Wang, E. M. N. Chirwa, *Proc. Mid-Atlantic. Ind. Hazard Waste Conf.,* **1997**, *29*, 32.
90. D. R. Lovley, *Ann. Rev. Microbiol.*, **1993**, *14*, 158.
91. H. Shen, Y. T. Wang, *Appl. Environ. Microbiol.*, **1995**, *61*, 2754.
92. M. S. Elovitz, W. Fish, *Environ. Sci. Technol.*, **1994**, *28*, 2161.

# 13

## OPTICAL AND REDOX SENSORS FOR METAL IONS

### 1. INTRODUCTION

Optical and redox sensors are materials that have a wide range of uses and applications in both medical and environmental situations. Sensors can be designed to make use of changes in either the wavelength or extinction coefficient of an optical sensing material, or in the redox potential of an electroactive compound. Alternatively for emissive materials, it is possible to use changes in the emission wavelength or intensities to monitor the presence or absence of chemical species. These chemical species for both and redox sensors can be cations, anions, or organic molecules. For a sensor to be useful it is necessary for the device to be selective for the specific chemical species of interest, and that the change in the property of the sensing material be responsive in a consistent manner to changes in concentration of the chemical species being detected or analyzed.[1-8] The sensing component is referred to as the reporter molecule. For optical reporter molecules the material needs to be photostable, and for redox reporters the oxidation-reduction step should be reversible. For metal ion sensors a binding site needs to be present that is selective for the targeted metal. In such sensors it is usual to employ chelate or macrocyclic ligands because they can be tailored to selectively complex a variety of different metal ions. For the detection of uncharged molecules a host will usually be selected such that its cavity matches the shape and size of the chosen guest. More recently, metal-containing optical sensors are being developed that can function as anion selective receptors, and again the receptor must be specifically designed to meet the binding requirements of the individual anions.[9]

Sensors obey mass law because the equilibrium constants of their interaction with analytes are low. As a consequence a plot of the sensor signal against the logarithm of the concentration of the analyte is "S-shaped" and not linear.[10] For a sensor to function efficiently the concentration of the analyte should fall within the dynamic range of this S-curve. An advantage of this mass law effect is that a lower concentration of a particular sensor can cover large concentration ranges of the analyte. For many metal ion sensors the system is designed with the donor atoms of the ligand system placed in such a manner that it can communicate with the fluorophore or the electroactive component of the reporter molecule. This arrangement is possible because the fluorophore or redox portion of the sensor is usually comprised of a conjugated $\pi$-system, and the ligand portion of a saturated $\sigma$-system. Sensors comprised of two such components have been called conjugate chemosensors.[11]

## 2. OPTICAL REPORTER MOLECULES

In order for a compound to function as an optical sensor for metal ions it is necessary for two components to be present. One is that a ligand binding site is present, and the other is that an optical reporter molecule is attached that changes its optical properties when a metal ion is bound into the ligand binding site. Optical reporter molecules can be ones having absorption bands in the visible region of the electronic spectrum which undergo a wavelength change when a metal ion is incorporated into the ligating site. Alternatively the emissive properties of a compound can be used in the design of optical reporter systems. These emissive properties can be changes in the emission wavelength of the reporter molecule upon complexation of the metal ion, or changes in the lifetime of the excited state. Excited state lifetimes can be sensitive to the presence of a metal ion because of a number of factors.[12, 13] One of these factors is that the presence of a metal ion changes the energetics involved in quenching processes that are controlled by photoelectron transfer reactions. Another factor is that the presence of a metal ion leads to differences in the relative energies of the singlet and triplet excited states, resulting in changes in the rates of intersystem crossing between them. A further factor that may be significant is that a complexed metal ion, especially one having a large atomic number, can cause quenching by spin-orbit coupling between the excited state and the metal ion. For the case of a flexible macrocyclic ligand, the coordination of a metal ion frequently leads to reduced flexibility of the entire molecule. Since the quenching rate of an excited state is strongly influenced by the conformational mobility of the molecule, any such changes are likely to be reflected in the excited state lifetime of the reporter group. Since it is important that the reporter group be both chemically stable and have a sufficiently long lifetime that the emission can be readily observed, it is usual for it to be an organic functionality. The optical properties of a series of potential reporter molecules are shown in Table 1.

**Table 1. Optical Properties for Emissive Organic Molecules**

| Compound | Extinction coefft at 313 nm | Quantum yield of fluorescence | Energy of excited state (nm) | Lifetime of singlet state (nsec) |
|---|---|---|---|---|
| Anthracene | $1.2 \times 10^3$ | 0.27 | 375[a], 681[b] | 4.9 |
| Pyrene | $1.5 \times 10^4$ | 0.58 | 372[a], 595[b] | 450 |
| Fluorene | $2 \times 10^1$ | 0.66 | 301[a], 421[b] | 10 |
| *trans*-Stilbene | $1.9 \times 10^4$ | | 329, < 572[b] | |

[a] Singlet state
[b] Triplet state

Fluorescent sensors have low intensities because of the short lifetimes of the excited state. The effectiveness of fluorescent chemosensors can, however, be improved if their sensitivity can be increased. One approach is to use conjugated polymers whose emission is frequently dominated by energy migration to local minima. For a conjugated polymer with a receptor attached to every repeating unit, the number of receptor sites is determined by the degree of polymerization. If energy migration is rapid compared to the fluorescence lifetime, the excited state samples every receptor in the polymer. As a result the occupation of a single binding site changes the entire emission. When a receptor site is occupied by a quenching agent, enhanced deactivation results. This concept has been verified by using the compound **1**, which is a good receptor for the electron transfer quenching agent $PQ^{2+}$. From the quenching constant obtained from a Stern-Volmer plot of these data there is a greater than 16-fold enhancement in the quenching resulting from the extended electronic structure.[14]

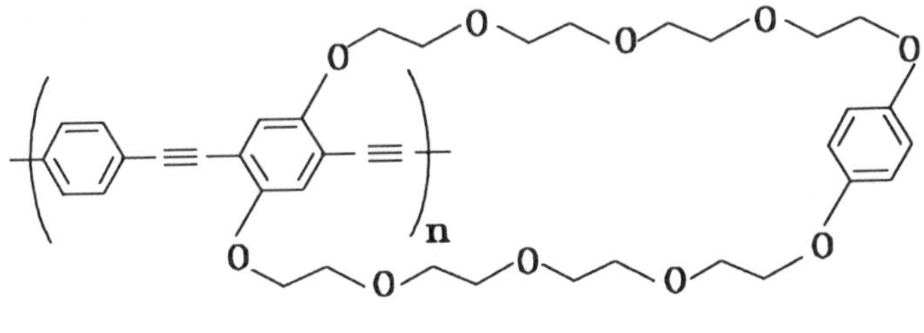

1

## 3. REDOX REPORTER MOLECULES

Redox chemosensors are similar to optical sensors except that they have an electroactive compound as the reporter molecule. Both organic and inorganic reporter molecules are possible. In addition to being reversible, it is advantageous if the oxidation-reduction process involves no chemical steps during the electron transfer process. In general it is also advantageous if the redox process involves the transfer of only a single electron. These considerations usually preclude the use of organic systems like the quinone-hydroquinone couple, and simple inorganic ions need to be bound to a ligand system if they are to be used as reporters. One such metal complex that is used is the ruthenium-bipyridyl complex, which undergoes electron transfer between a ruthenium(II) and (III) center. Other systems that are used as reporter molecules are the organometallic compounds *bis*-(cyclopentadienyl)iron (ferrocene) and *bis*-(cyclopentadienyl)cobalt (cobaltocene). The two oxidation states that are involved in these two redox systems are the (II) and (III) states.

## 4. CHEMILUMINESCENCE

One of the early observations that led the way to the development of sensors was the use of metal ion catalyzed chemiluminescence for analytical purposes. Two examples involve luminol oxidation. In one example trace amounts of iron(II) have been determined by measuring the iron(II)-catalyzed light emission from the oxidation of luminol by oxygen.[15] Since iron(II) is the only common metal ion to catalyze this reaction in aqueous solution, the method is selective for this metal ion. Nevertheless, some other first row transition metal ions do interfere. Another example involves the chloride and bromide ion enhancement of chemiluminescence of trace metal ion catalyzed luminol oxidation by hydrogen peroxide.[16,17] In the presence of bromide ion, enhancement of the chemiluminescence intensity is observed for chromium(III), iron(II) and cobalt(II). It is believed that the effect of halide ion is to increase the rate constant of the chemiluminescent pathway as compared to the non-emissive deactivation path.

A chemiluminescent flow-through sensor has been developed for copper(II) by co-immobilizing the reagents used for its analysis. These reagents, including luminol and cyanide, are immobilized on an ion exchange column, while the copper is retained by electrochemical preconcentration on a gold electrode in a flow cell. Injection of sodium hydroxide releases luminol and cyanide, which reacts with copper(II) stripped from the gold electrode to produce a chemiluminescence signal.[18]

# 5. CHELATE COMPLEXANTS

For sensors that are designed for metal ions it is important that the metal is bound strongly and selectively. This usually requires a multidentate binding ligand such as a chelate. The use of chelates for the development of fluorescent sensors for metal ions has a long history, since in 1867 it was reported that morin forms a strongly fluorescent chelate with aluminum(III).[19]

Energy transfer in aqueous solution between copper(II) ions and the dye rhodamine 800 has also been used to detect the metal in concentrations down to 110 ppb. By comparison, no significant quenching is observed with cobalt(II), nickel(II) or chromium(III).[20]

## 5.1. Nitrogen and Oxygen Donors

For transition metals that can exist in a range of different oxidation states, a photoinduced electron transfer mechanism may be involved in quenching the fluorescence from the reporter molecule. Such an example is found with a tetradentate amine which chelates to copper(II) with the

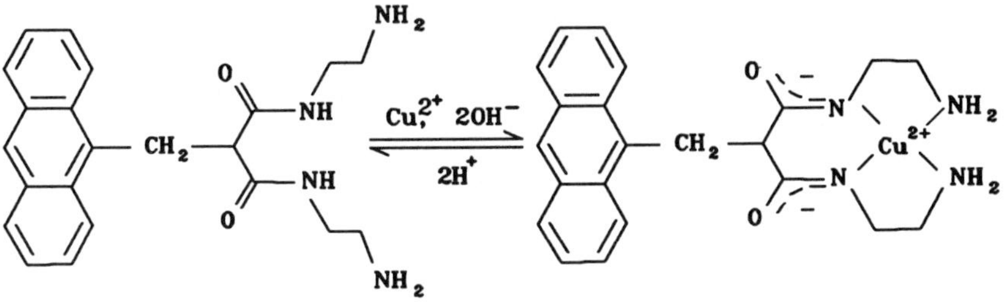

**Equation 1.**

simultaneous deprotonation of two of the amide groups equation 1.[21] In the absence of copper(II), the uncomplexed chelate shows the typical emission spectrum of anthracene, with no change over a pH range of 2-12. If, however, one equivalent of copper(II) is added to an acidic solution of the tetramine, followed by the addition of base, a progressive decrease of the fluorescence is observed from pH 5 until quenching is complete at approximately pH 7. Since excited states are both better oxidants and reductants than are the ground states from which they originate, this quenching of the excited state of anthracene by the complexed copper(II) is ascribed to the photoinduced electron transfer pathway shown in equation 2.[11]

$$An^* + Cu(II) \rightarrow An^- + Cu(III) \qquad (2)$$

A similar bidentate diamine ligand attached to an anthracene reporter has also been found to act as a fluorescent sensor. Again the chelate ligand is covalently bound to the anthracene reporter molecule. The non-fluorescent free ligand becomes fluorescent when it is complexed to two zinc chloride moieties (equation 3).[22] The observation of quenching in the complex is a

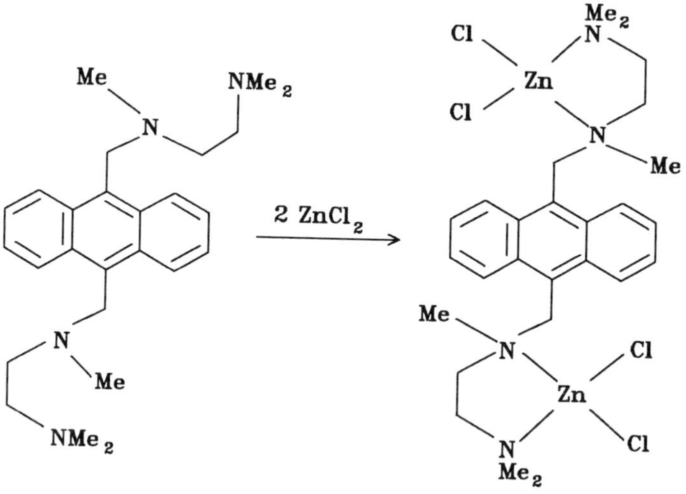

Me NMe 2 N N Me NMe 2

2 ZnCl$_2$ →

Cl Me$_2$ N Zn Cl N Me Me N Cl Zn Cl N Me$_2$

**Equation 3.**

consequence of the lone electron pairs on nitrogen being bound to zinc(II), thereby making the amine groups poorer reducing agents for the electron transfer quenching of the anthracene excited state. Another series of compounds **2-5** with polydentate amine groups attached to an anthracene reporter molecule act as fluorescent sensors for transition metal ions.[23]

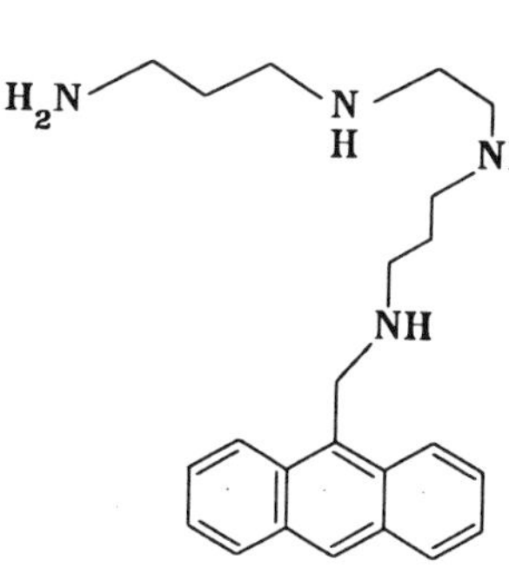

H$_2$N N H NH NH

2

H N N H NH$_2$ HN NH

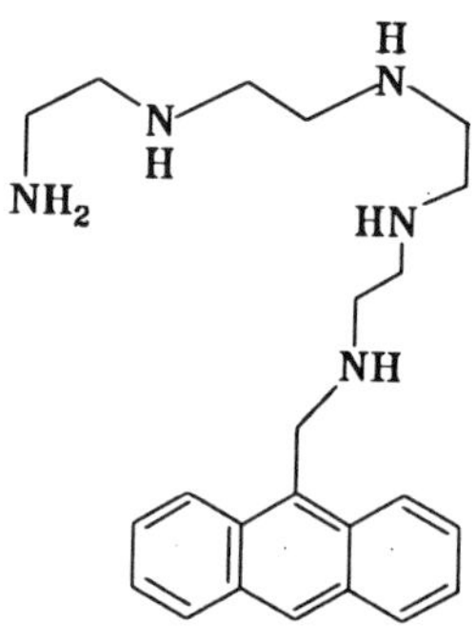

3

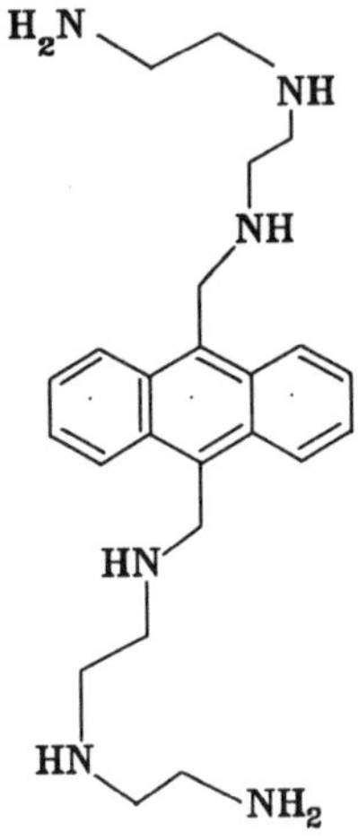

4 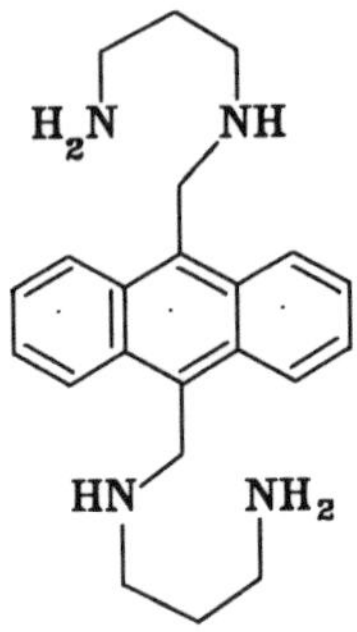 5

Liposomes enhanced with surface recognition groups have a high affinity for heavy metal ions and virus particles with unique fluorescent and colorimetric responses. These lipid aggregates can be immobilized in a silica matrix by the sol-gel method to give sensor materials. Fluorescent sensors prepared from *N*-[8-[1-octadecyl-2-(9-(1-pyrenyl)nonyl)-rac-glycerol]-3, 6-dioxaoctyl]iminodiacetic acid/ distearylphosphatidylcholine liposomes show a four to fifty-fold enhancement to metal ions.[24] Both ascorbate oxidase and alkyl phosphatase can be co-immobilized onto a porous polymer membrane and used in combination with flow-injection amperometry for the recognition of copper(II), zinc(II) and cobalt(II) ions.[25]

The fluoresence intensity of a nitrobenzaldehyde fluorophore attached to an amine receptor increases upon the addition of chromium(III). This seventy-fold enhancement is accompanied by a blue shift in the fluorescence band.[26] A tripodal ligand **6** having a dansyl chromophore as reporter has been used as a metal ion sensor. Addition of copper(II) or cobalt(II) results in a quenching of the emission. For zinc(II) and cadmium(II) the fluorescence maximum shifts to lower wavelengths, along with an increase in the fluorescence intensity. In solutions where the pH is greater than 4, the system is selective for copper(II).[27]

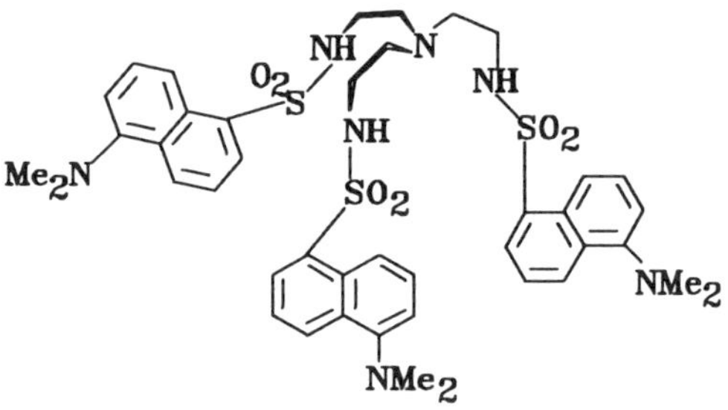

6

Optical control of metal ion binding is a potentially viable route to the design of metal sensors. Thus photochromic molecules with functional groups that bind metal ions open up the possibility of developing photoreversible sensors for trace metal ion detection. Three such systems have either a diphenylchromene **7**, a spiropyran **8**, or a spirooxazine **9** in their closed form. In the closed form these compounds do not strongly bind

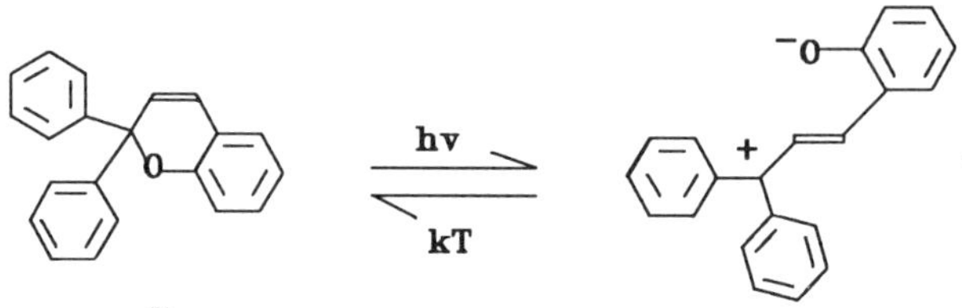

7

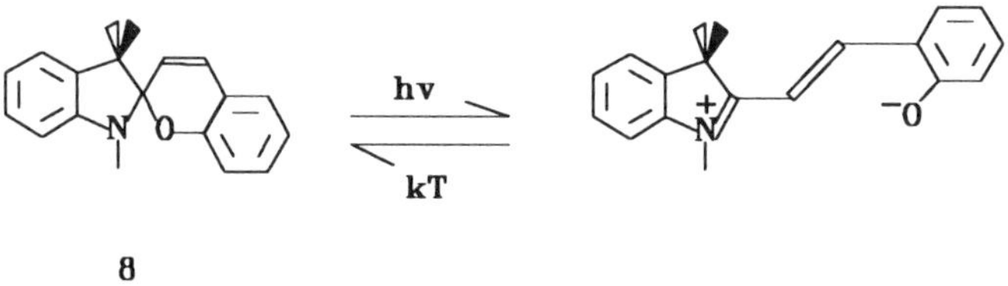

8

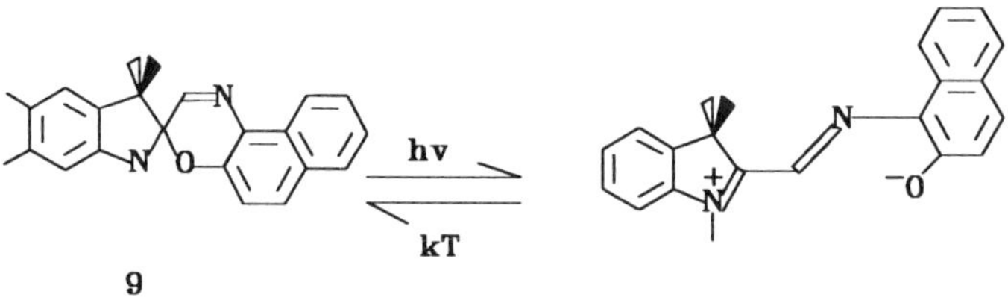

9

metal ions, but in their open zwitterionic form they bind metal ions such as lead(II) and zinc(II).[28, 29]

## 5.2. Conformational and Steric Effects

Another concept that can be used with chelating ligands is to make use of the conformational change that occurs upon complexation of the metal to change the distance between a donor and an acceptor molecule appended to the chelate, thereby modifying the electron transfer quenching rate between them. In equation 4 is shown a schematic as to how complexation of lead(II)

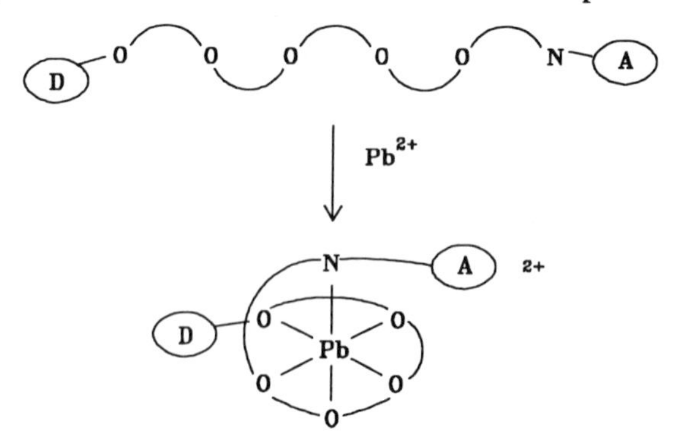


**Equation 4.**

with a sensor system having coumarins A and D bound to each end of a *penta*-(ethylene oxide) ligand can affect the intramolecular interactions, and thereby change the donor-acceptor quenching rate.[30, 31] The transfer efficiencies ($\Phi_T$) and Förster critical radii ($R_o$ in Å for the distance at which $\Phi_T = 0.5$) for this complex in acetonitrile solvent are shown in Table 2.

**Table 2. Transfer Efficiencies ($\Phi_T$) and Förster Radii ($R_o$) of the Free Ligand and the Lead(II) Complex**

| | $\Phi_T$ | $R_o$ (Å) |
|---|---|---|
| Free Ligand | 0.77 | 19.3 |
| Pb(II) complex | 0.89 | 18.3 |

## 5.3. Sulfur Donors

Sulfur donor ligands can also be used in the preparation of sensors for lead(II). The strong affinity of lead(II) for thiohydroxamic acids has been used in the preparation of a chemosensor system that shows a 13-fold enhancement in the fluorescence upon complexation of this metal (equation 5).[32] It is proposed that the enhanced fluorescence from the anthracene

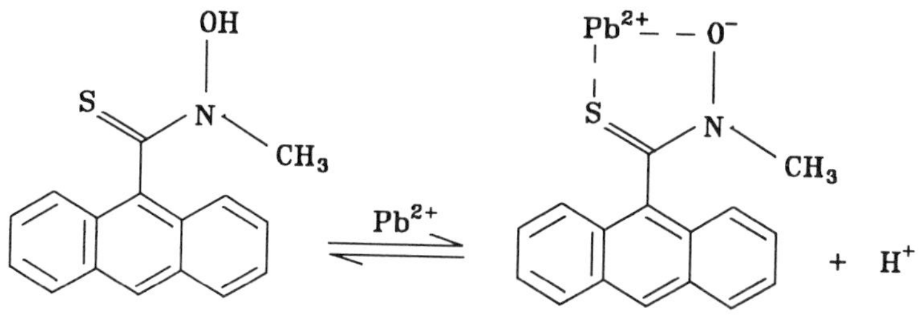

**Equation 5.**

reporter is observed because the approach of water as a quencher is hindered. A redox-switchable fluorescent probe that involves an open and a closed ring form has been designed for sensing heavy metal ions. The system involves an oxidized thiadiazole and a reduced iminoylthiourea combination with an anthracene reporter molecule equation 6. For the oxidized form there is an

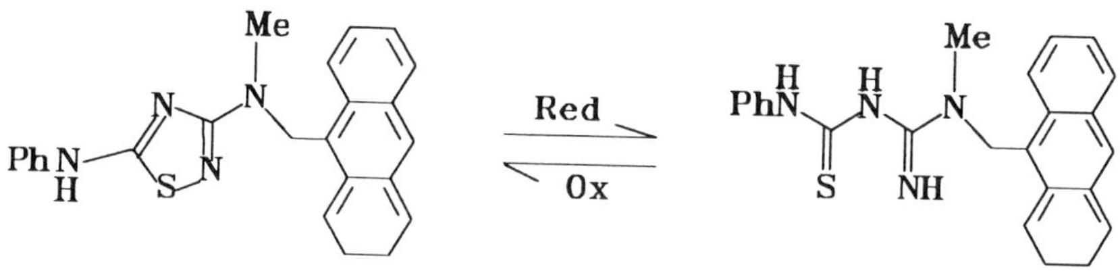

**Equation 6**

increase (in parentheses) in the anthracene fluorescence intensity upon the addition of copper(II) (2-fold), lead(II) (7.7-fold), and mercury(II) (44-fold), with the reduced form showing an increase for cadmium(II) (6-fold), and zinc(II) (3.6-fold).[33] Functionalized polythiophenes have been used as selective coating materials for quartz crystal microbalance sensors. These sensors selectively bind mercury(II) at the thiophene sulfurs, with no interference from iron(II), lead(II), cobalt(II) and chromium(III). Regeneration can be effected by the addition of EDTA.[34]

Sensors for heavy metal ions based on proteins have been developed, with a capacitive signal transducer being used to measure the conformational change following binding. The proteins GST-SmtA and MerR with binding sites for heavy metal ions are overexpressed in *Escheri-chia coli* which is immobilized onto a self-assembled thiol layer on a gold electrode. This assembly is then used as the working electrode in a potentiostat arrangement in a flow analysis system. The system has been used to detect copper(II), cadmium(II), mercury(II), and zinc(II) ions. Metal ions can be detected with both electrode types down to femtomolar concentrations. The metal binding sites are the cysteine residues in the proteins. The principle used to detect the metal ions is shown in Equation 7, where the conformational change of the

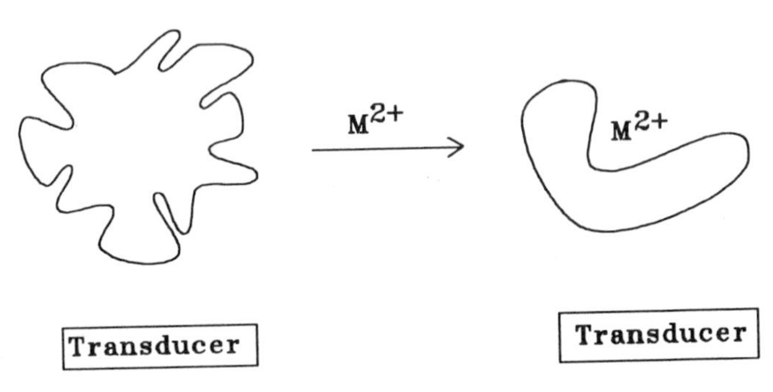


**Equation 7.**

protein in the presence of the metal ion is observed by capacitance measurements. The interface consists of a metallic conductor which is covered by nonconducting organic molecules, outside of which ions in an aqueous conducting liquid form a space-charge. The total capacitance is composed of a series of capacitances representing the self-assembled molecules, the protein, and the space-charge of ions. Since the inverse total capacitance is the sum of the inverse capacitances of each layer, it is important to make each capacitance as large as possible so that the changes caused by recognition dominate as much as possible. The electrodes can be regenerated with EDTA.[35]

## 5.4. Ruthenium Bipyridyl Complexes as Sensors

Chelate complexes of ruthenium(II) can be used as redox switchable fluorescent sensors. An example of such a sensor system is shown in equation 8 where a quinone/hydroquinone redox switch is incorporated

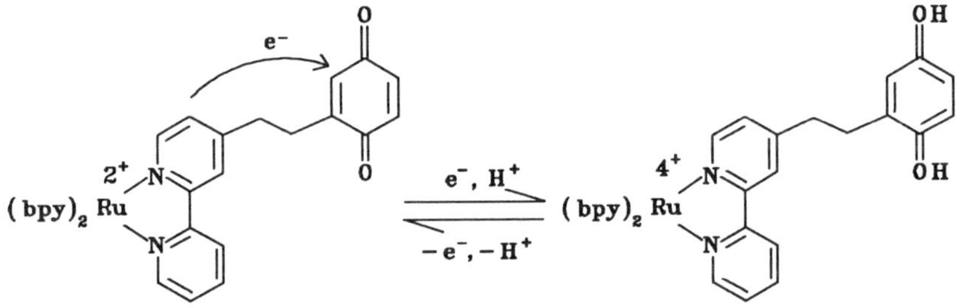

**Equation 8.**

onto a ruthenium bipyridyl complex.[36] In this sensor system the quinone fragment promotes electron transfer quenching from the excited state that is centered on the metal complex.[12] On reduction to the hydroquinone form this electron transfer is unfavorable, and the ruthenium complex is emissive. Similar chelate complexes of ruthenium(II) can be used as proton sensors if they have a site remote from the metal center that can undergo proton transfer. For example, the excited state of the ruthenium bipyridyl complex **10** is not quenched by the *tert*-amine nitrogens when they are protonated because their electron pairs are not available for electron transfer.[4]

Chelate complexes of ruthenium(II) can also be used as sensors for detecting the presence of small quantities of water. An example is the use of $Ru(bpy)_2dppz^{2+}$ (dppz = dipyrido-[3,2-9:23-c]-phenazine) as a probe of accessible water in Nafion. The emission of $Ru(bpy)_2dppz^{2+}$ in non-aqueous solvents is completely quenched by any water that is present.[37, 38] This quenching is due to coupling of the excited state of $Ru(bpy)_2dppz^{2+}$ with the O-H vibrational stretching mode in water.[39] Upon incorporation of the

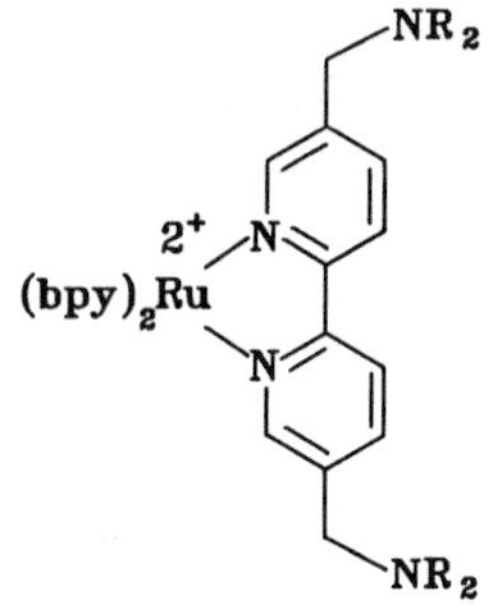

10

complex into Nafion, which is a perfluorinated ionomer, strong emission is detected.[40] This observation supports the postulate that the complex resides in the hydrophobic portions of the polymer films where the water concentration is low. The protruding phenazine moiety of the complex likely interacts with the hydrophobic fluorocarbon matrix, thereby interfering with hydrogen bonding to water molecules in the polymer film. As a consequence, emission intensities, lifetimes and lifetime distributions depend strongly on the loading level of the complex.

## 5.5. Calixarene Complexants

Calixarenes can be used as host molecules for synthesizing emissive sensors. Two types of photophysical reporter molecules have been used with calixarenes. The first type uses an emissive molecule attached to the ligand periphery, and the second uses an emissive lanthanide ion complexed into a metal binding site in the host. Electroactive ferrocene and cobaltocene reporter molecules have also been appended and the combination used as a redox sensor.

## 5.6. Optical Reporters on Calixarenes

Alkali metal ion sensors have been developed using the calixarene framework. A series of different binding atoms and reporter groups have been appended. One such chemosensor **11** has ketonic oxygens as the ligating groups and anthracenes as the reporter. Upon addition of lithium(I) or sodium(I) there is observed a marked decrease in the fluorescence intensity of the reporter molecule. For potassium(I) the emission intensity of the 418

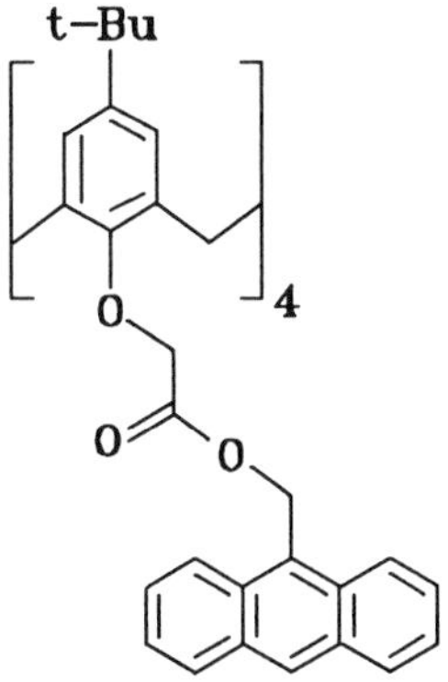

t–Bu

4

O

O

O

11

nm band decreases, but there is an increase of intensity at 443 nm. The system is iso-emissive at 432 nm.[41] A similar compound **12** having two

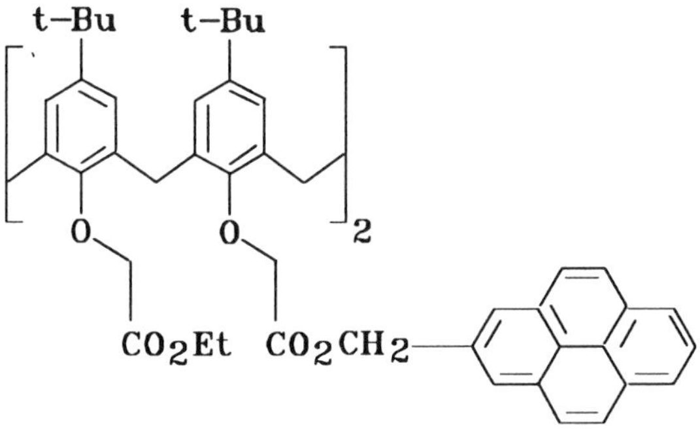

t–Bu t–Bu

2

O O

$CO_2Et$ $CO_2CH_2$

12

pyrene reporters shows a decrease in the excimer emission, and an increase in the intensity of the monomer emission upon addition of sodium(I).[42] A modified approach uses both a pyrene and a nitrobenzene quencher incorporated into the calix[4]arene molecule **13**. Upon addition of an alkali metal ion an increase in the fluorescence intensity is observed.[43] In the free state the pyrene and nitrobenzene quencher can rotate freely and come into close proximity in a statistical manner. When a metal ion is bound, the ester

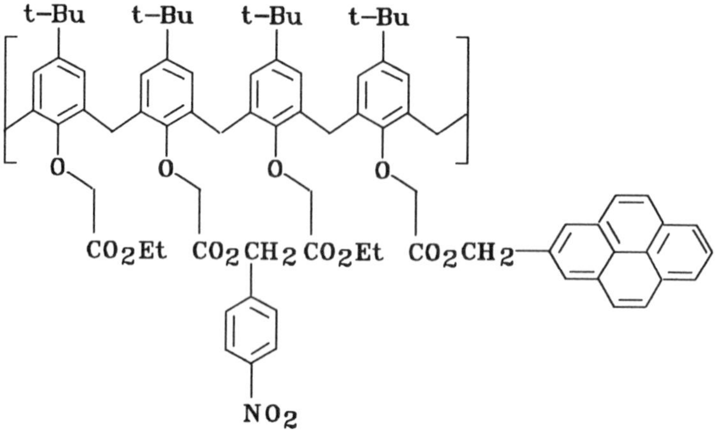

t–Bu t–Bu t–Bu t–Bu

$CO_2Et$ $CO_2CH_2$ $CO_2Et$ $CO_2CH_2$

$NO_2$

**13**

carbonyls become orientated inwards as the lone electron pairs on the oxygens are directed toward it, thereby reducing the collision probability between the pyrene fluorophore and the nitrobenzene quencher. This quenching pathway involves electron transfer from the photoexcited pyrene moiety to the nitrobenzene. Such electron transfer pathways are strongly dependent on the intermolecular distance between the fluorophore and the quencher, thereby making this a sensitive sensor system. This property of the calixarene framework to show conformational flexibility is particularly useful for such applications. An interesting extension would be to use larger ring size calixarenes since they exhibit even greater conformational mobility than the calix[4]arene used here. A calix[4]arene with two indoaniline reporters **14** shows such a selective calcium(II) induced color change. The

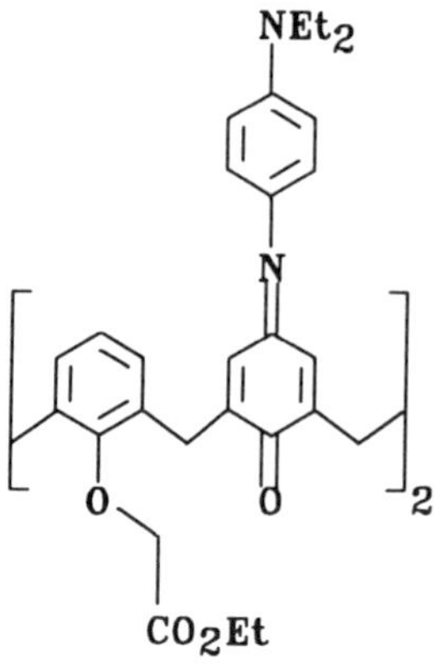

**14**

system has potential biochemical applications because it shows a high selectivity for calcium(II) over magnesium(II).[44] The calcium(II) ion is encapsulated in the cavity made by the distally located

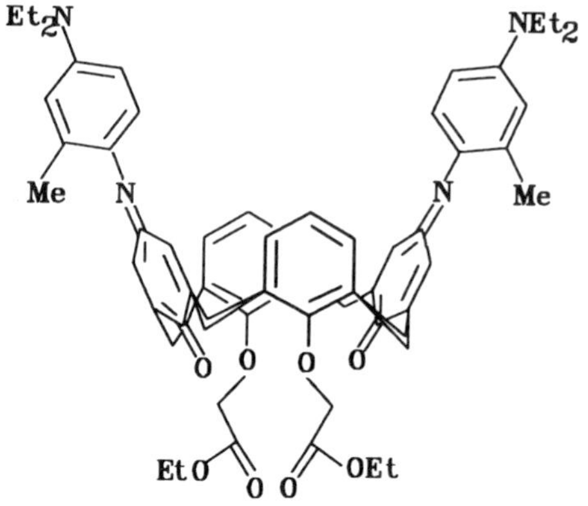

**14**

$OCH_2CO_2$ groups on the narrow rim of the calix[4]arene segment. The uncomplexed compound in ethanol solution is blue ($\lambda_{max}$ = 609 nm).

Addition of calcium(II) results in both a large bathochromic shift and an increase in absorption intensity. In the presence of a five-fold excess of calcium(II) the absorption band is observed at 724 nm. The addition of sodium(I), potassium(I) or magnesium(II) causes no bathochromic shift in the absorption band of this compound.[45] A further development of this concept incorporates an indoaniline receptor into a calix[4]crown **15**. Compound **15** also shows a bathochromic shift upon the addition of

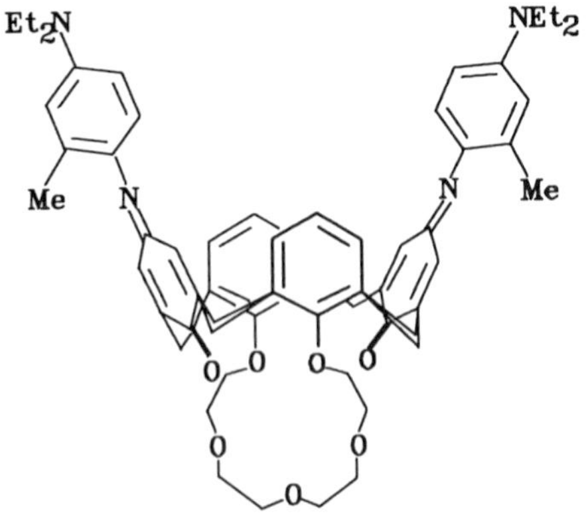

15

calcium(II), which is partially reversed by the addition of potassium(I). No reversal is observed with the addition of lithium(I), sodium(I) or cesium(I), leading to the conclusion that there is competition between calcium(II) and potassium(I) for the calixcrown binding site. The calcium(II)-induced blue to green shift is mainly attributed to efficient dipole interactions between the encapsulated calcium(II) and the quinone carboxyl groups of the indoaniline chromophores. The binding constant with calcium(II) has been estimated to be $1.8 \times 10^{6}$ $M^{-1}$. The large magnitude of the shift upon addition of calcium(II) can be explained on the basis of a conversion between resonance forms that have a neutral quinone or a zwitterionic phenolate structure (Equation 9).[46] This concept of incorporating the metal sensing function into the conjugation of the ligand host is a potentially useful way of incorporating large optical shifts upon metal binding. Such shifts can be particularly effective if the effect is transmitted across a large $\pi$-electron array involving multiple bonds and lone electron pairs. The effect of the metal ion is to act as an electron deficient center within this conjugated system, thereby perturbing its electron density distribution.

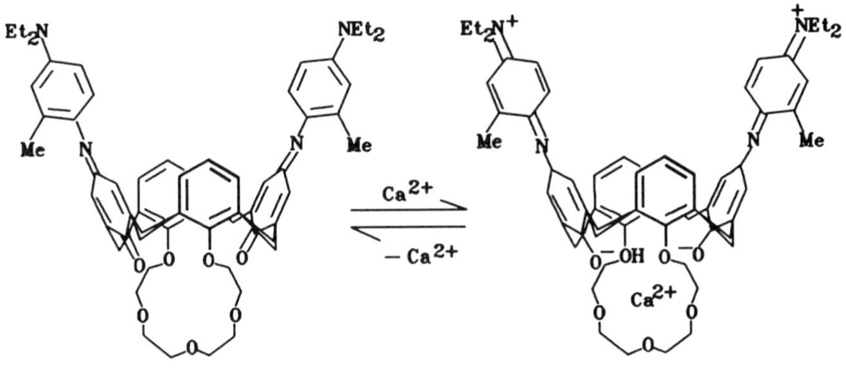

**Equation 9.**

The 4-sulfonic calix[6]arene derivatized with lissamine rhodamine B (**16**) is a fluorescent complexant for uranium(VI). When this complexant

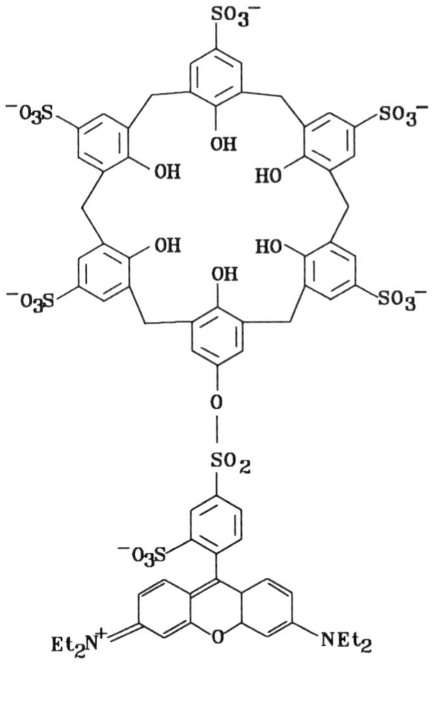

16

is applied to a capillary electrophoresis microchip it can be used for the selective detection of this species. This derivative has a high selectivity for complexing uranium(VI) in the presence of five transition metal, lanthanide, and actinide metal ion impurities, enabling the separation and detection of the uranium complex in under 40 sec on a "lab-on-a-chip" platform.[47]

## 5.7. Redox Reporters on Calixarenes

In addition to being emissive sensors, ruthenium(II) bipyridyl calix[4]arene receptors can also be used to electrochemically recognize dihydrogen phosphate, in the presence of excess bisulfate and chloride.[48]

## 5.8. Lanthanide Reporters on Calixarenes

A lanthanide complexed to a calixarene can act as the reporter group. The lanthanides that have been used are europium(III) and terbium(III). An advantage of using a calixarene as the complexant for a lanthanide reporter ion is that it can be readily modified to accommodate to the preferred high coordination number (8 or 9) of these trivalent ions.[49] For the case of lanthanides such as europium(III) and terbium(III) bound to a calixarene it is important to recognize that this host can bind either one or two such metal ions. Two lanthanide ions are, however, more likely to be coordinated to the larger calixarenes such as *p-tert*-butylcalix[8]arene. For the *bis*-europium complex, in addition to emission from the $^5D_0$ state, an Eu-Eu interaction occurs *via* the LMCT state, so that energy is rapidly lost by $^7F$(Eu)-$^8S$(LMCT) mixing.[50,51] Although europium(III) and terbium(III) can be excited by direct absorption, the absorption bands are weak because f-f transitions are forbidden by the Laporte rule. If these ions are to be effectively used in chemosensors therefore, it is necessary for their excited states to be accessed by energy transfer from excited ligand levels.[52] A particular advantage of using a ligand such as a calixarene is that water is hindered from entering the coordination sphere of the metal. This is particularly important because coupling of the O-H vibration in water with the excited state leads to its quenching, and consequent loss of its emissivity. The calixarene amide **17** is effective for allowing luminescence to be observed from europium(III) and terbium(III) in aqueous solution.[53-55]

However, higher emissivities have been achieved by incorporating three

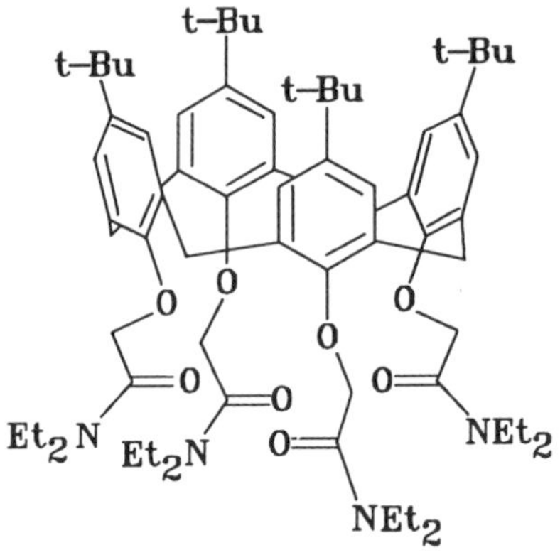

17

amide groups onto a calix[4]arene for metal binding, and one for an antennae group that can act as sensitizer **18**. The sensitizer S in this case is either a phenyl or a biphenyl group.[56]

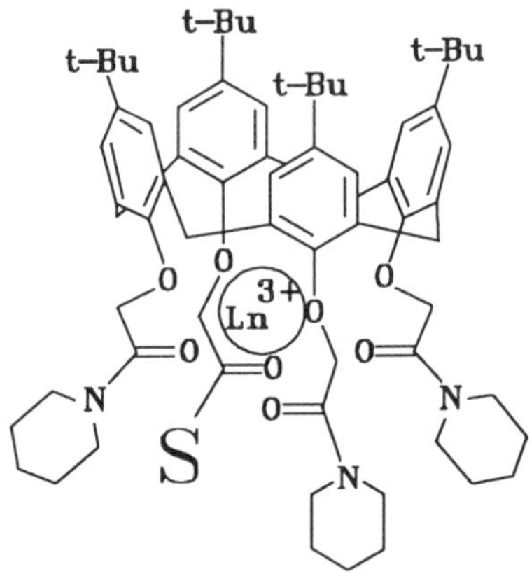

18

# 6. MACROCYCLIC COMPLEXANTS

Two types of macrocyclic ligand find common use. These are ones with a rigid ring structure such as porphyrins or phthalocyanines, and ones with a flexible ring structure such as crowns. The former type can be used as the complexing groups of a metallosensor because complexation of a metal ion within the cavity results in changes in the π-character of the unsaturated ring system, thereby leading to changes in the optical properties of the ring structure. By contrast, the latter type have a saturated ring structure and electron delocalization within it does not occur to any significant extent. These macrocycles have a flexible ring structure that is made more rigid upon complexation of a metal ion to the heteroatoms in the ring structure. These conformational changes that occur upon complexation of a metal can be transmitted to a reporter molecule that is present in the sensor system. Using such macrocyclic ligand systems sensors have been developed for a range of different metals.

## 6.1. *N*, *O*- and *S*-Donors

This group of chemosensors have an aliphatic macrocycle for metal complexation bound to either an optical or a redox reporter molecule. Such combinations can be used as sensors for metals such as lead(II). An example of such a macrocycle is **19**. This particular nitrogen containing macrocycle

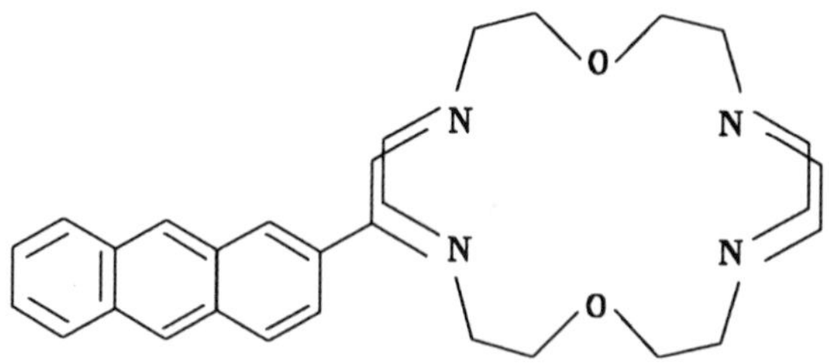

**19**

has good selectivity for lead(II). In methanol solution the lead complex shows a reduced fluorescence signal from the anthracene reporter group.[57] Because of the reversible redox couple between ferrocene $cp_2Fe(II)$ and the ferrocenium ion $cp_2Fe(III)^+$ ion, this system can be incorporated into an electrochemical sensor system. Four such compounds with either a macrocyclic or an acyclic coordination site for metals have been synthesized. The chemosensors **20**, **21**, and **22** electrochemically and selectively sense the presence of mercury(II), whereas maximum electrochemical shifts are produced in **23** in the presence of lead(II). The redox potential changes

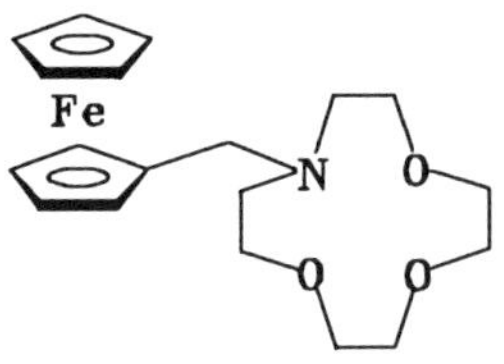

20

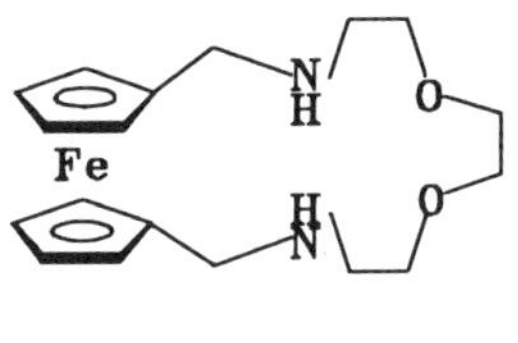

21

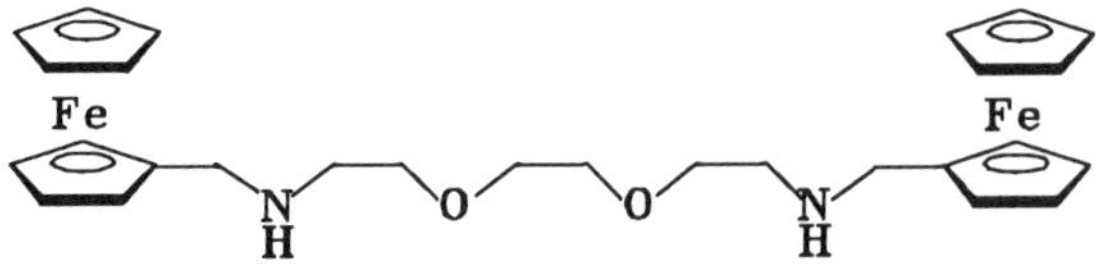

22

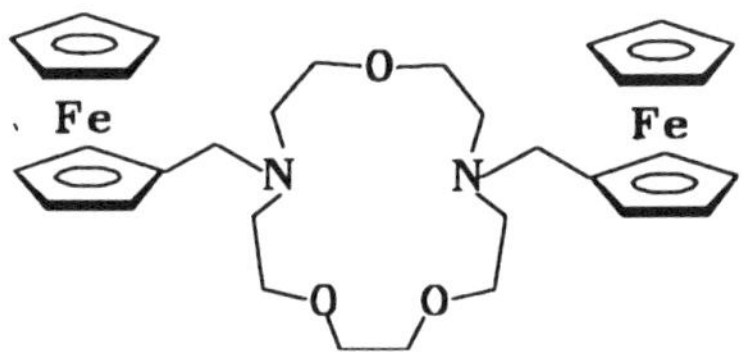

23

induced by metal ion complexation are dependent on the pH of the aqueous solution used in the measurements, with an anodic shift of 130mV being observed for the combination of **21** and mercury(II). The highest stability constants are found for mercury(II).[58] Redox switchable fluorescent chemosensors can use a macrocyclic complexant with an appended anthracene reporter. Once such system (**24**) uses the property of the

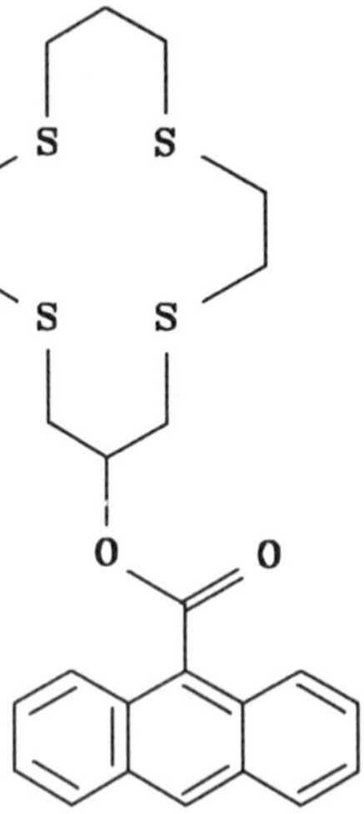

24

tetrathiamacrocycle to bind both copper(I) and copper(II), thereby allowing for the metal center to act as the switch, and the anthracene reporter to function as the fluorophore.[59] The copper(II) complex of this ligand is not fluorescent, but upon reduction the emissive copper(I) complex is formed

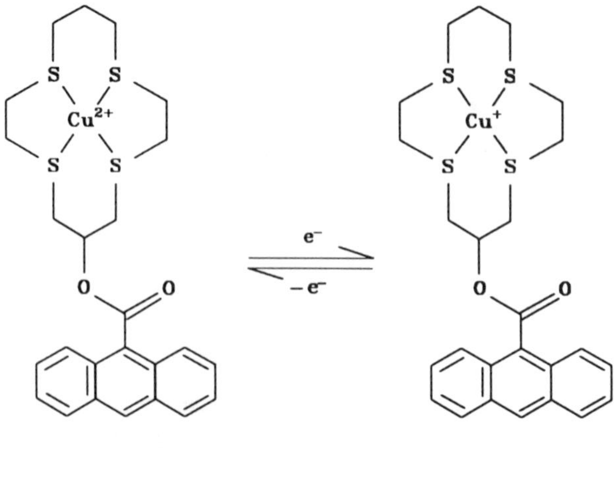

**Equation 10**

(equation 10). The fluorescence can be switched on and off by sequentially reducing the copper(II) to copper(I) at a potential of 200 mV (vs SCE), and then re-oxidizing the copper(I) back at 800 mV. The quenching of the excited state of anthracene by copper(II) (equation 11) is thermodynamically much more favorable than is the reverse electron transfer process that occurs with the copper(I) complex (equation 12).

$$An^{*} + Cu(II) \rightarrow An^{+} + Cu(I) \quad (11)$$

$$An^{*} + Cu(I) \rightarrow An^{-} + Cu(II) \quad (12)$$

Fiber optics is playing an increasing role in the design and operation of optical sensors, both for those that use absorption and those that use emission. Diarylthiocarbazone are a class of ligands that have been used in the preparation of absorptive sensors. These ligands develop an intense color in the presence of heavy metals. Such ligands have been supported on polystyrene by the coupling of nitroformaldehyde-β-naphthylhydrazone with diazotized aminopolystyrene. The polystyrene-supported final diarylthiocarbazone product is then obtained by treatment with aqueous ammonium sulfide, followed by oxidation. The resulting material has been used for the detection of lead(II) and mercury(II) in 0.005 M aqueous solutions.[60] The slow diffusion of heavy metals through such membranes can be problematic, however, and one approach to circumvent this problem uses flow injection analysis coupled with the flourescent sensor.[61]

Macrocycles having both a phenanthroline-crown binding site and a tetrathiafulvalene (TTF) redox reporter incorporated into the same ring (**25**, R = Me; hexyl) are sensors for lithium(I), silver(I), and copper(I). The

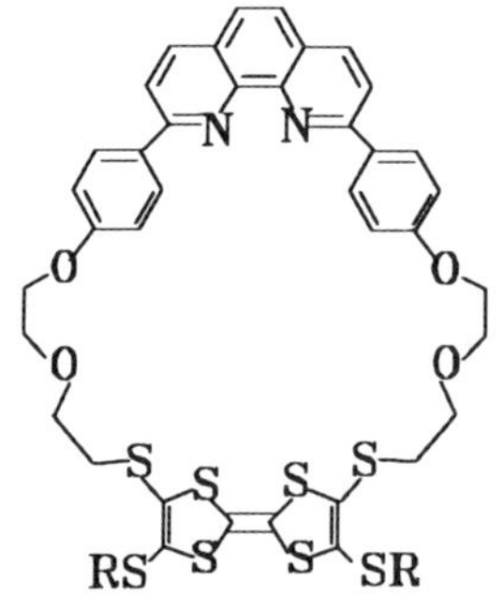

X⁻/X²⁻ cis/trans

25 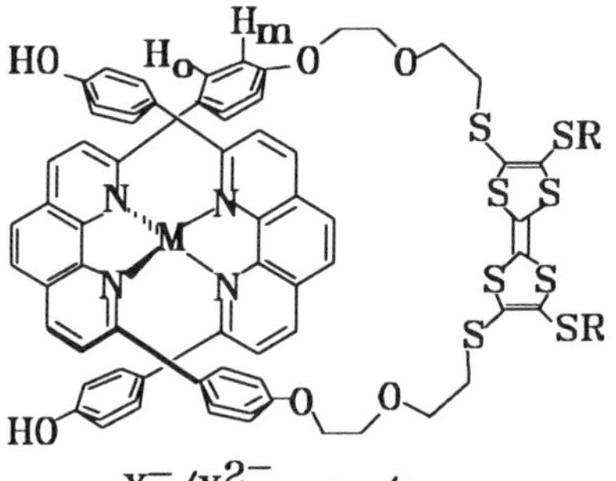 26

complexation occurs with the two phenanthroline nitrogens and the phenanthroline nitrogens of an added molecular thread such as a 4-phenol substituted phenanthroline (**26**). The copper(I) and silver(I) complexes show large shifts in the two TTF redox potentials.[62]

The hexathia-18-crown-6 (**27**) when present in PVC polymer results in a membrane that is a sensor for mercury(II) over a wide concentration range ($1.0 \times 10^{-3}$ to $4.0 \times 10^{-6}$ M). This chemosensor can be used in the pH range of 0.5-2.0., and the selectivity for mercury(II) over other metal ions is high.[63] Optical sensing of silver(I) and mercury(II) has also been accomplished with neutral dithiocarbamates attached to poly(vinyl chloride). The system shows good selectivity with a lower detection limit for silver(I) in the nM range.[64] A quinonoid polymer electrode prepared by the oxidation

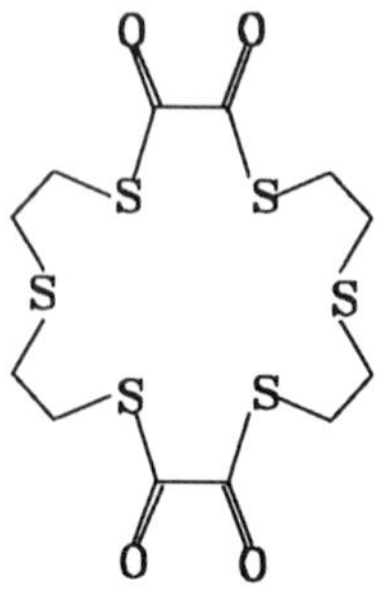

27

of mercaptohydroquinone shows a linear response for silver(I), mercury(II), copper(II), cadmium(II), and lead(II).[65]

## 6.2. Azamacrocycles

Azamacrocycles having a pendant anthracene reporter are selective chelation-enhanced fluorescent sensors for zinc(II) and cadmium(II) in aqueous solution. By contrast, copper(II) and mercury(II) cause quenching.[66] This series of azamacrocycles (**28**) are available for values of x ranging from 1 through 5. The pH dependence on the fluorescence lifetime of the reporter molecule correlates with a photoelectron transfer quenching

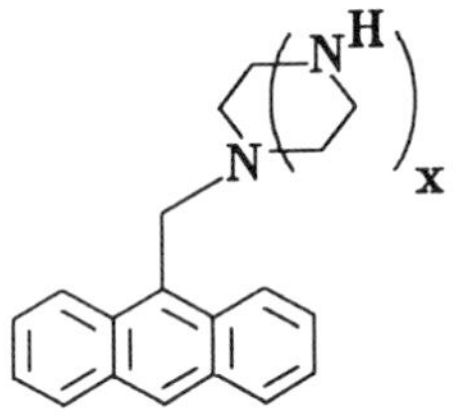

28

pathway *via* the lone electron pairs for the case of zinc(II) and cadmium(II). For the case of copper(II) and mercury(II), intracomplex quenching is the dominant pathway. Subsequently it has been proposed that the equilibrium process involves the presence of small amounts of a cyclometalated cadmium complex having a cadmium(II)-aryl bond.[67]

The metal ion induced phase changes in self-assembled membranes has been used as a sensor for metal ions such as copper(II). The systems **29** (m = 5, 10) and **30** (n = 5, 11) use single chain amphiphiles having an

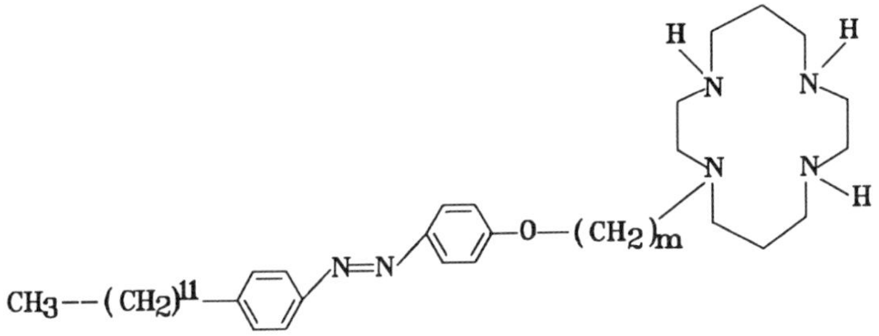

29

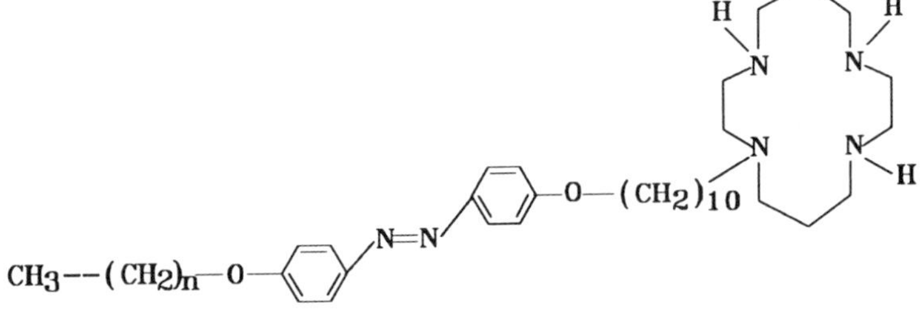

30

azobenzene chromophore in the alkyl chain, and the cyclic tetraamine cyclam as the headgroup. The azobenzene acts as the optical reporter, and the cyclam as the macrocyclic complexant. This cyclam terminated azobenzene chain aggregates in the bilayers of dihexadecyldimethyl ammonium bromide below its chain melting temperature (273 °C) to produce a blue shift (356 to 326 nm), and de-aggregates above this temperature as evidenced by the re-emergence of the 356 nm absorption (Figure 1). Metal ion binding stabilizesthe aggregate up to 45 °C, above which temperature the 356 nm absorption band for the lipid monomer is observed.[68]

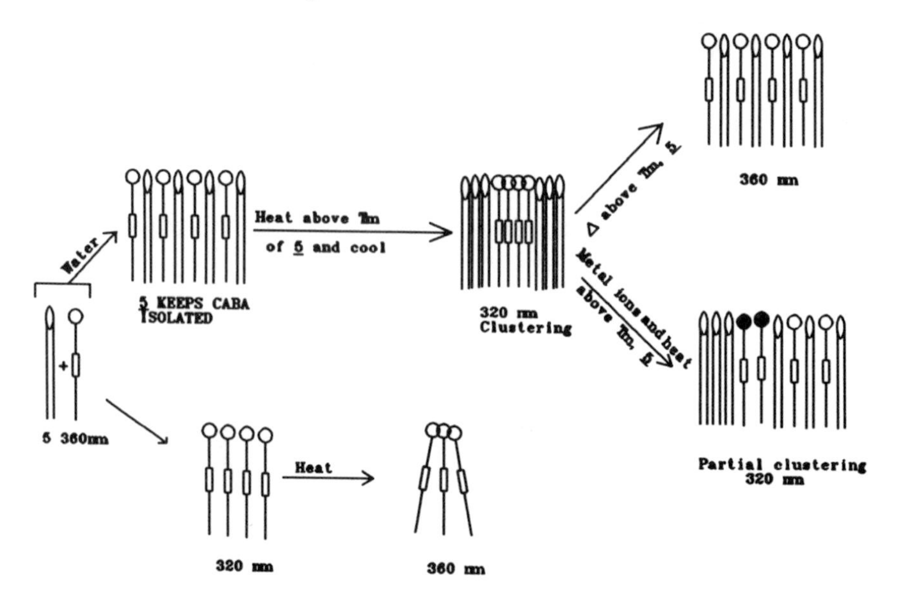


**Figure 1. Metal Ion Induced Phase Changes In Membranes**

## 6.3. Cryptands

Cryptands can be used as the complexant site for chemosensors for metals. In chelate and macrocyclic systems the presence of a transition metal usually leads to a partial quenching of the fluorescence of the reporter molecule. For the case of the cryptand **31** the presence of copper(II), nickel(II) and zinc(II) results in an enhanced fluorescence. The precise reason for this intensity reversal is presently unknown, but it is possibly due

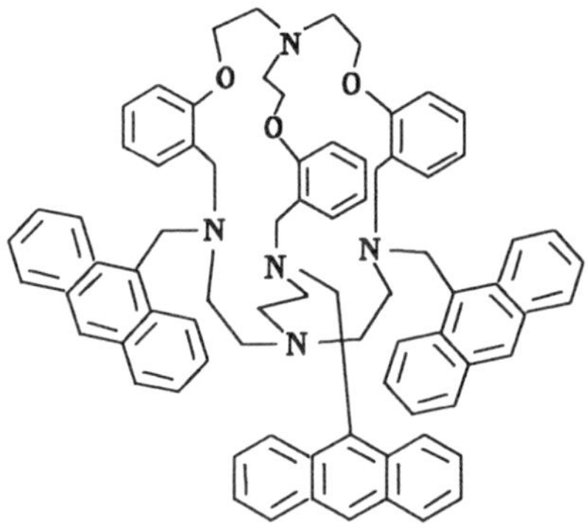

31

to the highly organized cryptate structure causing the redox activity of the metal ions to be suppressed.

Triamine and trianthrylamine cryptands of varying cavity dimensions **32-37** are complexants and emissive sensors. In uncomplexed form these cavitands are non-emissive because of photoinduced intramolecular electron transfer from the nitrogen lone pairs. Protonation or complexation of a transition metal ion causes a large fluorescence enhancement. Fluorescence enhancement is also observed in the presence of lead(II), europium(III) or terbium(III). No evidence is found for charge-transfer interaction between the anthryl groups and the bound metal, but also exciplex formation is not observed with the metal bound within the cavity.[69]

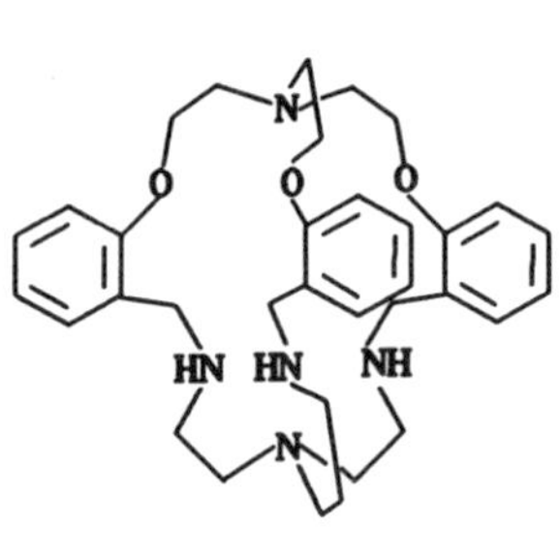

32

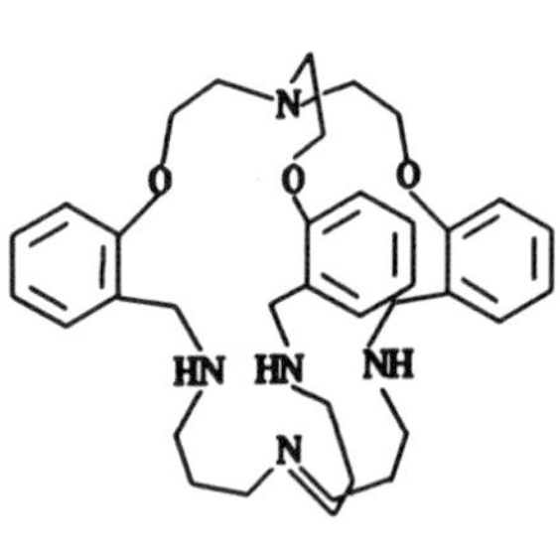

33

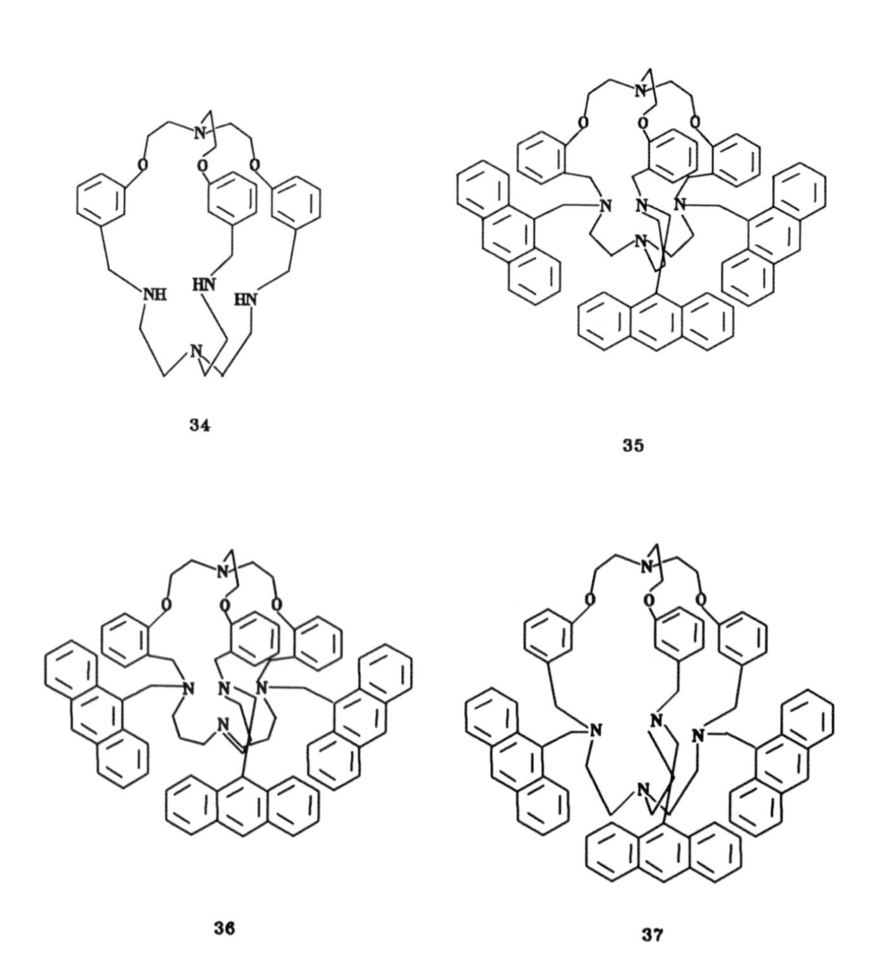


## 6.4. Porphyrins and Pyrroles

Metal porphyrin complexes have also been used as sensors. An example of this is the use of the *meso*-tetraphenylporphyrin Sn(IV) complex **38** as a proton sensor.[4] In this system the unprotonated amine on the ligand periphery acts as a quencher for the metalloporphyrin fluorophore. Protonation of the amine nitrogen eliminates this quenching pathway, and an increase in the emission is observed at higher solution acidities. A zinc

Ph

$NR_2$

Ph Sn(IV) Ph

N N N N

Ph

**38**

porphyrin complex that undergoes photoinduced electron transfer from its singlet excited state to a silver(I) bound closely to it within a tetracyclic-*N*,*O*-cryptate **39** can be used as a sensor for the silver(I). The photoreaction is

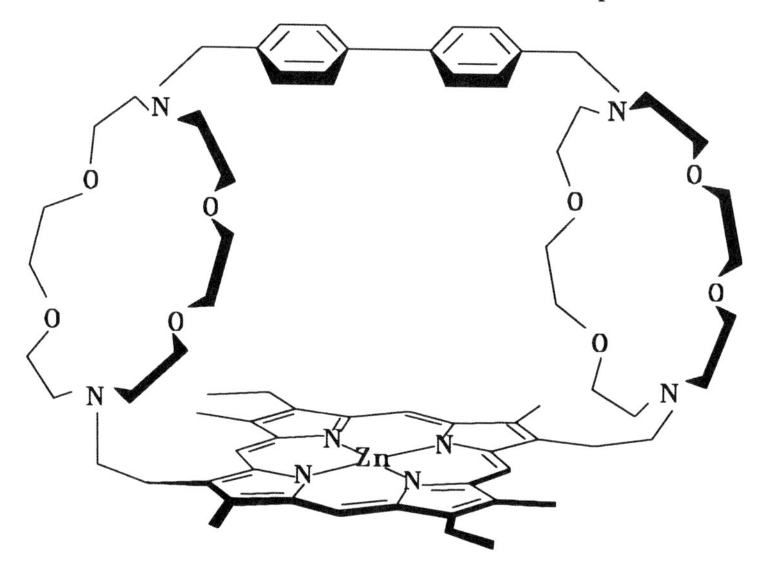


**39**

shown in equation 13, with the silver(I) being reduced by the excited state

$$ZnOEP^* + Ag^+ \rightarrow ZnOEP^+ + Ag \qquad (13)$$

of the zinc porphyrin.[70] A pyrrole-derived squaraine molecular wire **40** (m = 1, 2) that contains a flexible oxyethylene side chain (m = 1, 2) shows changes in its emission characteristics upon addition of lithium(I), but not

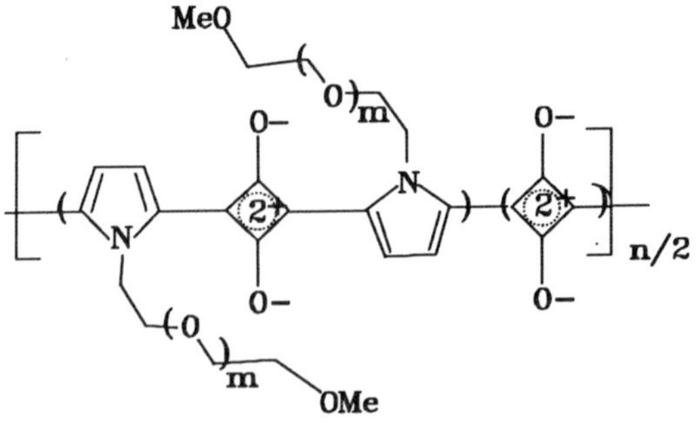

**40**

with sodium(I) or potassium(I). The change in the emission is believed to be due to a "flexible to a rigid" conformational change in the wire upon the binding of lithium(I).[71]

## 7. CROWN ETHERS AND CRYPTANDS FOR ALKALI AND ALKALINE EARTH CHEMOSENSORS

A class of saturated macrocycles having oxygen donor atoms are the crown ethers. These compounds are a particularly useful class for developing sensors for alkali and alkaline earth metals because they have high binding constants for these ions, and also by choosing ones that have the correct cavity size they can be made selective for the individual members of these series of metals. Upon complexation of an alkali or alkaline earth metal ion both the conformation and the flexibility of the crown ether changes, which is a useful feature because it may result in changes in the optical properties of the appended reporter molecules.

## 7.1. Naphthalene and Anthracene Reporters

The first example of a crown ether acting as a fluorescent sensor for an alkali or alkaline earth metal was reported by Sousa in 1977. This publication reported enhanced fluorescence from the attached napthalene chromophore reporter group of the compound 1,8-naphtho-21-crown-6 in the presence of an alkali metal.[72,73] This enhancement of approximately 60% in the fluorescence intensity is observed in an ethanol glass at low temperatures. This enhancement has been attributed to a cation-induced decrease in the triplet energy level relative to the fluorescent singlet state and the ground state of the reporter molecule. This strategy has been extended to the analogous 1,5-naphtho-22-crown-6. This particular compound is chosen because it has the additional property of having the facility to adopt a conformation where the crown ether moiety can hold the alkali metal quencher against the $\pi$-system of the naphthalene chromophore. As a result, a bound cesium(I) center with a large spin-orbit coupling term acts as a quencher because of its propensity to increase inter-system crossing from the fluorescent $S_1$ state to the non-fluorescent $T_1$ state of the chromophore. The lighter ion potassium(I) does not lead to such quenching. The system can therefore be used as a sensor system for mixtures of these two ions because of the dependence of the magnitude of the quenching effect on the two individual metal ions (equation 14).[73] The efficiency of such a sensor

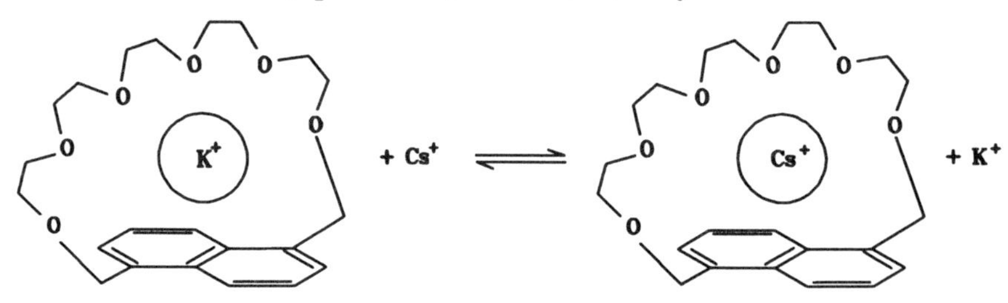


**Equation 14**

depends on the relative stability constants of the cesium(I) and potassium(I) complex.

In equation 4 an example was presented where the complexation of lead(II) resulted in a decrease in the separation of the donor and acceptor molecules at the ends of an open chain polyether ligand. For crown ethers, complexation of an alkali or alkaline earth ion can lead to greater rigidity in the polyether system, thereby resulting in increased separation of the donor and acceptor moieties (Scheme 1). Such a ligand system with a fluorescent

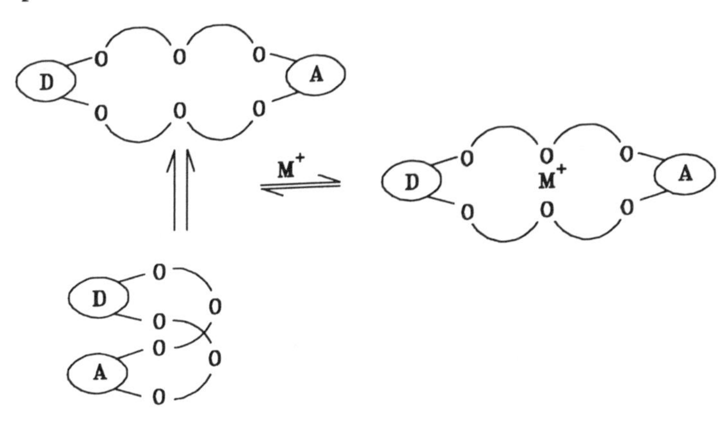


Scheme 1

naphthalene acceptor and aniline donor quencher **41** has been designed to test such a concept. It is estimated that complete complexation of potassium(I)

**41**

will lead to a 10% change in the intramolecular quenching rate. A complication to interpreting these data, however, is that complexation of potassium(I) leads to reduced quenching because the lone electron pair on nitrogen is a poorer reductive quencher for the photo-excited $\pi$-system of the condensed aromatic system. A system that shows such an increased fluorescence in the presence of added potassium ion is shown in equation 15.[4]

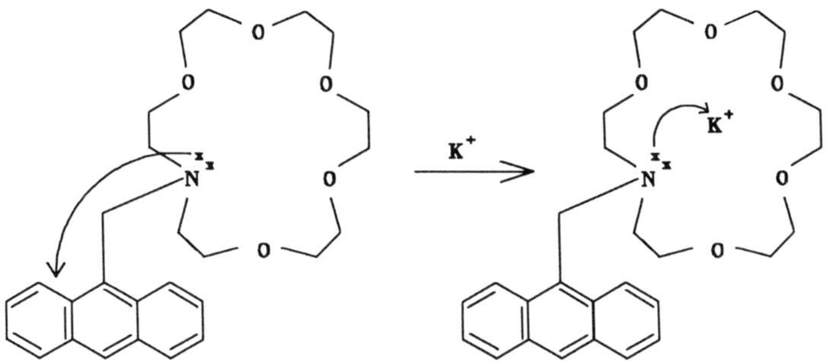

**Equation 15**

A similar potassium(I) enhanced fluorescence is also observed for the *bis*-(crown ether) **42**.[74] Another *bis*-(crown ether) sensor that has been used is **43**.[75]

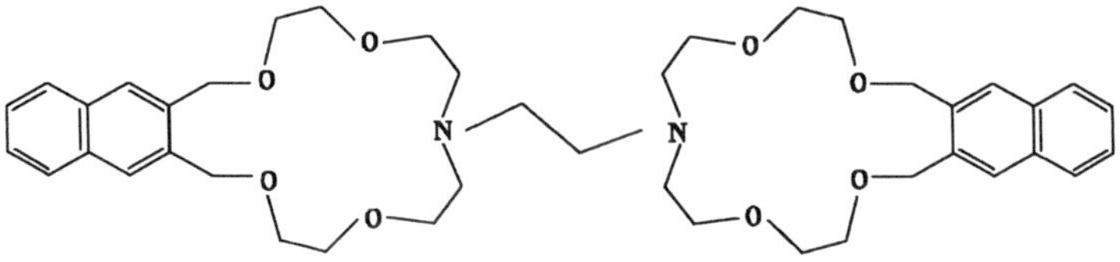

**42**

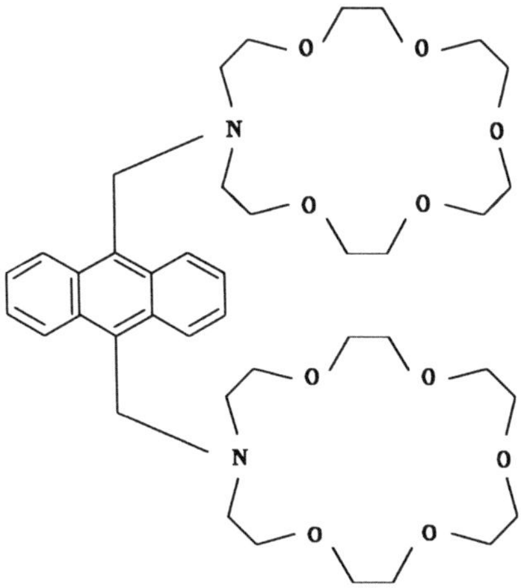

**43**

A crown ether ligand has been synthesized that incorporates two anthracene reporter molecules within the macrocycle. This crown can incorporate two sodiums into two separate 5-oxygen cavities, which results in a red shift of the excimer band. The proposed structural change upon complexation of sodium(I) is shown in equation 16.[76] Complexation results

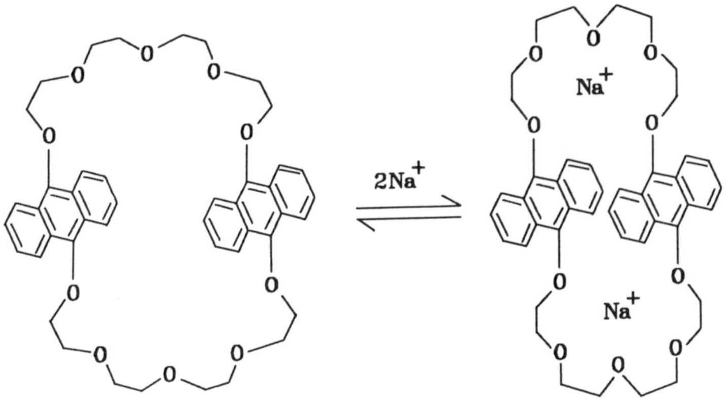

**Equation 16**

in a closer separation of the two anthracene reporter molecules in the macrocycle, and both structured monomer and unstructured excimer fluorescence emissions are observed in methanol solution. A similar two-site crown having anthracene reporters **44** also shows both monomer and excimer

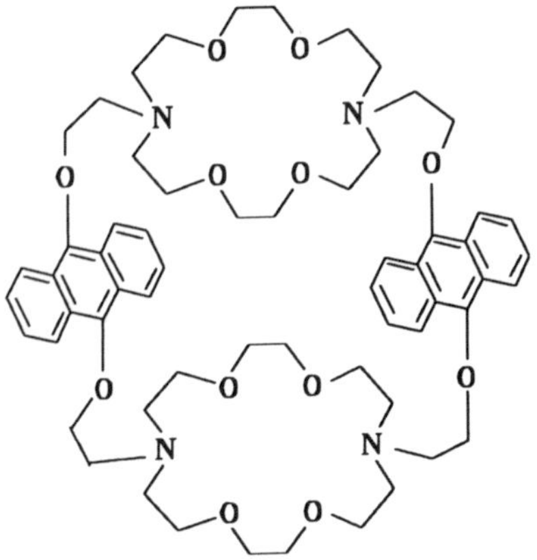

**44**

emissions which are modified upon binding rubidium(I).[77] Again the addition of rubidium(I) results in the observation of an intense red-shifted excimer emission. The effect is specific for rubidium(I) among the alkali metal cations.

Cryptands having anthracene reporters are also used as sensors for alkali metal ions. An example is **45** which is designed to minimize the proton

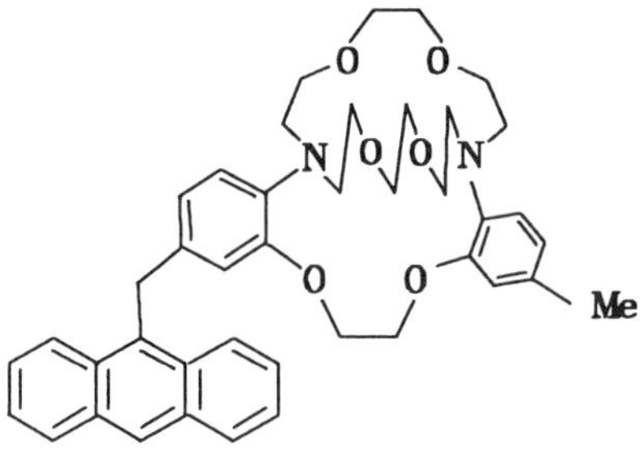

45

basicity of the ligand. This cryptand-anthracene combination shows up to a 10-fold increase in fluorescence intensity upon complexation of an alkali metal ion.[78] A similar supramolecular approach to metal ion recognition involves synthesizing conducting polymetallorotaxanes by the electrochemical polymerization of metallorotaxanes. These systems reversibly bind zinc(II) and copper(II), and show conductivity behavior characteristic of localized redox conductors (Equation 17).[79]

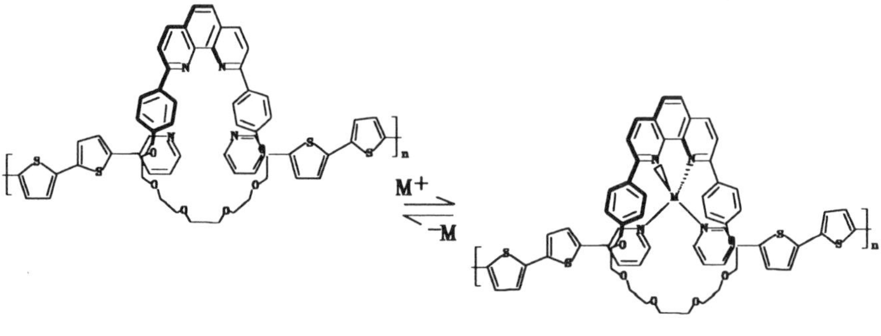

**Equation 17**

## 7.2. Iron Sulfur Cluster Reporters

Iron sulfur proteins are well known as redox active molecules in bioinorganic chemistry. One of the more common cores is the $Fe_4S_4$ cluster that undergoes 1-electron chemistry at low potential. These $Fe_4S_4$ clusters can be chemically attached to a crown ether functionality and used as the redox reporter. The addition of lithium(I), sodium (I), potassium(I), magnesium(II), or barium(II) results in an anodic shift in the reversible 2-/3- reduction wave. The magnitude of the effect can be correlated with the binding constant for the metal ion within the crown, and the nature of the redox active cluster core.[80]

## 7.3. Structural Features

The lowest excited triplet states have been studied in 1,5-naptho-22-crown-6 (**46**), 1,8-naphtho-21-crown 6 (**47**), and 2,3-naphtho-20-crown-6 (**48**), along with their sodium(I), potassium(I), cesium(I), silver(I), and .

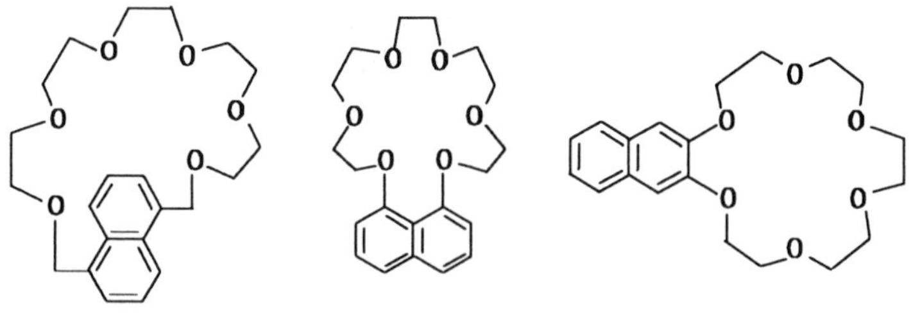

46 47 48

thallium(I) complexes Optical detection of magnetic resonance at 1.2 K in zero applied magnetic field indicates that the orientation of the heavy ion with respect to the naphthalene chromophore selectively influences both the sub-level radiative rates and their populating rates by intersystem crossing.[81] From these data, assumptions can be made as to whether the bound alkali metal lies within or above the plane of the macrocyclic oxygen donor.

# 8. SENSORS FOR BIOLOGICAL APPLICATIONS

An important use of sensors is in their application to living systems. Such applications need to lead to the non-destructive monitoring of the free concentrations of a range of different ions and molecules. *In vivo* concentration data can then be obtained for these species either in blood or in the intracellular state. Among the more important species that are targeted for monitoring are sodium(I), potassium(I), calcium(II), chloride ion and cyclic AMP.[82] For biological applications a number of additional features are required in a fluorescent sensor. One of these is that the material needs to be bound to an optical fiber with the resulting sensor being able to function in an aqueous medium. Another desirable feature is that the sensor be photostable in the presence of dissolved oxygen, and that it can discriminate between the chosen ion and a wide range of other species that are present in an *in vivo* environment. Emission wavelengths from the sensor should exceed 500 nm in order to reduce the interference from auto-fluorescence that occurs from reduced pyridine nucleotides that are present *in vivo.* However, the excitation wavelengths of the sensor should exceed 340 nm because shorter wavelengths are absorbed by *in vivo* nucleic acids and aromatic amino acids. For biological applications it is common to incorporate carboxylate groups so that the chemosensor is both water soluble and impermeable to passage through membranes. This latter feature is desirable so that once the chemosensor is introduced into cells, it does not readily migrate out again.

The biological applications of chemosensors will be a growth area over the next decade as increasing importance is placed on developing non-invasive techniques for medicinal diagnosis. Such systems need to be both selective and rapid, in addition to being compatible with living tissue.

The metal-induced dispersion of lipid aggregates can also be used as a selective and sensitive sensor system. Such a system uses a pyrene-labeled lipid functionalized with iminodiacetic acid. This system forms aggregates when placed in vesicles of distearoyl phosphatidylcholine, as evidenced by pyrene excimer formation in steady-state fluorescence measurements. Addition of divalent metal ions results in a rapid reversal of the ratio of emission intensities of excimer and monomer, which likely coincides with dispersion of the pyrene- lipid into the gel phase of the lipid. The system is selective for copper(II).[83]

## 8.1. Sodium

The measurement of intracellular sodium ion is important because its concentration gradient between the extracellular and intracellular state regulates a number of biological functions such as the transmission of electrical impulses. The most effective fluorescent indicator for sodium(I) is SBFI (**49**), a crown ether with two pendant arms and carboxylate groups appended to these arms.[84] This compound has good sodium(I) : potassium(I) selectivity.

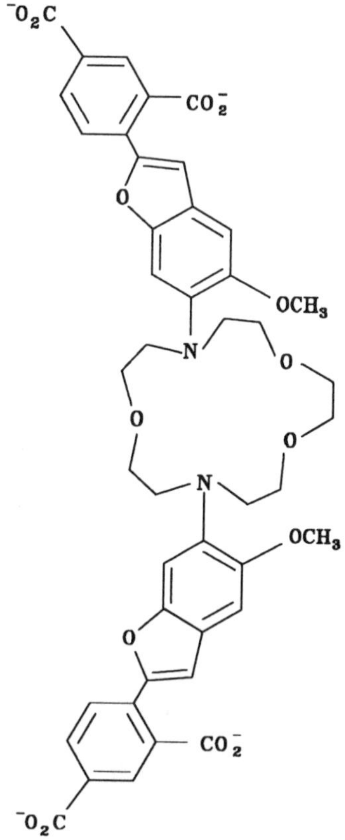

$^{-}O_2C$ $CO_2^-$ O $OCH_3$ N O O O N $OCH_3$ O $CO_2^-$ $^{-}O_2C$

**49**

## 8.2. Potassium

A compound that shows enhanced emission in aqueous solution in the presence of potassium(I) is the 4-methylcoumaro-[2,2,1]-cryptand **50**.[85] The excitation and emission peaks are at 340 nm and 420 nm respectively.

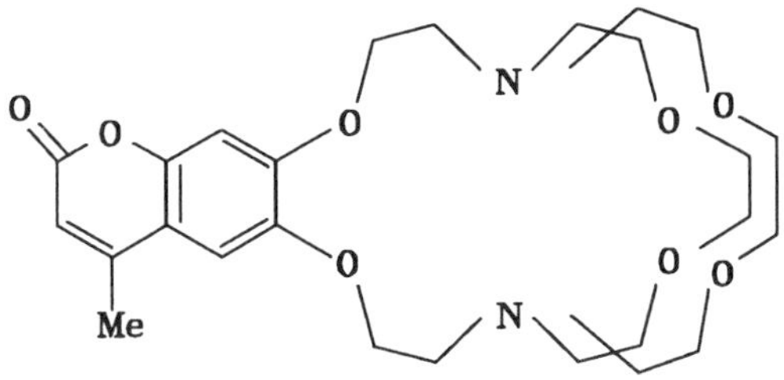

50

The addition of sodium(I), magnesium(II), calcium(II) and ammonium(I) does not change its fluorescence. The enhanced emission in the presence of potassium(I) is due to the reduced availability of the lone electron pairs on the cryptand nitrogens to become involved in photoelectron transfer quenching of the coumarin excited state.

## 8.3. Calcium

The measurement of intracellular calcium is important.[86] Local pulses of increased calcium(II) concentrations affect muscle control, and the majority of the cells of the immune system are influenced by its intracellular concentration. Since extracellular calcium(II) concentrations are some four to five orders of magnitude greater than intracellular ones, damage to the outer cell membrane causes the latter to rise. The fluorescent sensor must therefore enter the cell without puncturing the plasma membranes. Another problem to be overcome in developing an *in vivo* calcium(II) sensor is the achievement of selectivity against magnesium(II), which is present in cells at concentrations some four orders of magnitude higher than is that of calcium(II).

One strategy to develop an effective sensor is to design complexants that are based on EGTA, the most selective chelator available for calcium(II).[87,88] One such derivative, BAPTA (**51**), replaces the two -$CH_2CH_2$- groups between the nitrogens and ether oxygens in the EGTA backbone by *ortho* aromatic linkages.[86] BAPTA shows the necessary

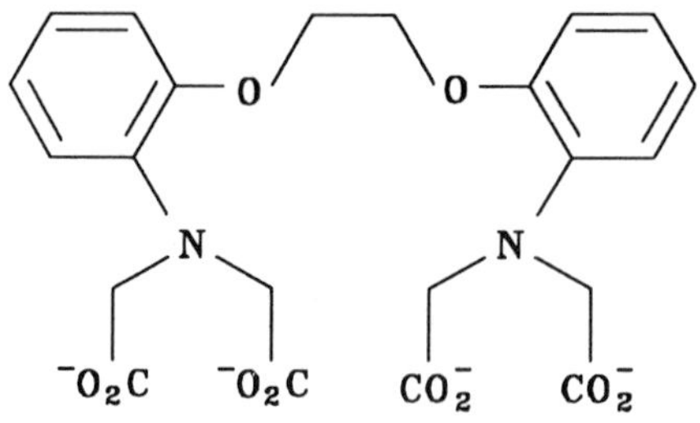

$^{-}O_2C$ $^{-}O_2C$ $CO_2^-$ $CO_2^-$

51

calcium(II) : magnesium(II) specificity because it retains both the dimensions and the large donor numbers appropriate for the larger calcium(II). The aromatic centers confer emissivity which is strongly affected by calcium(II) binding. Fluorescent chromophores have also been incorporated onto the BAPTA framework. Extension of the aniline chromophore in **51** to heterocyclics results in the derivatives FURA-2 (**52**) and INDO-1 (**53**). Both **52** and **53** show large spectral shifts upon

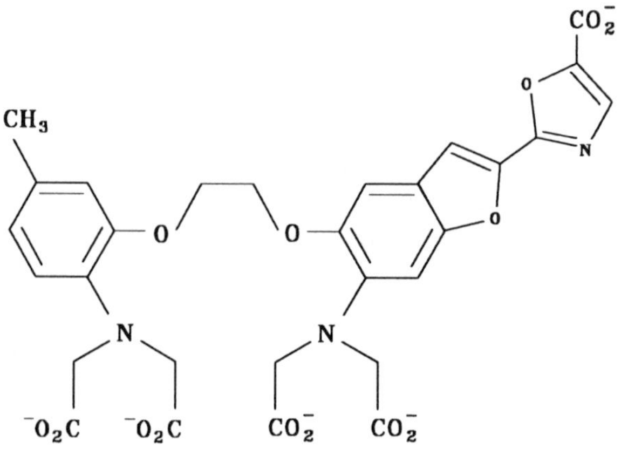

$CO_2^-$ $CH_3$ $^{-}O_2C$ $^{-}O_2C$ $CO_2^-$ $CO_2^-$

52

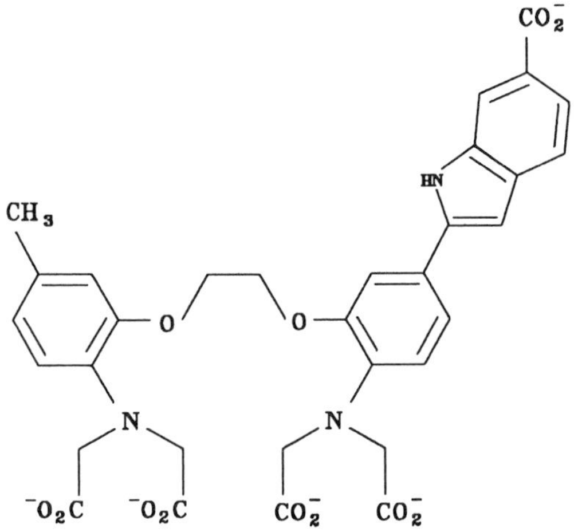

53

complexation with calcium(II).[89] Increasing the length of the conjugated chromophore has the effect of moving the absorption maximum further into the visible region, but a negative effect of such a strategy for changing the optical properties is that there is a corresponding increase in the hydrophobicity of the molecule, thereby decreasing its aqueous solubility.

A problem with the commercial use of **52**, however, is that within one hour it has migrated out of the cytoplasm and become localized within intracellular compartments. The sensor is still fluorescent, but it is now insensitive to changes in cytoplasmic calcium(II) levels. This problem has been solved by conjugating the derivatized **51** to dextran, a high molecular weight water soluble carrier. These conjugates have been used for neuronal tracing and ratiometric imaging of intracellular calcium.[90, 91]

## 8.4. Zinc

A review of peptide platforms for metal ion sensing has been published.[92] A fluorescent chemosensor based on a peptidyl ligating system has been used to detect nanomolar concentrations of zinc ion.[93, 94] In this system the metal binding induces protein folding, thereby shielding the fluorophore from the solvent and increasing the emission intensity.

Competing ions such as magnesium(II) and cobalt(II) have minimal effect on the effectiveness of the system. A similar fluorescent probe for zinc(II) has been developed by labeling the zinc finger consensus peptide CP with the fluorescent dyes Lissamine and Fluorescein.[95] Incorporation of the fluorescent dyes does not adversely affect the metal binding and folding properties of the zinc finger peptide, and since the absorbance and fluorescence of CP modified with the two dyes are not obscured by normal cellular components, the probe is a good candidate for *in vivo* applications.

The sensing of zinc in biological systems is important because of the presence in nature of these zinc finger peptides or domains. These zinc finger proteins have a tandem array of units with the amino and carboxyl termini far apart.[96] These proteins wrap around DNA, and interact primarily with the major groove. Their $N_2S_2$ coordination site allows for preferential binding of zinc(II) over both the first row transition metals, and also second row elements such as cadmium(II).[97] Fluorescence energy transfer measurements have been used to investigate the solution structure of a single domain zinc finger peptide.[98] The method uses two donor-acceptor-pair zinc finger peptides which incorporate a single tryptophan residue at the mid-point of the sequence. This functionality is the energy donor for two different acceptors. The acceptor at the amino terminus is a 5-(dimethylamino)-1-naphthalenesulfonyl group, and that at the ϵ-amino functionality of a carboxy-terminal lysine residue is a 7-amino-4-methyl-coumarin-3-acetyl group. The donor-acceptor distance distributions determined under both metal-free and zinc-bound conditions show a shorter distance when the metal is bound, and a longer distance with greater conformational flexibility when the metal ion is absent.

Because of their specificity for individual metal ions, enzymes can be used as the binding site for selected metal ions. Such biosensors have been developed for zinc. Carbonic anhydrase from mammalian erythrocytes, carbonate hydro-lyase, binds its zinc cofactor with high specificity. Only cobalt and manganese competitively bind, but then at reduced affinity. As a result a biosensor for zinc has been developed using the specific recognition for the metal ion by carbonic anhydrase, along with the fluorescent inhibitor dansylamide.[99] The 15-fold enhancement of the dansylamide fluorescence in the presence of zinc(II) allows for the metal to be detected at nanomolar concentrations. Transmission to the detector is effected through a single optical fiber. A limitation of this sensor is the requirement of exciting the dansylamide at 330 nm, and that it only has a moderate absorbance at that wavelength. An alternate approach (equation

18) involves having a fluorescent reporter bound to the carbonic anhydrase

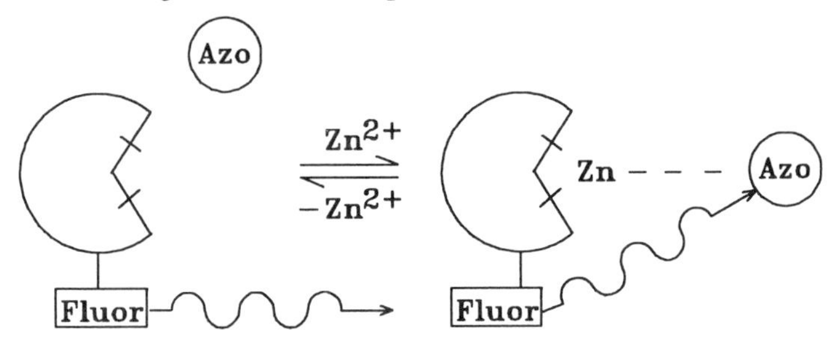


**Equation 18.**

that can undergo energy transfer to a receptor that binds to the complexed zinc.[100] A series of derivatized fluoresceins, rhodamines and coumarins are used as the fluorescent reporters (Fluor), and azosulfamide (Azo) is used as the quencher. The method functions because in the absence of zinc, azosulfamide does not bind to the enzyme, and the reporter then exhibits its normal fluorescence lifetime.

## 8.5. Lanthanides

The lanthanides europium and terbium have been used as emissive probes in biological systems. As an example, the complex between terbium and transferrin has been used as a label in the immunoassay of the antibiotic gentamicin.[101] The linkage of gentamicin to transferrin has been achieved using a carbodiimide reagent. Transferrin binds iron(III) in specific sites on the polypeptide chain, but in the absence of iron(III) the sites may be occupied by terbium(III) to form emissive complexes. Upon excitation at 295 nm, fluorescence enhancement from terbium(III) in the complex is observed. The complex shows good antibody recognition when titrated against anti-gentamicin. Another example is the use of a luminescent europium(III) to probe the metal-binding sites of bovine α-lactalbumin in $D_2O$ solution.[102] Addition of the bovine α-lactalbumin to a solution containing europium(III) results in a quenching of the protein luminescence, and an enhancement of that of europium(III). Other earlier examples have used iron(II), chromium(III), silver(I), and other transition metal ions as fluorescence quenchers for riboflavin.[103,104] The quenching effect is observed in aqueous solutions, and the presence of oxygen in the solution has no effect on the fluorescence intensities, both in the absence or presence of quenchers. These examples are part of a much larger effort focused in the general area of metalloimmunoassays.[105]

## 8.6. Signal Transmission

The use of emissive probes in biomedical applications requires a method to detect the changes in wavelength and intensity at an external site. One such method involves the use of fiber-optics.[106] Devices incorporating such technologies are relatively easily fabricated, and are of particular importance in biological applications where they can be used *in vivo*. For their application, however, it is important that the wavelength of the exciting radiation is not too low, or the photoluminescence from the optical fiber itself may be the dominant observed radiation. Other limitations of fiber optics for chemosensors are that they have limited dynamic ranges, and that mass transfer between the reagent and analyte may limit response time.[107] An alternate approach that is being explored is the use of a molecular photonic wire. One such wire has been synthesized that uses a boron-dipyrromethene dye as absorber at one end, a linear array of three zinc porphyrins as signal transmitters, and a free base porphyrin as emittor.[108] The system has a high yield of energy transfer, and shows no significant electron-transfer quenching. A chemical sensor has also been developed in which the ethylene-vinyl acetate polymer is used as a controlled release system to deliver reagents to the sensing region of an optical fiber for an immunoassay based on fluorescence energy transfer. A model system has been tested in which a fluoroscein-labeled antibody and Texas Red-labeled immunoglobulin G are used.[109] A fiber-optic magnesium and calcium ion sensor based on a natural carboxylic polyether antibiotic(**54**) has been developed. This

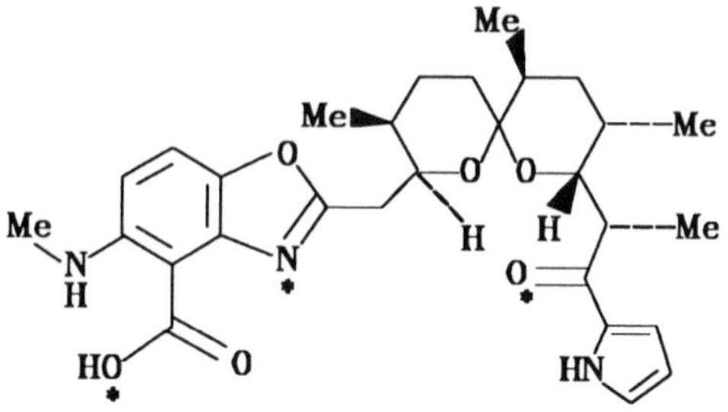

Me Me Me Me O O O Me N H N H H O HO O HN

**54**

compound selectively complexes magnesium(II) or calcium(II), leading to a decrease in the 430 nm fluorescence intensity. Since complexation occurs by replacement of a proton, the equilibrium can be controlled by the solution pH.[110]

Chemically facilitated Donnan dialysis in combination with continuous reagent flow has been used to develop a fiber optic colorimetric heavy metal sensor. The model describing the process relates the diffusion behavior of metal ions through a Nafion cation-anion exchange membrane with the stability constant of complexation, the ionic strengths of the receiving and sample solutions, the flow of the receiving solution, and the area : volume ratio of the membrane dialysis cell. This sensor is specific for lead(II) and cadmium(II) in aqueous solution, and uses sodium thiosulfate in the receiving solution to selectivity enhance the mass transport of these ions.[111, 112]

# 9. SOLID STATE SENSORS

The pulsed laser deposition of rhodium or iridium onto silicon results in the formation of polycrystalline films onto which microdrops of mercury can be electroplated. Applying Square Wave Anodic Stripping Voltammetry to these microelectrode array based sensors in the presence of trace amounts of either zinc(II), cadmium(II), or lead(II) shows a linear correlation with concentration over a range of 0.2-20 ppb. For a 5 minute preconcentration time, the detection limits for these ions are in the 0.2-0.5 ppb range[113] The *in situ* electrochemical determination of heavy metals in groundwater has been carried out using a microlithographically fabricated iridium ultramicroelectrode array sensor. Each array element has a mercury droplet coated onto its surface. The device has been used for the detection of copper(II), cadmium(II), lead(II), and zinc(II), although it is only stable for a few hours in an aqueous environment.[114]

Piezoelectric sensors have been prepared by immobilizing ligands onto the surface of a quartz crystal microbalance. The immobilization is achieved by reacting the ligand with the interfacial Si-OH groups of the high surface area silica particles, and then anchoring a monolayer of these particles onto the surface of the microbalance. The particles act as the interface by which ligand immobilization is achieved. Immobilization of phthalamic acid derivatives onto such hypersil silica particles results in a sensor system for the detection of uranium(VI).[115]

A magnesium selective electrode comprised of polymeric membranes doped with *N,N*-diheptyl-*N,N*-dimethylaspartamide favors magnesium(II) over calcium(II), sodium(I), and potassium(I) by factors of 300, 400, and 200 respectively. Since there is interference by protons, the use of this sensor is limited to use over a pH-buffered range of 8-9.[116] A polymer electrode prepared by the electrochemical oxidation of mercaptohydroquinone, followed by mercaptide modification, is a sensor for silver(I), mercury(II), cadmium(II), copper(II), and lead(II).[117] Metal chalcogenide thin films can be monitored by potentiostatic coulometry. Such electro-synthesized silver selenide films are sensors for gold(I) and mercury(II).[118] Semiconducting chalcogenide glasses based on arsenic selenide can be used as ion-selective electrodes for copper(II) and lead(II). The materials show Nernstian behavior over a wide concentration range.[119]

Phenolic polymers prepared *via* soybean hull peroxidase catalysis have been used as a sensor array. This sensor array that can be used for iron(III), copper(II), cobalt(II), and nickel(II) consists of fifteen phenolic homopolymers and copolymers generated from five phenolic monomers. The reporter unit is the intrinsic aqueous solution polyphenol fluorescence which is affected by the addition of metal ions. The response is dependent on the composition of the phenolic polymer used, thus the system has the potential to be widely useful.[120] Energy harvested from conjugated polymers can be used to amplify the output of analyte-sensitive fluorophores in chemosensors. A potentially more sensitive sensor is one of the "turn-on" type. Such a chemosensor has been developed using alternating layers of conjugated phenylene ethynylene polymer and a polyacrylate with the pH-sensitive dye fluoresceinamine covalently attached (**55**). Efficient energy transfer occurs

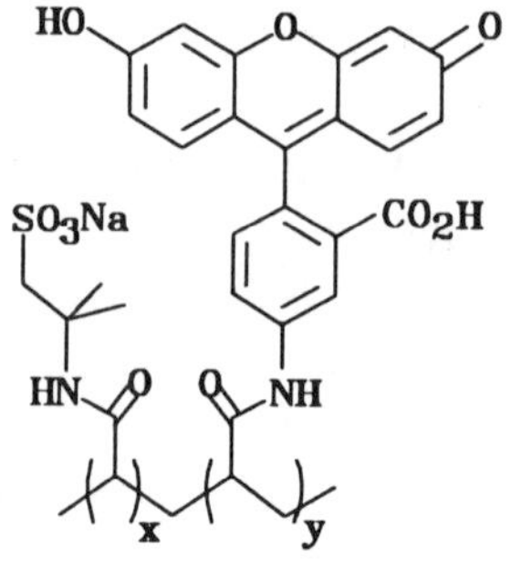

HO O O
$SO_3Na$ $CO_2H$
HN O O NH
x y

55

because of an overlay of the emission spectrum of the conjugated polymer with the absorption spectrum of the dye. Thus the fluorescence emission of the dye an increases 10-fold when the sensor is excited at the absorption maximum of the polymer rather than at the dye wavelength.[121]

An *in situ* electrochemical stripping sensor has been developed that involves the delivery of a complexant solution through a microdialysis sampling tube, and transport of the resulting complex to a downstream adsorptive stripping detector. The microdialysis sampling step minimizes the interference of surface-active macromolecules, and extends the linear dynamic range compared to conventional adsorptive stripping measurements.[122]

Langmuir-Blodgett film formation has been used as a technique for obtaining a specific orientation and concentration of a ligand on a solid support, and possibly on a fiber-optic device. An example is the monolayer of nonadecylpyridine, and substrates such as quartz, chromium, and polymers. The system can be used as a sensor for nickel(II) since this metal in the sub-phase induces a change in the properties of the monolayer due to its complexation. This complexation can be observed by changes in the absorption spectra of solutions containing this particular metal when the Langmuir-Blodgett film is immersed in the solution.[123]

# 10. SENSORS FOR ANIONS

The binding and recognition of oxyanions has been achieved using a variety of different complexant/reporter combinations. The complexant is a multidentate binding site that recognizes the anion, and the reporter changes its electrochemical or photophysical properties when an oxyanion binds to the host site. Alternatively, the chemosensor can be comprised of a complexant site that selectively recognizes both the cation and anion in the salt, thereby leading to the possibility of very high binding constants.

## 10.1. Acyclic Ligating Sites

Three oxyanion sensors (**56** and **57** (X = OMe, H)) with acyclic amide ligating sites have been used for the recognition of dihydrogen phosphate. In these systems the anion binds to the amide ligating sites, and

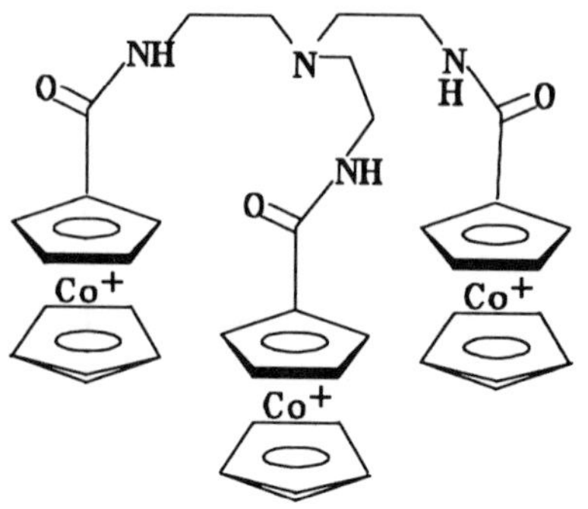

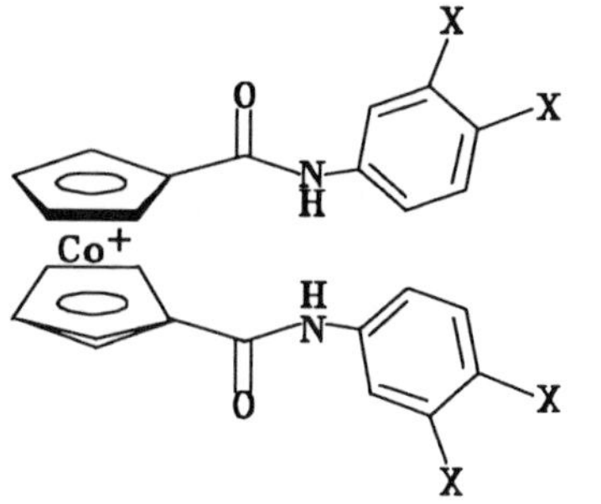

56

57

this event is reflected in a change in the redox potential of the cobaltocenium receptor.[124] Upon binding the dihydrogen phosphate anion, a shift in the potential of 200-240 mV is observed. Much smaller shifts are observed for the chloride ion. The stability constant for association of **56** with dihydrogen phosphate is particularly large. Other cobaltocenium reporters bound to amides are **58** and **59**, and these compounds have also been used as dihydrogen phosphate sensors.[125, 126] Similar amide types have been synthesized where the reporter molecule is of the ruthenium(II)-bipyridyl type (**60**). In this case oxyanion binding results in changes in the fluorescence of the reporter.[127] In addition to amides, amines have been used

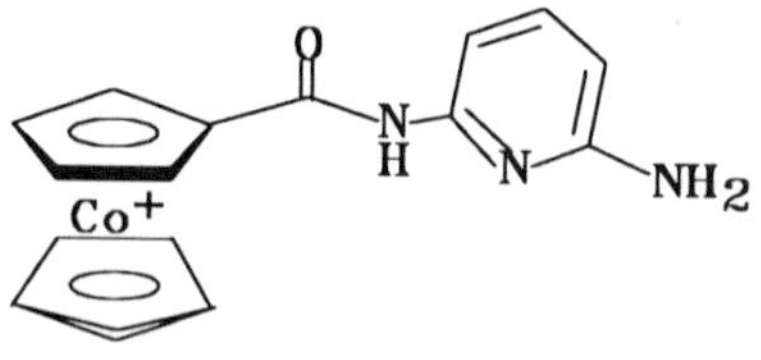

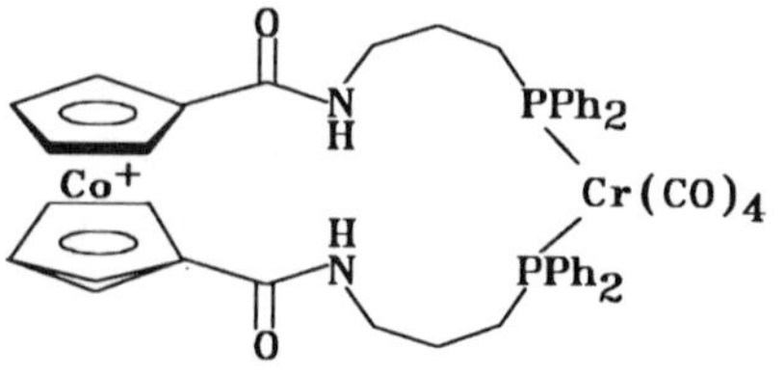

58

59

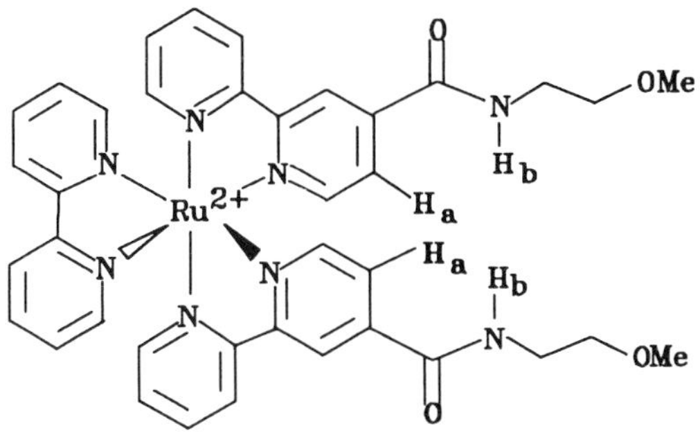

**60**

as luminescent sensors for oxyanions. The example shown in **61** has a ruthenium-bipyridyl type reporter bound to the acyclic amine.[128] This chemosensor can be used for hydrogen phosphate and dihydrogen phosphate.

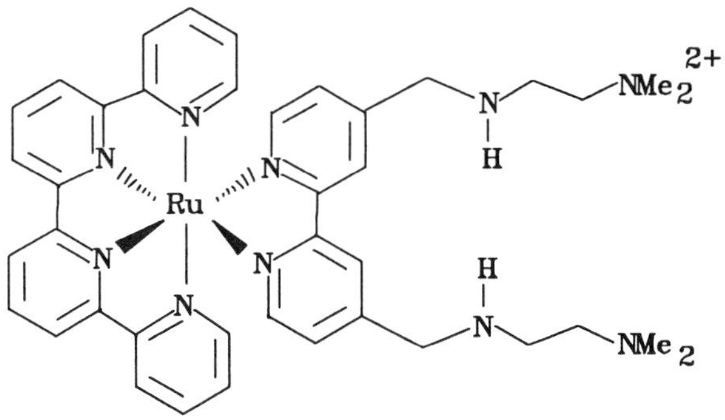

**61**

## 10.2. Calixarenes

Calixarenes have recently been used for anion recognition. At present much of this research centers on non-metallic anions such as chloride or phosphate, however, since there are examples of the liquid-liquid extraction of metalloanions, it is likely that chemosensors will be developed. These calixarene systems employ either an electrochemical or an optical reporter molecule for sensing the anion binding. A series of such calixarenes are **62**, **63**, and **64** (X = H, Y = Tos; X = Tos, Y = H). The compounds with

X = H and Y = tosylate show very high binding selectivities for dihydrogen phosphate over chloride. This selectivity is likely due to the bulky tosyl groups *para* to the amide substituents causing a favored receptor conformation in which the Lewis acid groups are rigidly held in close proximity to each other. The ruthenium(II) and rhenium(I) systems **63** and **64**, are potentially optical sensors, and the cobalt system **62** is a potential redox sensor.[124, 127-129] Compound **62** recognizes both chloride and dihydrogen phosphate anions, and **64** with X = H and Y = Tos has a much

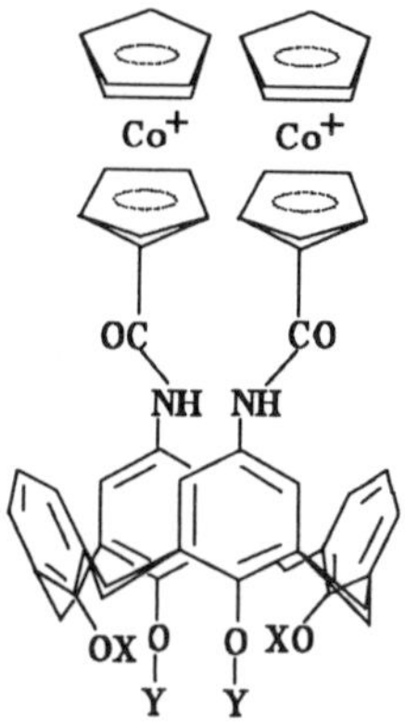

**62**

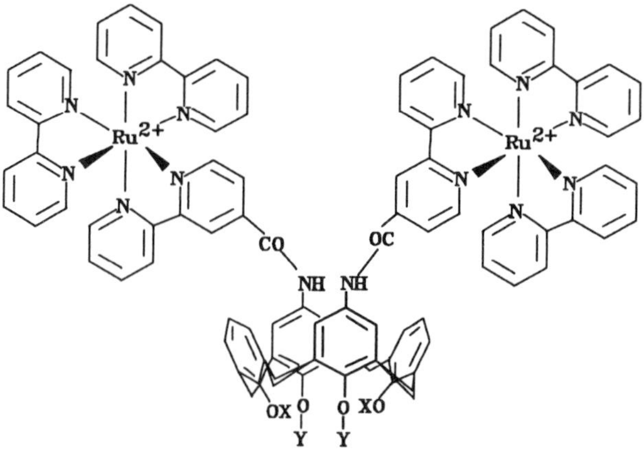

**63**

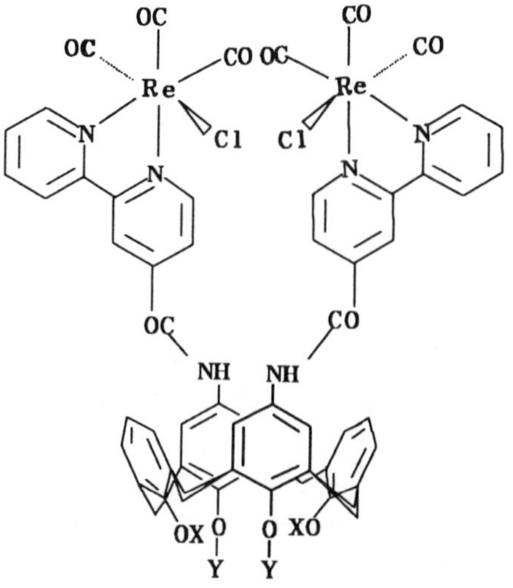

**64**

higher stability constant for dihydrogen phosphate than chloride

A calixarene with an emissive bipyridyl ruthenium(II) reporter molecule appended **65** is an overall dication. This system can therefore act as an emissive sensor for anions.[130] Although halide ions cause redox shifts, they are significantly smaller than those observed with dihydrogen phosphate.

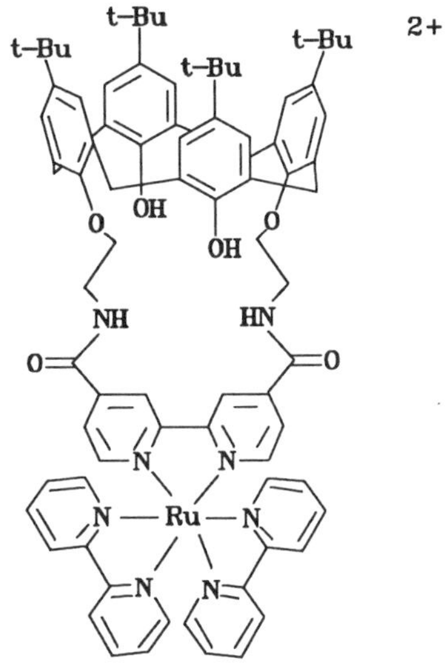

65

# 11. MOLECULAR COMPUTATION

Optical sensors for metal ions are also being considered for use in molecular-scale computation systems. One such system uses chemical inputs and optical outputs to add up to two. The system uses two related molecules, one containing a fluorophore, and the other a chromophore, that carry out AND and OR logic operations, respectively (Figure 2). Both molecules contain two receptors that are sensitive to the same inputs, protons and calcium(II) ions. In computing language each cation, when present in sufficiently high concentration, corresponds to the number 1 in the decimal system. When both inputs are present the AND molecule fluoresces at 419 nm, but the XOR molecule does not respond. This output, 01 in binary code, represents 1 in the decimal system.[131] The system operates in wireless mode in water and responds to physiological levels of protons and calcium ions. Logic systems with ionic inputs and fluoresence outputs can in principle be "wired" together through intermediate units that have optical inputs and ionic outputs.

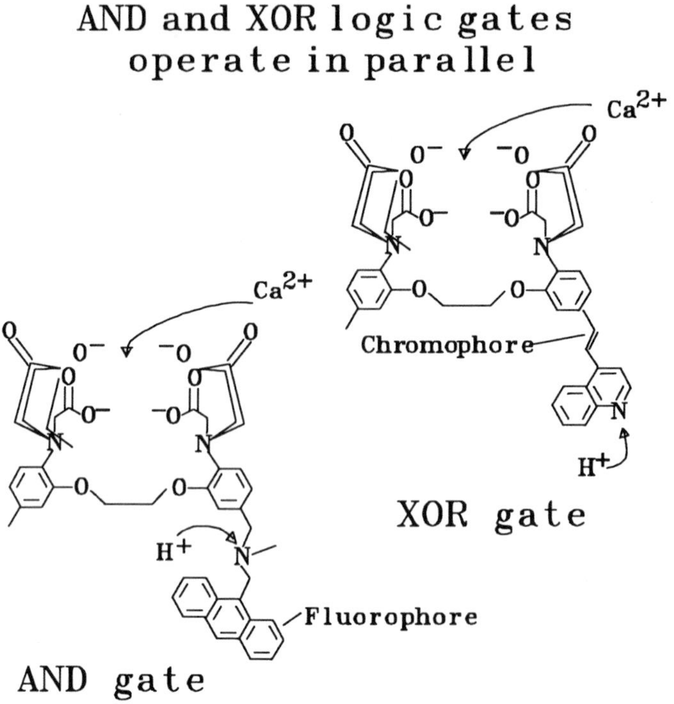

**Figure 2. Optical Sensors in Molecular Computing**

# REFERENCES


1. A. W. Czarnik, ed., *Fluorescent Chemosensors for Ion and Molecule Recognition*, ACS Sympos. Ser., No. 538, 1993.
2. D. Schuetzle, R. Hammerle, J. W. Butler, eds. *Fundamentals and Applications of Chemical Sensors*, ACS Sympos. Ser., No. 309, 1986.
3. T. E. Edmonds, *Chemical Sensors*, Chapman and Hall, New York, 1988.
4. R. A. Bissell, A. P. de Silva, H. Q. N. Gunaratne, P. L. M. Lynch, G. E. M. Maguire, K. R. A. S. Sandanayake, *Chem. Soc. Rev.*, **1992**, *21*, 187.
5. A. P. de Silva, S. A. de Silva, *JCS, Chem. Comm.*, **1986**, 1709.

6. L. Fabbrizzi, A. Poggi, *Chem. Soc. Rev.*, **1995**, *24*, 197.
7. J. Janata, A. Bezegh, *Anal. Chem.*, **1988**, *60*, 62R.
8. J. Janata, *Anal. Chem.*, **1990**, *62*, 33R.
9. B. Dietrich, *Pure & Appl. Chem.*, **1993**, *65*, 1457.
10. R. Narayanaswamy, *Anal. Proc.*, **1985**, *22*, 204.
11. A. W. Czarnik, *Accs Chem. Res.*, **1994**, *27*, 302.
12. D. M. Roundhill, *Photochemistry and Photophysics of Metal Complexes*, Plenum, New York, 1994.
13. H. Hennig, D. Rehorek, *Photochemische and Photokatalytische Reaktionen von Koordinations-verbindungen*, Akademie-Verlag, Berlin, 1987.
14. Q. Zhou, T. M. Swager, *J. Am. Chem. Soc.*, **1995**, *117*, 7017.
15. W. R. Seitz, D. M. Hercules, *Anal. Chem.*, **1972**, *44*, 2143.
16. C. A. Chang, H. H. Patterson, *Anal. Chem.*, **1980**, *52*, 653.
17. R. Escobar, Q. Lin, A. Guiraum, F. F. de la Rosa, *Analyst,* **1993**, *118*, 643.
18. W. Qin, Z. Zhang, H. Liu, *Anal. Chem.*, **1998**, *70*, 3579.
19. F. Goppelsröder, *J. Prakt. Chem.*, **1867**, *101*, 408.
20. D. J. S. Birch, O. J. Rolinski, D. Hatrick, *Rev. Sci. Instrumen.*, **1996**, *67,* 2732.
21. M. Kodama, E. Kimura, *JCS, Dalton Trans.*, **1979**, 325.
22. M. Huston, K. Haider, A. W. Czarnik, *J. Am. Chem. Soc.*, **1988**, *110*, 4460.
23. S. Y. Hong, A. W. Czarnik, *J. Am. Chem. Soc.*, **1993**, *115*, 3330.
24. S. A. Yamanaka, D. H. Charych, D. A. Loy, D. Y. Sasaki, *Langmuir*, **1997**, *13*, 5049.
25. I. Satoh, *Ann. N. Y. Acad. Sci.*, **1996**, *799*, 514.
26. B. Ramachandram, A. Samanta, *Chem. Phys. Letts.*,**1998**, *290*, 9.
27. L. Prodi, F. Bolletta, M. Montalti, N. Zaccheroni, *Eur. J. Inorg. Chem.*, **1999**, 455.
28. M. T. Stauffer, D. B. Knowles, C. Brennan, L. Funderburk, F-T Lin, S. G. Weber, *JCS, Chem. Comm.*,**1997**, 287.
29. M. T. Stauffer, S. G. Weber, *Anal Chem.*, **1999**, *71*, 1146.
30. B. Valeur, J. Bourson, J. Pouget, M. Kaschke, N. P. Ernsting, *J. Phys. Chem.*, **1992**, *96*, 6545.
31. B. Valeur, J. Mugnier, J. Pouget, J. Bourson, F. Santi, *J. Phys. Chem.*, **1989**, *93*, 6073.
32. M.-Y. Chae, A. W. Czarnik, *J. Fluoresc.*, **1992**, *2*, 225.
33. G Hennrich, H. Sonnenschein, U. Resch-Genger, *J. Am. Chem. Soc.*, **1999**, *121*, 5073.
34. S. C. Ng, X. C. Zhou, Z. K. Chen, P. Miao, H. S. O. Chan, S. F. Y. Li, P. Fu, *Langmuir,* **1998**, *14*, 1748.
35. I. Bontidean, C. Berggren, G. Johansson, E. Csöregi, B. Mattiason, J. R. Lloyd, K. J. Jakeman, N. L. Brown, *Anal Chem.*, **1998**, *70*, 4162.
36. V. Goulle, A. Harriman, J.-M. Lehn, *JCS, Chem. Comm.*, **1993**, 1034.
37. E. Amouyal, A. Hamsi, J.-C. Chambron, J. -P. Sauvage, *JCS, Dalton Trans.*, **1990**, 1841.
38. Y. Jenkins, A. E. Friedman, N. J. Turro, J. K. Barton, *Biochemistry,* **1992**, *31*, 10809.
39. J. Fees, W. Kaim, M. Moscherosch, W. Matheis, J. Klima, M. Krejcik, S. Zális, *Inorg. Chem.*, **1993**, *32*, 166.

40. E. Sabatani, H. D. Nikol, H. B. Gray, F. C. Anson, *J. Am Chem. Soc.*, **1996**, *118*, 1158.
41. C. Pérez-Jiménez, S. J. Harris, D. Diamond, *JCS, Chem. Comm.*, **1993**, 480.
42. T. Jin, K. Ichikawa,T. Koyama, *JCS, Chem. Comm.*, **1992**, 499.
43. I. Aoki, T. Sakaki, S. Shinkai, *JCS, Chem. Comm.*, **1992**, 730.
44. Y. Kubo, S. -I. Hamaguchi, A. Niimi, K. Yoshida, S. Tokita, *JCS Chem. Comm.*, **1993**, 305.
45. Y. Kubo, S. Tokita, Y. Kojima, Y. T. Osano, T. Matsuzaki, *J. Org. Chem.* **1996**, *61*, 3758.
46. Y. Kubo, S. Obara, S. Tokita, *JCS, Chem. Comm.*, **1999**, 2399.
47. Q. Lu, J. H. Callahan, G. E. Collins, *JCS, Chem. Comm.*, **2000**, 1913.
48. P. D. Beer, Z. Chen, A. J. Goulden, A. Grieve, D. Hesek, F. Szemes, T. Wear, *JCS, Chem. Comm.*, **1994**, 1269.
49. D. M. Roundhill, *Progr. Inorg. Chem.*, **1995**, *43*, 533.
50. J.-C. G. Bünzli, P. Froidevaux, J. M. Harrowfield, *Inorg. Chem.*, **1993**, *32*, 3306.
51. P. Froidevaux, J.-C. G. Bünzli, *J. Phys. Chem.*, **1994**, *98*, 532.
52. J.-C. G. Bünzli, P. Froidevaux, C. Piguet, *New J. Chem.*, **1995**, *19*, 661.
53. N. Sabbatini, M. Guardigli, A. Mecati, V. Balzani, R. Ungaro, E. Ghidini, A. Casnati, A. Pochini, *JCS, Chem. Comm.*, **1990**, 878.
54. M. F. Hazenkamp, G. Blasse, N. Sabbatini, R. Ungaro, *Inorg. Chim. Acta*, **1990**, *172*, 93.
55. E. M. Georgiev, J. Clymire, G. L. McPherson, D. M. Roundhill, *Inorg. Chim. Acta*, **1994**, *227*, 93.
56. N. Sato, S. Shinkai, *Workshop on Calixarenes and Related Compounds*, Fukuoka, Japan, **1993**, Abstr. PS/B-13.
57. M.-Y.Chae, X. M. Cherian, A. W. Czarnik, *J. Org. Chem.*, **1993**, *58*, 5797.
58. J. M. Lloris, R. Martinez Mañez, M. E. Padilla-Tosta, T. Pardo, J. Soto, P. D. Beer, J. Cadman, D. K. Smith, *JCS, Dalton Trans.*, **1999**, 2359.
59. G. de Santis, L. Fabbrizzi, M. Licchelli, C. Mangano, D. Sacchi, *Inorg. Chem.*, **1995**, *34*, 3581.
60. R. B. King, I Bresinka, *ACS Sympos. Ser.*, **2000**, *762*, 23.
61. J. Zhang, H. Prestol, A. Gahr, R. Niessner, *Proc. SPIE-Int. Opt. Eng.*, **2000**, *4077*, 32.
62. K. S. Bang, M. B. Nielson, R. Zubarev, J. Becher, *JCS, Chem. Comm.*, **2000**, 215.
63. A. R. Fakhari, M. R. Ganjali, M. Shamsipur, *Anal. Chem.*, **1997**, *69*, 3693.
64. M. Lerchi, E. Reitter, W. Simon, E. Pretsch, D. A. Chowdhury, S. Kamaka, *Anal. Chem.*, **1994**, *66*, 1713.
65. G. Arai, A. Fujii, I. Yasumori, *Chem. Lett.*, **1985**, 1091.
66. E. U. Akkaya, M. E. Huston, A. W. Czarnik, *J. Am. Chem. Soc.*, **1990**, *112*, 3590.
67. M. E. Huston, C. Engleman, A. W. Czarnik, *J. Am. Chem. Soc.*, **1990**, *112*, 7054.
68. A. Singh, L.-I. Tsao, M. Markowitz, B. P. Gaber, *Langmuir,* **1992**, *8*, 1570.
69. P. Ghosh, P. K. Bharadwaj, J. Roy, S. Ghosh, *J. Am. Chem. Soc.*, **1997**, *119*, 11903.
70. M. Gubelmann, A. Harriman, J.-M. Lehn, J. L. Sessler, *JCS Chem. Comm.*,

**1988**, 77.
71. C. R. Chenthamarakshan, J. Eldo, A. Ajayaghosh, *Macromolecules*, **1999**, *32*, 5846.
72. L. R. Sousa, J. M. Larson, *J. Am. Chem. Soc.*, **1977**, *99*, 307.
73. J. M. Larson, L. R. Sousa, *J. Am. Chem. Soc.*, **1978**, *100*, 1943.
74. L. R. Sousa, B. Son, T. E. Trehearne, R.W. Stevenson, S. J. Ganion, B. E. Beeson, S. Barnell, T. E. Mabry, M. Yao, C. Chakrabarty, P. L. Bock, C. C. Yoder, S. Pope, *ACS Sympos. Ser.*, **1993**, *538*, 10.
75. A. P. de Silva, K. R. A. S. Sandanayake, *Angew. Chem., Int. Ed. Engl.*, **1990**, *29*, 1173.
76. H. Bouas-Laurent, A. Castellan, M. Daney, J.-P. Desvergne, G. Guinand, P. Marsau, M.-H. Riffaud, *J. Am. Chem. Soc.*, **1986**, *108*, 315.
77. F. Fages, J.-P. Desvergne, H. Bouas-Laurent, J.-M. Lehn, J. P. Konopelski, P. Marsau, Y. Barrans, *JCS, Chem. Comm.*, **1990**, 655.
78. A. P. de Silva, H. Q. N. Gunaratne, K. R. A. S. Sandanayake, *Tetrahedron Lett.*, **1990**, *31*, 5193.
79. S. S. Zhu, P. J. Carroll, T. M. Swager, *J. Am. Chem. Soc.*, **1996**, *118*, 8713.
80. R. J. M. Klein Gebbink, S. I. Klink, M. C. Feiters, R. J. M. Nolte, *Eur. J. Inorg. Chem.*, **2000**, 253.
81. S. Ghosh, M. Petrin, A. H. Maki, L. A. Sousa, *J. Chem. Phys.*, **1987**, *87*, 4315.
82. R. Y. Tsien, *Ann. Rev. Biophys. Bioeng.*, **1983**, *12*, 94.
83. D. Y. Sasaki, D. R. Shnek, D. W. Pack, F. H. Arnold, *Agnew. Chem. Int. Ed. Engl.*,**1995**, *34*, 905.
84. A. Minta, R. Y. Tsien, *J. Biol. Chem.*, **1989**, *264*, 19449.
85. D. Masilamani, M. E. Lucas, *ACS Sympos. Ser.*, **1993**, *538*, 162.
86. R. Y. Tsien, *Biochemistry*, **1980**, *19*, 2396.
87. C. R. Schauer, O. P. Anderson, *J. Am. Chem. Soc.*, **1987**, *109*, 3646.
88. C. K. Schauer, O. P. Anderson, *Inorg. Chem.*, **1988**, *27*, 3118.
89. G. Grynkiewicz, M. Poenie, R. Y. Tsien, *J. Biol. Chem.*, **1985**, *260*, 3440.
90. D. M. O'Malley, S. M. Lu, W. Guido, P. R. Adams, *Neuroscience*, **1992**, *18*, 14.
91. S. Gilroy, R. L. Jones, *Proc. Nat. Acad. Sci. USA*, **1992**, *89*, 3591.
92. Z.B. Imperiali, D. A. Pearce, J.-E. Sohna Sohna, G. Walkup, A. Torrado, *SPIE-Int. Soc. Opt. Eng. I*, **1999**, *3858*, 135.
93. G. W. Walkup, B. Imperiali, *J. Am. Chem. Soc.*, **1996**, *118*, 3053.
94. A. Torrado, B. Imperiali, *J. Org. Chem.*, **1996**, *61*, 8940.
95. H. A. Godwin, J. M. Berg, *J. Am. Chem. Soc.*, **1996**, *118*, 6514.
96. J. M. Berg, *Acc. Chem. Res.*, **1995**, *28*, 14.
97. B. A. Krizek, D. L. Merkle, J. M. Berg, *Inorg. Chem.*, **1993**, *32*, 937.
98. P. S. Eis, J. R. Lakowiez, *Biochemistry*, **1993**, *32*, 7981.
99. R. B. Thompson, E. R. Jones, *Anal. Chem.*, **1993**, *65*, 730.
100. R. B. Thompson, M. W. Patchan, *Anal. Biochem.*, **1995**, *227*, 123.
101. N. J. Wilmott, J. N. Miller, J. F. Tyson, *Analyst*, **1984**, *109*, 343.
102. J.-C. Bünzli, J.-M. Pfefferlé, *Helv. Chim. Acta*, **1994**, *77*, 323.
103. A. W. Varnes, R. B. Dodson, E. L. Wehry, *J. Am. Chem. Soc.*, **1972**, *94*, 946.
104. G. Weber, *Biochem. J.*, **1950**, *47*, 144.
105. M. Cais, S. Dani, Y. Eden, O. Gandolfi, M. Horn, E. E. Isaacs, Y. Josephy, Y.

Saar, E. Slovin, L. Snarsky, *Nature,* **1977**, *270*, 534.
106. J. I. Peterson, G. G. Vurek, *Science,* **1984**, *224,* 123.
107. W. R. Seitz, *Anal. Chem.,* **1984**, *56*, 16A.
108. R. W. Wagner, J. S. Lindsey, *J. Am. Chem. Soc.,* **1994**, *116*, 9759.
109. S. M. Barrard, D. R. Walt, *Science*, **1991**, *251*, 927.
110. K. Suzuki, K. Tohda, H. Ohzora, S. Nishihama, H. Inoue, T. Shirai, *Anal. Chem.*, **1989**, *61*, 382.
111. Z. Lin, L.W. Burgess, *Anal.Chem.*, **1994**, *66*, 2544.
112. Z. Lin, K. S. Booksh, L.W. Burgess, B. R. Kowalski, *Anal. Chem.*, **1994**, *66*, 2552.
113. B. Le Drogoff, M. A. El Khakani, P. R. M. Silva, M. Chaker, G. G. Ross, *Appl. Surf. Sci.*, **1999**, *152*, 77.
114. J. Herdan, R. Feeney, S. P . Kounaves, A. F. Flannery, C. W. Storment, G. T. A. Kovacs, R. B. Darling, *Environ. Sci. Technol.*, **1998**, *32*, 131.
115. R. Cox, D. Gomez, D. A. Buttry, P. Bonneson, K. N. Raymond, *ACS Sympos. Ser.*, **1994**, *561*, 71.
116. M. V. Rouilly, M. Badertscher, E. Pretsch, G. Suter, W. Simon, *Anal Chem.*, **1988**, *60*, 2013.
117. G. Arai, A. Fujii, I. Yasumori, *Chem. Lett.*, **1985**, 1091.
118. M. T. Neshkova , V. D. Nikolova, V. Petrov, *J. Electroanal. Chem.*, **2000**, *487*, 100.
119. A. E. Owen, *J. Non-Cryst. Solids*, **1980**, *35-36*, 999.
120. X. Wu, J. Kim, J. S. Dordick, *Biotechnol. Prog.*, **2000**, *16*, 513.
121. D.T. McQuade, A. H. Hegedus, T. M. Swager, *J. Am. Chem. Soc.*, **2000,** *122*, 12389.
122. J. Wang, J. Lu, D. Luo, J. Wang, M. Jiang, B. Tian, K. Olsen, *Anal. Chem.* **1997**, *69*, 2640.
123. G. Caminati, E. Margheri, G. Gabrielli, *Thin Solid Films*, **1994**, *244*, 905.
124. P. D. Beer, *JCS, Chem. Commun.*, **1996**, 689.
125. P. D. Beer, M. G. B. Drew, A. R. Graydon, *JCS Dalton Trans.*, **1996**, 4129.
126. J. E. Kingston, L. Ashford. P. D. Beer, M. G. B. Drew, *JCS, Dalton Trans.*, **1999**, 251.
127. P. D. Beer, M. G. B. Drew, D. Hesek, M. Shade, F. Szemes, *JCS, Chem. Comm.*, **1996**, 2161.
128. P. D. Beer, J. Cadman, *New J. Chem.*, **1999**, *23*, 347.
129. P. D. Beer, D. Hesek, K. C. Nam, M. G. B. Drew, *Organomet.*, **1999**, *18*, 3933.
130. P. D. Beer, P. A. Gale, D. Hesek, M. Shade, F. Szemes, Abstr. 3rd Int. Calixarene Conf., Fort Worth, TX, May 1995, Abstr. LI-8.
131. A. P. de Silva, N. D. McClenaghan, *J. Am. Chem. Soc.*, **2000**, *122*, 3965.

# Index


Acetylacetone, 166, 210
Actinides
storage, 1
ion exchange, 21
polymer filtration, 24
ELECTROSORB, 57
carbamoylphosphonates, 204
calixarenes, 205
carboxylates, 207-209
β-diketones, 211
phosphoramides, 222
acetamidophosphine oxide, 224
biotreatment, 300
lanthanum phosphate, 301
Higher coordination, 193
phosphates and phosphine oxides, 193, 194
tributyl phosphate (TBP), 194, 195, 210
tetravalent actinides, 196
Acyclic polyethers, 166, 183
Adsorbent
extraction and recovery wheel, 2
solid support, 10, 20
solid phase adsorbent, 10, 25
chemical or thermal treatment, 23
iron(III) hydroxides, 27
titanium oxides, 28
silicates, 22, 28
thiol, 28
chitosan 29,
managanese(IV), 40
organic, 115
recovery, 126
polymers for uranium, 204
calixarene amide, 216

Alfalfa, 292, 294-296
Alkali metal ion sensors, 320
Alkyl-8-hydroxyquinolines, 75
Alkyl phosphatase, co-immobilized, 314
Amberlite LA-2, 225
Amides
lipophilic, 70
disubstituted, 134
for gold, 180
thioamides, 183
dialkylamides, 199
diamides, 200-202
1,3-diamides, 201
picolinamides, 201
succinamides, 202, 215
for lanthanides, 214
phosphoramides, 222
calixarenes, 240-243
cobaltocenium, 356
Amines
picrates, 19
quaternary amines, 21
ethylenediamine, 109, 115
propylenediamine, 109
polyamines, 116, 117, 143, 238
open-chain amines, 143

cyclic amines, 143
for cadmium, 153
for copper, 164, 165
for silver and gold, 179
long chain, 180
for uranium(VI) from (IV), 196
for actinides, 204
for lanthanides, 214
for oxyanions, 238-242
for alkali metals, 274
rhodamines, 351
as oxyanion sensors, 356
Amino Crowns, 264
Ammonium molybdophosphate, 110
Anion binders, 238
Anthracene reporters, 339, 342, 343
Applying Square Wave Anodic Stripping Voltammetry, 353
Aqueous biphasic systems, 15, 132
Arbuscular mycorrhizal (AM) fungi, 291
Arsenazo III (AZ), 222
Ascorbate oxidase, 314
Atomic absorption spectroscopy
technique, 41, 67
mercury, 114
lead and cadmium, 136
lead, 137
chelate extraction, 149, 234
cadmium, 156
Avicel, 115
Azamacrocycles
for mercury(II) and copper(II), 84
for mercury, 116
for copper, 175
for silver and gold, 184
with anthracene reporters, 332
Barley, 294
Battery, 3
Bermuda grass, 292
Binschedler's green, 110
Bioaccumulation, 3, 107
by seafood, 107
Biodegradation, 2, 194
of citrate, 194
Biomethylation
of mercury(II), 136, 146, 148
Bioremediation
by *zoogloea ramigera*, 237
challenge, 289, 290
categories, 297
engineering aspects, 298
for cadmium, 300
for chromium, 301
Bis-(2-ethylhexyl) phosphoric acid, 166, 169
Bis- and tetra-capped clefts, 253
Bis-(benzocrown ether)s, 248
Cadmium
accumulation, 3
with carbon, 26
with kaolinite, 29
with derivatized chitin, 30
electrodeposition, 37
monoclonal antibody, 42
chelators, 44
electrokinetics, 53-55
with xanthates, 67
with EDTA, 69
with Aliquat 336 and Titan Yellow, 70
with alizarin, 74
with salicylic acid and TAP, 75
with multidentates and podands, 78-81
with azamacrocycles, 86, 87
with pendant macrocycles, 90
with calixarene phosphine, 99
with oxyethylene carboxylic acids, 108
with sulfur donors, 109
with inorganic sulfides, 113
cyanide complex, 114
with macrocycles, 117, 119
with porphyrins, 122
with calixarenes, 122, 123
occurrence, 129
with PEG, 132
with chelates, 133-136
with phosphonic acids, 138
with acyclic crowns, 140
with macrocycles, 142-144

with calixarenes, 145, 146
with dithiocarbamates and carbon dioxide, 147
toxicity and occurrence, 151
carbodithioates for therapy, 152
metallothionein, 151, 156
tobacco smoke, 151
with nitrogen and oxygen donors, 152-154
with sulfur donors, 154-157
with cyclodextrins, 157
in food chain, 158
with *O*-donor chelates, 166
with oxime, 168
with *N*-donor chelates, 172
with *S*-donor chelates, 174
with azamacrocycles, 175
with calixarenes, 177
biohydrometallurgical process, 178
*thlaspi caerulescens*, 291, 293
phytoremediation, 293-296
bioremediation, 299-300
fluorescence sensor, 314, 318
redox sensor, 332, 333
solid state sensor, 353, 354

Calcon, 70

Calixarenes
with polyamines, 30
cyclic condensation products, 64
introduction, 95-97
cyclic oligomers, 96, 145
*P*- and *S*-donors, 122-124
*N*-donors, 124-125
for lead, 145
for lead, copper, silver and cadmium, 157
for copper, 177
for silver, 187
for actinides and lanthanides, 205, 215, 222
anion recognition, 231, 240
for alkali metals, 247
as metal ion sensors, 320-327
for anion recognition, 357-359

Calixcrown binding site, 324

Camphor-3-thioxo-2-oxime, 138

Capillary electrophoretic method, 113

Capillary electrophoreisis microchip, 326

Capillary Tube isotachophoresis, 114, 179, 207
with EDTA, 219

Carbamoylphosphonate type, 204

Carbon dioxide
applications, 5
chelate solubility, 80
with fluorochelates, 81
with fluorinated dithiocarbamate, 111
for mercury, 117, 121
for lead, 146
with dithiocarbamates, 147
for cadmium, 156
for copper, 177
for uranium, 195
for lanthanides, 210, 211 243
see also Supercritical carbon dioxide

Carbon materials, 26

Cavitand
pyridine for silver, 186
for actinides, 204
resorcinarene for europium, 205
for anions, 238
for alkali metals, 273
for emissive sensors, 335

Chelates
protonated chelates, 11
polymer chelates, 38
metal chelate, 42
chelate assisted pressurized liquid extraction(CAPLE), 43
in electrokinetics, 49
common chelates, 69
oximes, 72
preorganized chelates, 78
for use in carbon dioxide, 80
*O, O*- and *N*-donor, 110

for mercury, 109-112
for *in vivo* therapy, 133, 152
for lead, 134
pendant chelates, 143
fluorinated chelates, 147
copper chelates, 166
oxime chelates, 169
*N*-donor, 172-174
sulfur-donor chelates, 174
*N,O*-donor, 181, 197
*S*-donor, 183
*O, O*-donor, 211
EDTA and analogues, 217, 294
aromatic, 219
*N* and *O*-donor, 311
Chelate assisted microwave extraction, 44
Chelate effect, 5
Chelating Resins, 21, 236
Chelation therapy, 130, 134
Chemiluminescence, 310, 311
Chitosan, 29, 291
Chromium
chromium hydroxide, 27, 235
adsorbed on glass, 28
adsorbed on lignin, 30
Soil washing, 37, 38
electrokinetic extraction, 53-56
chromium(VI), 58-61, 70, 99
potential chromium oxidation score, 59
with 1,5-diphenyl carbohydrazide, 64
with chelates, 68-70
with oximes, 73
with salicylic acid and TAP, 75
with multidentates, 76
with *N,N*-chelates, 81
with a HFM, 169
with *N,O*-chelates, 182
chromium(VI) structure, 231
chromium(VI) toxicity, 233
chromium(VI) analysis, 234
chromium-diphenylcarbazone complex, 234
chromium(VI) precipitation, 235
chromium(VI) adsorption, 236
chromium(VI) complexation, 237
chromium(VI) phytoremediation, 237
chromium(VI) with amines, 238
chromium(VI) from soils, 243
with PEG, 243
chromium phytoremediation, 292
hyperaccumulated chromium, 292
with alfalfa, 295, 296
with creosote bush, 295
with microorganisms, 298
bioremediation, 301
chemiluminescence, 311
fluorescent sensor, 314
quartz microbalance sensors, 318
with lanthanide reporters, 351
Chromogenic and Fluorogenic crown ethers, 121
Chromogenic crowns, 266
Cobaltocenium reporters, 356
Copper
as adsorbent for metals, 25
with carbon, 26
on ferric oxide, 27
on glass, 28
copper-doped manganese, 28
on polymers, 29
electrokinetic extraction, 54
with chelates, 68-76
with podands, 78-80
with aza- and thiamacrocycles, 85-87
with crowns, 89-90
with calixarenes, 97, 99
with *S*-donors, 109
with cyanide, 114, 164
with acids, 131
with ionophores, 134
with *O*-donors, 140

with *N*-donors, 142
adsorption, 150
with *N*-donors and *O*-donors, 153
copper-thionein proteins, 156
occurrence, 163
Wilson's and Menkes' disease, 164
with thiocyanate, 164
with chelates, 166-175
with macrocycles and
calixarenes, 175-177
carbon dioxide, 177
with pyridine podands, 182
with picolylamine polymer, 236
phytoremediation, 291, 293-296
bioremediation, 299
rhodamine sensor, 311
chelate sensor, 312
liposome sensor, 314
redox sensor, 317
protein sensor, 318
macrocyclic sensor, 330-333
membrane sensor, 334
cryptand sensor, 334-336
supramolecular sensor, 343
lipid sensor, 345
solid state sensor, 353
phenolic polymer sensor, 354

Coronands, 91
Creosote Bush, 295
Criterion Maximum Concentration (CMC), 232
Cryptands
for heavy and precious metals, 81
problems in their use, 96
selectivity, 100
for alkali metals, 255, 273
fluorinated cryptands, 272
for chemosensors, 334, 335, 338
with anthracene reporters, 343
Cyclic Thioether, 144
Cyclodextrins, 157
Dansylamide, 350
Debye screening length, 50
Detergent extraction, 181
Di-2-ethylhexyl phosphoric acid (HDEHP), 195
Diethylenetriaminepentaacetic acid (DTPA), 69, 218
Diethyldithiophosphate, 137
Dihydroxycrownophanes, 87
Diphonix, 21, 22, 236
Diphonix Resins, 21, 22
Distearoyl phosphatidylcholine, 345
Distribution Ratio, 11, 164, 225
Dithiocarbamates
as chelates, 44
as multidentates, 65-67
for mercury(II), 111
for lead(II), 135
for extraction therapy, 152
for cadmium(II), 154, 155
for heavy metals, 174
for silver and gold, 183
for silver and mercury, 332
Dowex 50W-X8 ion exchange resin, 217
DPKBH, 74
EDTA
with metal hydroxides, 27
for lead from soils, 37-39
for cadmium, 42-45
for metals, 69
for mercury, 114
with Avicel, 115
for lead, 133
for lead and mercury, 135
for copper, 166
for uranium(IV), 197
for lanthanide(III) oxides, 208
for lanthanide(III) ions, 214, 217-219
for chromium(VI), 292
with phytoremediation, 294
for regeneration, 318, 319
E. Coli cells, 237, 302
Electrodeposition, 37, 38
Electrokinetic extraction, 35, 43, 47-49, 53-58, 243
Electromigrations, 48

Electroosmosis, 48, 50, 53, 55
Electroosmotic flow, 53-55
Electroosmotic water flow, 52
Electrophoretic transport model, 50
Electrosorb, 57
Emissive sensors, 320, 326, 335
Emulsion liquid Membranes, 18, 118
Enzymes, 3, 350
Ethyleneglycol bis-(2-aminoethylether) tetraacetic acid (EGTA), 44, 69, 294, 348
EXAFS (Extended X-ray Absorption Fine Structure), 283
Fast binding, 6
Fiber Optics, 331, 352
Fluorescent sensors, 309, 311, 313, 314, 319, 332
Genetically engineered microbes, 297
Glass microbeads, 115
Gibbs free energies of hydration, 248
Goethite, 27, 235, 236
Gold
    uses, 4, 163
    for mercury, 25
    with macrocycles, 90, 93
    with dithiocarbamoyls, 99
    for back extraction, 112
    with macrocycles, 123
    non-toxicity, 164
    extractants, 178-184
    binding to alfalfa, 296
    detection by silver selenide, 354
Hardness, 7
Hemispherands, 273
Heterocyclic amines, 239
High pressure liquid chromatographic (HPLC), 67
High stability, 6, 26, 138, 218
HINAP, 74
Hollow fiber membrane extractors, 239
Human food chain, 158, 290
Humic Acid, 39, 131, 132, 165, 291, 295
Hydrometallurgical, 4
Hydration energies, 13
Hyperaccumulator plant, 293
Hyperthermophilic microorganisms, 298
Hyphan I, 72
IDICPMS analysis, 155
Immunoassay methods, 42
*In situ* stabilization
    by cementation, 36
    as insoluble sulfide, 39, 40,
Incineration, 3
Indian mustard, 292-294
Indoaniline reporters, 322
Inductively coupled plasma-atomic emission spectroscopy, 41, 67, 234
Ion exchange resins, 5, 20-22, 27, 204
    polymeric beads, 20
    gellular and macroreticular, 21
Itai-Itai disease, 151
Kaolin, 44, 60, 61
Kelex
    as a chelate for copper, 166
    structure, 167
    on a support for copper(II), 169
    for lanthanides, 219
Lab-on-a-chip, 326
Langmuir-Blodgett film, 355
Lariat crowns, 248, 263, 264, 274
Laser-induced breakdown spectroscopy penetrometer system, 41
Lead
    occurrence, 3
    elemental lead removal, 25
    with carbon, 26
    with ferric oxide, 27
    with clays, 28, 29
    with lignin, 30
    with EDTA, 37
    with Metaset-Z, 38
    *in situ* stabilization, 40
    immunoassay, 42
    electrokinetic extraction, 54
    lead hydroxide, 55
    with aminocarboxylic acids, 68, 154
    with oximes, 73

with polyethers, 76, 166
with preorganized chelates, 78
onto polymers, 80
with carbon dioxide, 81
with macrocycles, 86, 87, 90, 91
with calixarenes, 97, 98
with *S*-donors, 109, 184
with organic adsorbents, 115
with macrocycles, 117, 121
with calixarenes, 125, 177
toxicity and occurrence, 129
trimethyllead, 131
with polyethylene glycol, 132
*in vivo* therapy, 133
with *N*-donors and with *O*-donors, 134
with dithiocarbamates, 135
analytical methods, 136
with *S*-donors, 137, 156
with macrocycles, 139, 141-144, 175
lead-210 isotope, 141
mono-and bi-nuclear lead complexes, 143
with liquid carbon dioxide, 146
analytical methods, 148
sediment analysis, 149
phytoremediation, 291, 293-296
bioremediation, 299-300
conformational sensing, 316, 340
*S*-donor sensors, 317
quartz crystal microbalance sensors, 318
macrocyclic chemosensors, 328, 329
diarylthiocarbazone sensors, 331
quinonoid sensor, 332
cryptand sensor, 335
solid state sensors, 353
membrane sensor, 354
Lettuce, 290, 291, 293
Lipophilic and Oligomeric polyethers, 76
Liquid-Liquid Extraction
types, 5
for radioactive metals, 10
preferred conditions, 15
picrate method, 19,
with concave hydrocarbons 64
with dithiocarbamates, 66, 136
with macrocycles, 84
with tetraazamacrocycles, 86
with jojoba oil, 109
with *N*-donor chelates, 110
for mercury, 113
for lead(II) and cadmium(II), 133, 147
for lead(II) and copper(II), 134
with dithiophosphates, 137
for lead, 138, 140, 141, 143-145
for cadmium(II), 152
for cadmium(II) with amines, 153
for copper, 166
with Acorga P50, 169
rates, 170
for copper(II) with carbon dioxide, 177
for silver(I) and gold(III), 183
for actinide, 193, 197, 198, 199, 205
for uranium(VI), 196
for europium, 206
for lanthanides, 210-213, 217, 219, 224,
with acylpyrazolones, 221
with calixarene phosphoramides, 222, 223
with heterocyclic amines, 239
for alkali metal ions, 247, 264, 277
with crown ethers, 248, 255
for alkali metals with crowns, 254, 259, 266
with lariat crowns, 263
for sodium(I) and potassium(I), 276

thermodynamics, 280
for cesium(I), 281, 283
Liquid membranes, 10, 16, 18, 19, 24, 30, 118
Lissamine rhodamine B, 325
LIX 64N, 168
LIX Reagents, 168
LMCT state, 326
Macrocycles
concept, 64
metallomacrocycles, 65
*O*-donor macrocycles, 82, 118, 140
azamacrocyles, 84, 116, 141, 175, 332
thiamacrocycles, 84, 116, 176, 184
molecular modeling, 93
*N*-donor macrocycles, 117, 141
*S*-donor macrocycles, 117, 143, 176
*N,O*-donor macrocycles, 119, 213
chromogenic and fluorogenic, 121
for lead and cadmium, 139
lariat crowns, 263
furan macrocycles, 271
fluoro, 272
*N,O*-donor chemosensors, 328
*N,S,O*-donor chemosensors, 331
alkali metal chemosensors, 338
Macrocyclic effect, 6, 64
Malonamides, 198-202, 214, 215
tetrahexylmalonamides, 201
Manganese oxide, 150, 279
Membrane methods, 23
Menkes disease, 164
merApe9, 296, 299
Mercuric ion reductase, 296, 298
Mercury
sulfhydryl binding, 3
elemental mercury, 25, 29, 107, 108, 125, 126
With carbon, 26
mercury vapor, 27, 106, 113, 126
with derivatized silicates and polymers, 28-30
soil washing, 37
electrodeposition, 38
antibody affinity, 43
sulfides, 53
electrokinetics, 54
with Aliquat 336 and Titan Yellow, 70, 79
with hydroxyquinolines, 75
with acrylamide resin, 80
with fluorochelates, 81
azamacrocycles and thiamacrocycles, 84
*N,O*-donor macrocycles, 88
*S*-donor macrocycles, 89
*O*-donor macrocycles, 91
aromatic macrocycles, 93
dithiocarbamoyl calixarenes, 99
toxicity, 106
*O*-donor complexants, 108
*N*-donor and *S*-donor complexants, 109
*N*-donor and *S*-donor chelates, 110
dithiocarbamates, 111
techniques, 113
organomercury, 113, 298
column materials, 114
phenylmercury, 114
radioactive mercury, 114, 125
macrocycles, 116
porphyrins, 121
calixarenes, 122
soft ligands, 130
thiocyanate, 132
with carbohydrates, 157
in food chain, 158
with isooctyl thioglycolate, 178
with *N,S*-donor chelates, 183
with *S*-donor chelates, 183
with *N*-donor macrocycles, 186

with hydrocarbons, 187
phytoremediation, 291
transgenic plants, 296
bioremediation, 298
fluorescence sensors, 317
protein sensors, 318
redox sensors, 329
thiacrown sensor, 332
silver selenide sensor, 354
Metal oxides, 25, 27, 279
Metal porphyrin complexes, 336
Metallacarboranes, 284
Metallomacrocycles, 65
Metallophytes, 291
Methylmercury(II), 105, 107, 112, 118, 126, 187, 298
Micelles and vesicles, 164, 165
Microbial reduction of chromium(VI), 235, 302
Microbially-enhanced chemisorption, 300
Microelectrode array based sensors, 353
Microemulsions, 18, 19, 113
Mining, 1, 3, 9, 49, 163
Molecular modeling, 44, 75, 93
Molecular photonic wire, 352
Molecular-scale computation, 359
Nafion, 319, 320, 359
*N*-dodecyl-hexa-(ethylene glycol) ether, 171
*N*-(hydroxymethyl)thioamide resin, 110
*N*, *N*-ethylene-bis-(salicylaldimine), 167
*N,N,N,N*-tetrabutylsuccinylamide, 201
Oat, 294
Optical reporter, 307, 308, 334, 357
Palladium
uses, 4
with chitosans, 30
with XAD-4, 67
with PAN, 73
with thiophosphonyls, 77
with selenocrowns, 89
with *S,O*-macrocycles, 90
with a dipyridyl amide polymer, 110
with a calixarene ketone, 187
palladium complex as extractant, 278
PAN, 73
PAQH, 79
Pertechnetate, 233, 239
Phosphate
tributyl phosphate, 10
for ion exchange, 22
phosphate host matrix, 40
dithiophosphate, 44
triisoamyl phosphate, 75
molybdophosphate, 110
tributyl phosphate-plasticized zinc dithizonate foam, 115
stannic phosphate, 115
diethyldithiophosphate (DTP), 137
dialkyldithiophosphates, 137, 156
ammonium hydrogen phosphate, 156
methyldiphenyl phosphate, 166
cerium(IV) phosphate, 180
for actinides, 193-195
TEHP, 204
for lanthanides, 207, 209-212
metaphosphate, 210
extraction of, 231-232, 240
analysis, 234
structure, 237
with calixarene crowns, 241
glycerol 2-phosphate for uranium, 300
organophosphate, 301
dihydrogen phosphate, 326, 355-359
Photocontrol, 267
Photoinduced intramolecular electron transfer, 335
Photoreversible sensors, 315
Phytoextraction, 289, 291, 294
Phytoremediation, 2, 237, 244, 289-296, 302
Plating, 3, 46, 163
Platinum, 4, 25, 30, 67, 73, 99, 180
Picrate method of Analysis, 19

Podant ionophores, 181
Polyethylene glycol
as aqueous biphasic system, 15, 132
use, 16
for chromium(VI), 243
for sodium(I), 254
Polymer FiltrationTM, 24
Polymeric supports, 108
Polymetallorotaxanes, 343
Polythiophenes, 318
Portland cement, 40
Potential chromium oxidation score, 59
Proteins
cadmium-thionein and copper-thionein proteins, 156
metallothionein, 151, 156
for sensors, 318
iron sulfur, 344
zinc finger proteins, 350
"Proton Ionizable" crowns, 263
PUREX process, 10
Purolite S940, 154
Pyrene reporters, 322
Pyrometallurgical, 4
Quartz crystal microbalance, 318, 353
Quinone/hydroquinone redox switch, 319
Radioactive
isotopes, 2
metals extraction, 10, 21
metals by electrokinetics, 48, 53, 57
mercury, 114, 125
silver, 179
with ROMP resin, 208
pertechnetate, 233
signature, 234
alkali metal ion, 247
cesium, 247, 281
Recombinant cells, 300
Reverse osmosis, 5
Reversible complexation, 6
Rhizofiltration, 289
Ruthenium(II)-bipyridyl type, 356
Salicylic acid, 75
Sediment Analysis, 149
Selenacrown ethers, 89
Selenium, 3, 63, 70, 231, 232
Silicates, 22, 28
Silicon surfaces, 26
Silver
precious metal, 4
with mercury, 25
with aminimide, 71
*N,S*-chelate, 77, 174
*O,S*- and *S,S*-chelate, 77
mercaptoamide resin, 80
*O*-macrocycles, 82
with azamacrocycles, 86, 87
with lariat crowns, 90
molecular mechanics, 93
with calixarenes, 98, 99
cyanide, 114
with organic adsorbents, 115
with crowns, 121
with 8-quinolinol and acetylacetone, 137
with azacrowns, 142
with calixarenes, 145, 157, 177
extraction, 178-188
iodide, 181
with crowns, 251
with cryptands, 274
with *N,O,S*-macrocycles, 331
with hexathiacrown, 332
sensor with *N,O*-cryptate, 337
sensor with zinc porphyrin, 338
sensor with crowns, 344
sensor with lanthanides, 351
selenide films, 354
SME 529, 166, 168
Smelting, 3, 151
Softness, 7
Solid phase microextraction (SPME), 150
Stern-Volmer plot, 309
Sulfhydryl, 3, 30
Sunflower, 292
Supercritical carbon dioxide
with fluorinated extracts, 80
with dithiocarbamate, 111
with a dibenzo-*bis*-triazole

crown, 117
with a dithiocarbamate extractant, 147, 156
with hexafluoroacetylacetone, 177
for uranium and plutonium, 195
with phosphine oxide extractants, 196
with tributyl phosphate, 210
with fluoroketone extractants, 211
Surfactant flooding, 36, 59
Surfactants, 12, 20, 202, 299
Switch grass, 292
Tetraazamacrocycles, 86, 141, 175
Tetrahexylmalonamides, 201
Tetraphenylporphyrin complexes, 121
Tetrathiafulvalene (TTF) redox reporter, 331
TEVA-spec, 196
Thiamacrocycles
for mercury, 84, 116
for copper, 176
for silver and gold, 184
Thiohydroxamic acids, 317
Toxic effects, 3, 106, 164, 232
Toxicological effects, 106, 163, 233
Triamine and Trianthrylamine cryptands, 335
Tributyl phosphate
PUREX, 10
for mercury, 115
for actinides, 194, 195
for lanthanides, 210
for cesium(I), 283
Trioctylphosphine oxide, 194
Triton X-100, 222
U/TEVA-spec, 196
Vitrification, 44
Water Hyacinth
for chromium, 237
for chromium(VI), 292, 302
for cadmium, 293
Wilson' s disease, 164
Xanthates, 44, 67, 111, 183
Zeolites
anionic, 3
Ion exchangers, 22
natural, 23, 28
in soils, 35
for chromium(VI), 236
Zinc finger peptides, 350